Chromosomes
in
Evolution
of
Eukaryotic Groups

Volume II

Editors

Arun Kumar Sharma
Ghosh Professor and Program Coordinator
Center of Advanced Study in Cell and Chromosome Research
Department of Botany
University of Calcutta
Calcutta

Archana Sharma
Professor of Genetics and Head
Department of Botany
University of Calcutta
Calcutta

CRC Press
Boca Raton, Florida

Library of Congress Cataloging in Publication Data

(Revised for vol. 2)
Main entry under title:

Chromosomes in evolution of eukaryotic groups.

Bibliography: v. 1, p.   ; v. 2, p.
Includes indexes.
  1. Chromosomes—Evolution. 2. Eukaryotic cells—
Evolution. 3. Evolution. I. Sharma, Arun Kumar,
1923- . II. Sharma, Archana Mookerjea. [DNLM:
1. Cells. 2. Chromosomes. 3. Evolution. QH 371.C5
C557]
QH371.C5 1983   574.87'322   82-9449

ISBN 0-8493-6496-5 (v. 1)
ISBN 0-8493-6497-3 (v. 2)

Direct all inquiries to CRC Press, Inc., 2000 Corporate Blvd., N.W., Boca Raton, Florida, 33431.

International Standard Book Number 0-8493-6496-5 (Volume I)
International Standard Book Number 0-8493-6497-3 (Volume II)

Library of Congress Card Number 82-9449
Printed in the United States

FOREWORD

Chromosomes of higher organisms exhibit myriads of forms of wide diversity, but their mechanism of origin is still a controversial issue. Their prokaryotic ancestry, so for considered as undisputed, has run into bad weather, due to the discovery of intervening sequences associated with the property of splicing. This theory, envisaging a gradual evolution of complexity in the chromosome from the genophore, is facing replacement by the concept of independent origin of eukaryotic chromosomes parallel to that of the prokaryota.

The chromosomes of higher organisms have undergone diverse modifications both during and after evolution. Several structural features are indeed of phylogenetic and evolutionary significance. At one end of the broad spectrum is the peculiar chromosome of the Dinophyceae with very little histone and absence of functional differentiation of segments. The other extreme is represented by the complex chromosome structure of primates, having even reverse-banded segments associated with special genetic attributes.

Structural features of chromosomes with high phylogenetic potential include, among others, the centromere — diffuse, polycentric, or localized; number and nature of nucleolar constrictions; sex chromosomes with special reference to their multiple mechanisms; early and late replication, heterochromatin, repeated DNA and banding pattern. The nature and position of repeats in introns and even exons, as brought out lately, may all reveal facts of fundamental significance to a evolutionist.

Aspects of chromosome behavior often stressed in the study of evolution are heteroploidy, fragmentation, and translocation. The association of such factors with the evolution of species is undeniable. However, all three mechanisms, excepting the relative infrequency of polyploids in animal systems, are widespread in eukaryota, though not restricted to any particular level of taxonomic hierachy. In fact, in plant systems, such changes occur with equal prominence within and between species, as well as between genera and families. As such, their universal incidence at all levels of taxonomic hierarchy disqualifies them for consideration as specific parameters of phylogenetic significance. This understanding is of supreme importance in the application of chromosome science to the study of evolution.

The amount of data on the importance of chromosome structural analysis in the study of evolution is indeed enormous. The analysis of such a compendium of information is a task of tremendous magnitude. This exercise is also complicated as the paths of evolution are divergent. A definite pattern of chromosome evolution in the eukaryotic system is yet to be established, even in structural features of chromosomes of established phylogenetic significance. As a result it is difficult to draw a single undisputed phylogenetic tree for a plant or animal system as a whole, based solely on chromosomal characteristics.

It was initially desired to present the pathways of evolution of chromosomes indifferent groups of eukaryota according to their sequence in an established progression of complexity. Such a desire, however pious it may be, could not be fulfilled due to various reasons beyond our control for which we offer our apologies in advance. Limitations of space and time also did not permit a discussion of all the groups which may have to be taken up later in detail.

THE EDITORS

Arun Kumar Sharma, D.Sc., F.N.A., F.A.Sc., F.N.A.Sc., is Sir Rashbehari Ghosh Professor and Programme Coordinator (Director), Centre of Advanced Study on Cell and Chromosome Research, Department of Botany, University of Calcutta and the President, Indian National Science Academy, New Delhi (1983 to 1984). He obtained his M.Sc. (1945) and D.Sc. (1955) degrees from the University of Calcutta and was Head of the Department of Botany from 1969 to 1980. He has made significant contributions on diffe rent aspects of chromosome research; and has built up one of the largest centers of chromosome research in the world. His works cover cytotaxonomic studies on angiosperms, particularly mono-cotyledons, speciation in asexual plants, and development of techniques of chromosome analysis from both meristematic and differentiated nuclei. Among his more than 350 papers, recent works include theory of chromosome dynamism, demonstration of the variability of chemical components of the chromosome during organogensis, additional genetic elements in chromosomes, dynamic DNA and stufy of the chromosome involving in vitro nutagenasis.

Professor Sharmaa has been the General President of the Indian Science Congress Association (1980), Vice-President of the Indian National Science Academy, Chairman of the Indian National Committee of IUBS and Co-Chairman, Global Seminar on Role of Scientific Societies (AAAS/INSA/ISCA). He has been a member of the council of all three Academies of India and President of several societies, including the Indian Botanical Society, Indian Society of Cell Biology, Genetic Association of India, Society of Cytologists and Geneticists, and others. His numerous Awards include the Shanti Swarup Bhatnagar Award of the Council of Scientific and Industrial Research, the J. C. Bose Award of the University Grants Commission, the Silver Jubilee Medal of the Indian National Science Acadamy, the Birbal Sahni Medal of the Indian Botanical Society, the FICCI, and the Jawaharlal Nehru Fellowship.

He has been visiting lecturer to different centers of the world and led the Indian delegation several times to international conferences including the International Genetics Congress, the Hague (1963); the International Botanical Congress, Leningrad (1975); the International Cell Biology Congress, Berlin (1980); and the IUBS General Assembly, Bangalore (1975), and Ottawa (1982). He is the founding Editor of the international journal, *The Nucleus,* and member of the editorial board of several journals. He is co-author, with Archana Sharma of *Chromosome Techniques — Theory and Practice,* a standard reference and textbook.

Archana Sharma, Ph.D., D.Sc., F.N.A., F.A.Sc., F.N.A.Sc., is Professor of Genetics and Head of the Department of Botany, University of Calcutta (1980 to 1982). She obtained her M.Sc. (1951), Ph.D. (1955), and D.Sc. (1960) degrees from the University of Calcutta and specialized in cytogenetics and human genetics. She has made outstanding contributions to cytotaxonomy, the cause of polyteny in differentiated nuclei, and the development of new techniques for the study of chromosome structure. Her group is actively engaged in the study of chromosomal and genetic polymorphisms in normal and pathological human populations, differentiating patterns in the human fibroblast, and genetic polymorphisms in relation to environmental mutagenesis and genetic diseases. Other significant research includes studies of the effect of metallic pollutants on genetic systems, both antagonistic and synergistic. She has more than 150 papers and several books to her credit.

Professor Sharma is a Fellow of all three Academies of India, member of the council of the Indian National Science Academy, member of the Science and Engineering Reseach Council of the Government of India, and General Secretary of the Indian Science Congress Association, with which she has been involved for nearly two decades. As official delegate of the Government of India, she has participated in several international conferences, including the IUBS General Assembly Sessession at Helsinki, and the International Cell Biology Congress at Berlin, and has been the Visiting Scientist in the U.S.S.R. under the

Government of India exchange program and a member of the delegation from the Academy to the people's Republic of China. She is the recipient of the Shanta Swarup Bhatnagar Award of the Council of Scientific and Industrial Research (1976) and the J. C. Bose Award of the University Grants Commission, and National Lecturer, University Grants Commission. She is the Editor of *The Nucleus,* and a member of the Editoral Board of a number of other journals.

CONTRIBUTORS

John W. Bickham
Associate Professor
Department of Wildlife and Fisheries
 Sciences
Texas A & M University
College Station, Texas

Keith Jones
Professor
Jodrell Laboratory
Royal Botanic Gardens, Kew
Richmond, Surrey
England

Ann Kenton
Jodrell Laboratory
Royal Botanic Gardens, Kew
Richmond, Surrey
England

G. K. Manna
Professor of Zoology and
 National Fellow, UGC
Department of Zoology
University of Kalyani
Kalyani, West Bengal
India

Susumu Ohno,
Ben Horowitz Chair of Distinguished
 Scientist
Beckman Research Institute of the City of
 Hope
Duarte, California

Archana Sharma
Professor of Genetics and Head
Department of Botany
University of Calcutta
Calcutta
India

Arun Kumar Sharma
Ghosh Professor and Program
 Coordinator
Center for Advanced Studies in Cell and
 Chromosome Research
Department of Botany
University of Calcutta
Calcutta
India

Geeta Talukder
Research Associate
Human Genetics Unit
Center for Advanced Studies in Cell and
 Chromosome Research
University of Calcutta
Calcutta
India

A. C. Triantaphyllou
Professor
Department of Genetics
North Carolina State University
Raleigh, North Carolina

Niilo Virkki
Cytogeneticist
Department of Crop Protection
Agricultural Experiment Station
University of Puerto Rico
Río Piedras, Puerto Rico

Trevor G. Walker
Senior Lecturer
Department of Plant Biology
University of Newcastle upon Tyne
Newcastle upon Tyne
England

TABLE OF CONTENTS

Chapter 1

CONSERVATION OF LINKAGE RELATIONSHIPS BETWEEN GENES AS THE UNDERLYING THEME OF KARYOLOGICAL EVOLUTION IN MAMMALS*

Susumu Ohno

TABLE OF CONTENTS

*   This work was supported in part by an NIH grant and by research grants from the Bixby Foundation and the
     Wakunaga Pharmaceutical Co., Ltd.

I. INTRODUCTION

Monitoring the state of Arkansas judicial battle waged between creationists and evolutionists, I came to realize that one hundred years after the death of its creator, Darwinism too has acquired a distinctly religious tinge. No one would seriously dispute the obvious fact that man as a species has an absolute biological need to believe in a supreme being of one kind or another, for I know not a single tribe or race in human history which has not invented its own version of a god or gods. The tragedy of modern man, then, is found in his disaffection with traditional religions. He has to seek a surrogate religion elsewhere. Thus, for modern biologists, natural selection came to represent the infallible "All Mighty" that does no wrong. This religious belief in natural selection is absurdly anti-Darwinian, for the inherent ingredient of the Darwinian world of evolution by natural selection is the eventual doom of all species ever created. It follows then that every evolutionary strategy ever devised must necessarily contain an *a priori* flaw that presages final destruction.

The current attitude of evolutionists reminds me of the attitude of historians in the heat of conflicts who are then inclined to find the infallible generalship of their victorial commanders in nearly all their victories. Fortunately, time cools emotions and historians have become enlightened and realize that most battles were won in spite of a number of serious blunders committed by the victors. I shall give only one example. At the battle of Waterloo, the generalship of neither Wellington nor Blücher was altogether brilliant. On the contrary, they committed at least two nearly fatal blunders. First of all, they failed to really appreciate Napoleon's genius in mobilizing and deploying his troops with lightning speed. Thus, Blücher's army was struck before it managed to unite with Wellington's army, and suffered a rout. Were it not for the unaccountable tardiness of Grouchy, to whom Napoleon entrusted the task of pursuing Blücher's defeated army, the dramatic rescue of Wellington by Blücher on the evening of the battle of Waterloo would not have taken place. While Napoleon correctly identified the vital cross-road "Quatre-bras" as the key to the battle of Waterloo, Wellington failed to appreciate its significance. This blunder was minimized by the Prince of Saxe-Weimar, on Wellington's side, who occupied Quatre-Bras on his own initiative. Yet Ney, Napoleon's bravest marshal, should have had no trouble in dislodging them if he had initiated his attack as promptly as ordered.

It is the very fact that the modern biologists in general, evolutionists in particular, thus far have failed to adapt the enlightened approach already practiced by modern historians in dealing with evolutionary events that makes me suspect a religious aura emanating from the concept of natural selection. Take our wisdom teeth, for example — only a deep-rooted, unshakably religious faith in the might of natural selection would prompt one even to contemplate a possible selective advantage offered by the possession of so useless a tooth characterized by miss-directed erruptions. What was the selective advantage in losing the ability to synthesize vitamin C *de novo* within our own bodies? It is granted that while our ancestors lived in tropical forests as brachiating apes, the abundant supply of fresh fruits and greens rendered this function unnecessary. Yet, had natural selection been endowed with even a trace of foresight, we would have kept this ability as did most of the greens-eating mammals. Once our ancestors inhabited the temperate zone with long winters, this genetic defect became a serious disadvantage. Some years ago, I thus arrived at the view that the successful emergence of a new species need not represent the culmination of a series of brilliant strokes written by natural selection. More often, such an event may have occurred in spite of a series of evolutionary blunders committed in the past, because of the more recent acquisition of certain redeeming traits.

Viewed in this light, it is no longer necessary to attach a forced significance either to the situation of extensive speciation in spite of the karyological conservation seen in certain mammalian families and genera or to the opposite situation of extensive karyological diversification accompanied by a minute amount of evidence of adaptive radiation seen in other mammalian families and genera. The fact is that, in either case, the linkage relationship between genes has been conserved to a surprising extent.

Excluding a small number of exceptional species that are characterized by either the superabundance or extreme paucity of constitutive heterochromatin, the genome size of eutherian as well as marsupial mammals has remained fairly constant, the haploid set of chromosomes containing roughly 3.2×10^9 base pairs of DNA.[1] In view of the fact that the diploid chromosome number of eutherian mammals may be as low as six and seven, as in the red muntjac deer *(Muntijacus muntijak vaginalis)*,[2] or as high as 92, as in the fish-eating rodent *(Anotomys leander)* of Peru,[3] this genomic constancy should be viewed as somewhat of a surprise. Here, I shall concentrate on the general topic of the surprisingly rigid conservation of the linkage relationship between genes in spite of an apparently extensive karyotypic diversification.

II. EVOLUTIONARY INSIGNIFICANCE OF TRANSPOSABLE ELEMENTS AND JUMPING GENES PRODUCED BY RETROVIRUS REVERSE TRANSCRIPTASE

In recent years, it was found that certain repeated DNA sequences of dubious functional significance (transposable elements) may change their positions within the genome with considerable rapidity. This observation prematurely prompted the revival of the once largely discarded notion, extreme evolutionary fluidity of the genome. Such indeed is a dangerously misleading notion, for the fact is that as far as linkage relationships between coding sequences are concerned, conservation is the rule of mammalian evolution. It follows then that if these so-called transposable elements contributed significantly to mammalian evolution, their roles have been confined to the following two aspects:

1.   By inserting itself in the midst of a gene, thus disrupting the reading frame of coding sequences, it might have silenced that particular gene.
2.   By inserting itself upstream of a gene, it might have modified the expression of a gene, e.g., from a steroid hormone-dependent inducible expression to a constitutive expression.

Nothing more needs to be said of these transposable elements.

Another hitherto little-recognized cause of disruption of the preexisting linkage relationship between genes is the back incorporation of a cDNA copy of the functional *messenger* RNA (mRNA) into a new position of the genome. The reverse transcriptase specified by omnipresent retrovirus genomes is inherently capable of performing the above. In the mouse *(Mus musculus domesticus)* genome, a pair of genes for adult hemoglobin α-chain and one gene for embryonic hemoglobin α-chain form a 25,000 base-pair-long cluster in the region that is reasonably close to the centromere of chromosome 11. In addition, however, another unspecified chromosome contains one more hemoglobin α-chain coding sequence. This sequence is entirely homologous to that of its mature cytoplasmic mRNA, thus lacking two intervening sequences — an obvious work of reverse transcriptase.[4] The evolutionary significance of such a daring experiment of nature is diluted by the fact that such a cDNA copy of the mature cytoplasmic mRNA cannot function as a gene.

III. CYTOLOGICAL EVIDENCE OF LINKAGE GROUP CONSERVATION IN SPITE OF EXTENSIVE CHROMOSOMAL REARRANGEMENTS

As with all things biological, a superficial glance of karyological evolution in mammals yields no apparent rule. At one extreme various bears of the family Ursidae maintain very similar karyotypes, and the same can be said of the domestic cat, lion, tiger, and leopard, of the family Felidae. The extreme of extremes in evolutionary karyological conservation, however, is found in the order Cetacea. Dolphins, killer whales, and sperm whales are toothed whales, belonging to the suborder Odontoceti, while the blue, gray, and sei whales are baleen whales, belonging to the suborder Mysticeti. These two suborders already existed as distinctly separate lineages by the late Oligocene (some 28 million years ago) at the latest and most likely much earlier. Yet the common dolphin *(Delphinus delphis)*, belonging to the former, and the sei whale *(Balaenoptera borealis)*, belonging to the latter, presented the $2n = 44$ karyotypes that were nearly indistinguishable even after the G-banding of individual chromosomes.[5] In these instances, the evolutionary conservation of linkage groups is all too obvious. At the other extreme, each species belonging to the family Equidae is characterized by its own unique karyotype, i.e., the Przewalski wild horse ($2n = 66$), the domestic horse ($2n = 64$), the somali wild ass ($2n = 62$), the onager wild ass ($2n = 54$), the Grevy's zebra ($2n = 46$), the common zebra ($2n = 44$), and the mountain zebra ($2n = 32$).

How many disturbances of linkage relationships have been caused by such extensive chromosomal rearrangements? The answer, apparently, is surprisingly little. The muntjac, a small deer living in Asia, has remained in many respects similar to the original deer that first emerged in Miocene times (26 to 16 million years ago). The two neighboring species or subspecies, the red or Indian muntjac *(Muntijacus muntijak vaginalis)* and the Reeves or Chinese muntjac *(Muntijacus muntijak reevesi)* are so closely related with each other that viable but sterile hybrids between the two can readily be obtained. Yet the former exhibits the lowest diploid chromosome number of all mammalian species ($2n = 6$ for females and $2n = 7$ for males due to an X-autosome translocation)[2] whereas the latter possesses the standard cerbid karyotype made of 46 acrocentric chromosomes. The detailed comparative study of parental haploid sets coexisting in interspecific hybrid individuals revealed that chromosome 1 of the former corresponds to eight acrocentric autosomes of the latter that have been united by one Robertsonian and six tandem fusions. Chromosome 2 of the former also corresponds to another eight acrocentric autosomes of the latter, whereas the third autosome (and so-called Y_2) of the former, that later fused with the X, represents the remaining six autosomes of the latter.[6] Such Robertsonian and tandem fusions serve to conserve the original linkage relationships between autosomal genes, instead of disrupting them.

The evolutionary conservation *in toto* of the X-linkage group by all eutherian as well as marsupial species originally proposed in 1964[7] and subsequently expanded in 1967[8] has now been established as an inviolable law. Yet, the X chromosomes of certain mammalian species are metacentrics, while those of others are acrocentrics, thus revealing occasional internal rearrangements within the X-linkage group. The interesting point here is the essential conservation of the G-banding pattern by individual X chromosomes of divergent mammalian species which makes it possible to discern the nature of internal rearrangements that changed the metacentric X to an acrocentric X or vice versa.[9] Particularly impressive is the strikingly similar G-banding pattern exhibited by the metacentric human X on one hand and the metacentric X chromosome of the wood lemming *(Myopus schisticolor)* on the other — one a primate, the other a rodent. We shall come back to this later.

IV. CONSERVATION *IN TOTO* OF THE X-LINKAGE GROUP AND A PAUCITY OF SEX-DETERMINING GENES AMONG THE X-LINKED GENES

Inasmuch as the evolutionary conservation *in toto* of the X-linkage group by all eutherian mammals, originally proposed in 1964,[7] has now been firmly established as law,[10] I shall refer only to the latest of the numerous reports on this subject. Utilizing cattle-mouse somatic

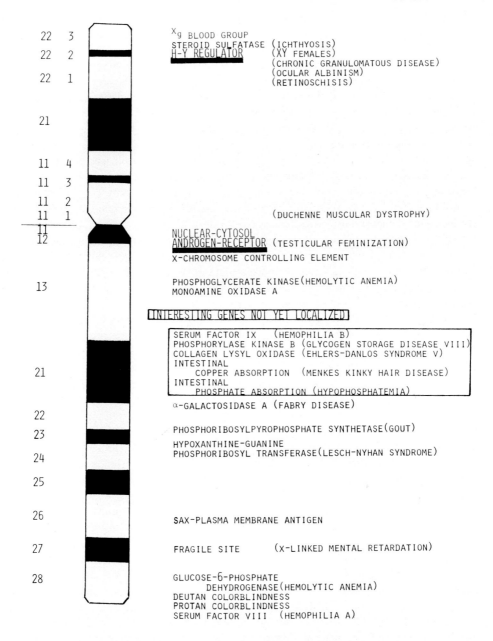

FIGURE 1.   The G-banded human X-chromosome is schematically illustrated at the left. The 20 X-linked genes, whose approximate location on the X are known, are aligned on their roughly corresponding positions at the right. In addition, five other X-linked genes of special interest but of unknown locations are presented in the box under the heading of "Interesting Genes Not Yet Localized". (For further details, refer to Reference 10.)

hybrids, the X linkage of glucose-6-phosphate dehydrogenase, phosphoglycerate kinase, α-galactosidase, and hypoxanthine-guanine phosphoribosyl transferase has been established in cattle, *Bos taurus*.[11] The first two of the four enzymes noted above are also X linked in kangaroos and other marsupials.[12,13] Thus, the conservation extends to marsupial mammals as well.[8]

Of more than 100 X-linked gene loci identified in man,[10] Figure 1 lists some of those that have been localized with variable accuracy to certain regions of the human X. Although their locations are not yet known, five other X-linked genes of particular interest are also included in Figure 1. With regard to nearly all the human X-linked genes shown in Figure 1, the X linkage has been confirmed in at least one other mammalian species.

The first glance at Figure 1 reveals the paucity of genes that are directly involved in the act of sex determination and differentiation. This is because only a very small number of genes are required for the mammalian primary (gonadal) and secondary (extragonadal) sex determination.[14] The embryonic plan of mammalian embryos is inherently feminine, and the role played by the Y chromosome is strictly limited to cause testicular organogenesis by diverting the inherent tendency of embryonic indifferent gonads to organize the ovary. The secondary sex determination is solely due to the presence or absence of testosterone and the fact is that XY and XX embryos are equally sensitive to testosterone.[15] Thus the X-linked testicular feminization locus that specifies the one and only species of nuclear-cytosol androgen-receptor protein emerges as the master of secondary sex determination.[14] This receptor demonstrates a two-fold preference toward 5α-dihydrotestosterone over testosterone, and the *Tfm* locus in man is thought to be located very near the centromere on the long arm side of the human X (Figure 1). Interestingly, the gene for 5α-reductase is autosomally inherited,[16] and so are the nuclear-cytosol receptors for other steroid hormones (e.g., hydrocortisone) as well as gonadotropins. Even the genes for hypothalamic-releasing factors of gonadotropins are apparently on autosomes.[17] The autosomal location of most of the sex-related genes is no surprise, for sex differences are very subtle at the molecular level, most of the sex-related genes being utilized by males and females alike.

The small Y chromosome is largely a dummy, a great part of it being readily dispensable. As to the acrocentric Y chromosome of both man and the mouse, the male-determining gene that governs the expression of testis-organizing H-Y antigen[18] apparently resides either on the extremely short second arm or immediately below the centromere on the long arm side.[19] Wisely, the Y of the mouse pairs with the X at its long arm end during male meiosis, thus avoiding a danger of transferring the critical male-determining gene from the Y to the X by cross-overs. Unaccountably, man chose to court constant danger by pairing the critical short arm of the Y with the very tip of the short arm of the X. The transfer by a crossing-over of the male-determining gene from the Y to the X should produce XX males and XY females as reciprocal products. Observing Figure 1, it should be noted that the X_g blood group gene is situated near the tip of the short arm of the human X chromosome.[10] There is some evidence that in exchange for acquiring the male-determining gene, one of the X chromosomes of XX males has donated the X_g gene to the Y. XX males produced by the mating between an X_g^+/Y father and X_g^-/X_g^- mother are expected to be X_g^-/X_g^+ heterozygotes, thus typing as X_g^+. The fact is that they almost invariably type as X_g^-.[20,21] It would be of extreme interest to type XY women with gonadal dysgenesis, produced by the same mating type, for X_g blood group antigen. If they type as X_g positive, the defect clearly lies in their Y which gained the X_g^+ gene from the paternal X in exchange for the male-determining gene. Such XY women and XX men can then be truly regarded as reciprocal products of meiotic cross overs in male germ cells. While the failure of XY women to organize testes can in some instances be attributed to their defective Y chromosomes, in other instances the fault is clearly in their X chromosomes.[19] In such instances, the defect is mapped again close to the very tip of the X chromosome short arm (Figure 1). It is of interest to note that the defect localized in the roughly corresponding region of the metacentric X chromosome of the Scandinavian rodent *Myopus schisticolor* also produces XY females, this time fertile.[22] There can be no greater demonstration than the above in dramatizing the evolutionary conservation of the mammalian X chromosome. The wild-type gene of this region may be concerned with the expression of H-Y antigen. Conversely, it might be specifying the specific plasma membrane receptor for H-Y antigen. Successful testicular organogenesis is the result of the interaction between the male specific, but ubiquitously expressed, H-Y plasma membrane antigen and its specific receptors expressed only by somatic elements of the gonad, but of both sexes.[14]

The very fact that some mammalian X chromosomes are metacentrics, while others are acrocentrics, reveals that there had been frequent internal rearrangements within the confine of *in toto* evolutionary conservation. Indeed, when the genetic map of the human X, which is a metacentric (Figure 1), is compared that of the acrocentric X of the mouse (*Mus musculus domesticus*), one readily detects conserved as well as rearranged linkage relationships between individual genes. For example, in both species, the *testicular feminization* (androgen-receptor) locus and the phosphoglycerate kinase locus are in close proximity of each other. However, one sees a sign of an inversion in relative positions of the *testicular feminization* locus and the inactivation center locus of two species in relation to their respective centromeres. *A priori*, random inactivation of one or the other X early in female embryonic development of eutherian mammals, originally envisaged by Lyon in 1961,[23] requires the presence in each individual X of one major center from which the inactivation spreads toward both ends. In the mouse, such a center has been defined as the *Xce* locus, and is mapped roughly six cross-over units upstream of the *testicular feminization* locus.[24] In the human X, the corresponding center is thought to reside in its long arm,[25] but it apparently maps downstream of the *testicular feminization* locus. In Figure 1, this locus is shown as the X-chromosome controlling element, the original term employed by Cattanach.[24]

While there are numerous examples in mammals of functionally related genes existing as closely linked clusters (e.g., genes for hemoglobin β-like chains and genes for various classes of immunoglobulin heavy chains), such a clustering is merely a consequence of these genes arising from tandem duplicates of one ancestral gene. More significant is the close linkage of genes involved in sequential steps of the same metabolic pathway, for the observed coordinate regulation of these genes in prokaryotes is often the direct consequence of their close linkage enabling them to constitute an operon. Such close linkages of potentially functional significance are extremely rare in mammals. The more typical situation is exemplified by genes of the pentose phosphate shunt; the first enzyme (glucose-6-phosphate dehydrogenase) is X linked, while the second enzyme (6-phosphogluconate dehydrogenase) is autosomally inherited. Thus, it is somewhat of a surprise to find two solid examples and abundant hints of the X linkage of two members of the same metabolic pathway. The X-linkage of both antihemophilic factor VIII (hemophilia A) and factor IX (hemophilia B) are the oldest known examples, although the approximate location on the human X is known only for the former. The two enzymes of the purine-scavenging pathway, hypoxanthine-guanine phosphoribosyltransferase and phosphoribosylpyrophosphate synthetase, are not only X linked but also in close proximity of each other on the human X (Figure 1). It should be noted, however, that the latter is also involved in other aspects of purine and pyrimidine metabolism. When the products of the deutan and protan colorblindness loci are identified in the future, they might be shown to be involved in sequential steps of forming trichrome vision.

The X linkage of two genes for intestinal absorption of copper and phosphate (Figure 1) may merely reveal their common ancestry, the two having arisen from duplicates of the same ancestral gene. The comparison of Menkes kinky hair disease of man with a *Mottled* series of mutations of the mouse, on the other hand, strongly indicates that the mammalian X chromosome carries more than one gene involved in the copper transport. Consistent with the defect confining itself in the intestine, heterozygous mothers of sons with Menkes kinky hair disease are not visible mosaics in their external appearance. By contrast, female mice heterozygous for the *Mottled* series of mutations are clearly mosaics, their otherwise wild-type coat showing patches of affected areas characterized by near-white hairs. The wild-type allele of the *Mottled* locus is clearly not involved in the intestinal absorption of copper. Rather it is likely to be involved in subsequent steps of the uptake of circulating copper by individual cells. However, I was informed of an especially dark-skinned woman who was an obligatory heterozygote, having produced a son affected by Menkes kinky hair disease.

She presented a mottled reminiscent of $M^o/+$ heterozygous female mice. In this connection, it is also of interest to note the X-linked inheritance of two copper-dependent enzymes, lysyl oxidase and monoamine oxidase A. As the former catalyzes the first step in the cross-linking of collagen, the connective tissues are invariably affected in both Menkes kinky hair disease of man and *Mottled* mutations of the mouse. Not only X-linked monoamine oxidase but also autosomally inherited tyrosinase are copper dependent; this accounts for neurological effects as well as for dilution of hair color caused by copper transport mutations of the mouse and man. Tyrosinase is involved not only in the synthesis of dopa, one of the neurotransmitters, but also of melanin.

The X linkage of phosphorylase kinase B has been established in the mouse.[26] While the mildest form of glycogen storage diseases which is X linked in man is thought also to be due to a mutational deficiency of the same enzyme, there is a growing suspicion that the enzyme deficient in the case of human patients is an activating enzyme of phosphorylase kinase.[10] If this is proved in the future, the mammalian X may carry not only the gene for phosphorylase kinase B but also the gene for its activating enzyme.

As to the question of the evolutionary conservation *in toto* of the mammalian X, no satisfactory answer has been provided. Development of the unique dosage compensation mechanism that depends upon the inactivation of one entire X does not provide a satisfactory answer, for the avian Z chromosome has also been conserved *in toto* with equal stringency, in spite of the apparent absence of the dosage compensation mechanism in this class of vertebrates.[8] This point, made in 1967,[8] has recently been confirmed on the cytoplasmic form of acotinase.[27] While this enzyme is autosomally inherited in all the mammalian species, residing at the short arm tip of chromosome 9 in the case of man,[10] the Z linkage of this enzyme has been shown in five species of birds belonging to three diverse orders. Furthermore, on a per cell basis, ZZ males demonstrated nearly twice as high activity of this enzyme when compared to ZW females.[27] The abundance of pairs of genes involved in sequential steps of diverse metabolic pathways on the mammalian X, or the avian Z for that matter, might really be the reason for their evolutionary conservation *in toto*.

V. EVOLUTIONARY CONSERVATION OF AUTOSOMAL SEGMENTS

Unlike the X and the Y chromosomes, *a priori* there is no sure way of singling out counterparts of a given autosome of one species in another. This then is the first difficulty in attempting to assess the degree of evolutionary conservation of the autosomal linkage group. On the surface, the diversity of diploid chromosome numbers among mammals from a high of 92 to a low of 6, as already noted, appears to rule out the extensive evolutionary conservation of autosomal linkage groups. Yet, if a large autosome, as a rule, arose by simple tandem fusions of smaller ones, or vice versa, as indicated by the comparative study of the two muntjac deer species,[6] the evolutionary conservation, to a surprising degree, of autosomal linkage groups then is expected.[28] Indeed, there are numerous examples of evolutionary conservation of small autosomal segments, e.g., the segment containing genes for various classes of immunoglobulin heavy chains and that containing genes for major histocompatibility antigens. Thus, we shall confine our attention only to the conservation of large autosomal segments.[29,30] By far the most impressive is the homology between the human chromosome 21 and the mouse chromosome 16. Aside from genes for 18S and 28S ribosomal RNAs (rRNAs) residing within the nucleolar organizer, the above-noted human and mouse autosomes share the genes for interferon receptor, superoxide dismutase, phosphoribosyl-glycinamide synthetase, and phosphofructokinase. The short arm of human chromosome 1 is also homologous in part to a portion of the mouse chromosome 4; the homology involves the genes for phosphoglucomutase isozyme, adenylate kinase isozyme, enolase isozyme as well as for 6-phosphogluconate dehydrogenase, the already noted second enzyme of the pentose phosphate shunt. In man, the serum albumin locus resides in the long arm of chromosome 4 while its short arm contains the genes for another phosphoglucomutase isozyme and peptidase isozyme. In the mouse, the gene for another phosphoglucomutase isozyme resides only 16 cross-over units away from the serum albumin locus in chromosome 5, thus revealing an internal rearrangement within the conserved autosomal segment separating the mouse from man. Nevertheless, in the case of the mouse, the gene for peptidase isozyme also resides on chromosome 5. In all mammalian species studies, the glyoxylase I locus resides in the autosome that also carries the major histocompatibility antigen gene (MHC) complex (chromosome 6 of man and chromosome 17 of the mouse). In this instance, one can readily discern an inversion within the MHC complex itself which sets rodents apart from all other mammalian species, including man. Aside from genes for C4 and other complements, the MHC complex contains the two classes of genes for plasma membrane antigens. The class I genes specify ubiquitously expressed polypeptides 348 amino acid residues or so long that invariably form dimers with β_2-microglobulin (H-2K1, H-2K2, H-2D, and H-2L antigens of the mouse and HLA-A, HLA-B, and HLA-C antigens of man). Antigens specified by class II genes, on the other hand, are expressed only by lymphocytes and macrophages, and the two chains that make up these antigens are less than 290 amino acid residues long each (various Ia antigens of the mouse, and various HLA-DR antigens of man). Within the MHC complex of man and other primates, as well as of the dog and the pig, the genes for class II antigens are clustered on the side of the glyoxylase I locus, while those for class I antigens are clustered on the other side, this arrangement apparently representing the mammalian archetype. Only in the mouse and the rat do the genes for class II antigens cluster in the middle of the MHC complex, flanked on both sides by gene for Class I antigens.[31] The evolutionary significance of such a small-scale inversion affecting only rodents is not very obvious. It probably represents one of those frozen accidents of evolution.[28]

VI. CLOSE LINKAGES BELIE IMMENSE DISTANCES SEPARATING INDIVIDUAL GENES

It has always been obvious that natural selection cannot afford to survey all the 3.2×10^9 base pairs of DNA contained in the haploid set of the mammalian genome, for so ambitious a surveillance would surely cause the extermination of every mammalian species from an unbearable mutation load. Of the mammalian genomic DNA, 95% or more is thought to be ignored by natural selection. The net consequence of this is an immense distance separating neighboring genes residing in the euchromatic region of mammalian chromosomes. Thus, the mammalian euchromatic region can be viewed as a barren stretch of desert in which still-functioning genes are scattered as though oases.[32] This prediction made in 1972 has been verified by the recent advent of DNA cloning. In prokaryotes with the extremely tidy genome, two adjacent genes may be separated by a spacer as short as 20 or so base pairs.[33] A difference of two orders of magnitude exists in mammals, for even a pair of very recent tandem duplicates such as the genes for human hemoglobin γ^{ala} and γ^{gly} are still separated by the spacer 3000 base-pairs-long. The average distance between neighboring mammalian genes was estimated as 35,000 base pairs.[32] The recent study on a very long stretch of mammalian DNA containing genes for various classes of immunoglobulin heavy chain constant regions certainly supports the above estimation. Distances separating C 3, C 1, C 2b, and C 2a genes of the mouse were determined as 34,000, 21,000, and 15,000 base-pairs.[35] It should be realized that these immunoglobulin heavy chain constant region genes are relatively recent tandem duplicates of each other; thus, they should be forming an exceptionally tight cluster. In the event of two genes of independent origins becoming neighbors, the distance separating them is expected to be considerably greater.

The reason for desertification of the euchromatic region is found in the extreme inefficacy of the mechanism of gene duplication as the sole means of generating new genes with previously nonexistent functions.[1] The two alternative fates await a redundant copy of the already existing gene. By accumulating all the randomly sustained mutations, it may emerge triumphant as a new gene with a hitherto nonexistent function. But its far more likely fate is degeneracy to join the rank of *junk* DNA. Mammalian ancestors at the stages of fish and amphibians had an option of utilizing the mechanism of polyploidization as a means to experiment with a redundant copy of every gene locus.[1] Nevertheless, the study on tetraploid fish species of variable antiquity has revealed that after 50 million years, about half of these redundant copies became degenerate and joined the rank of junk DNA, while a few managed to become new organ-specific isozyme loci. Mammals have long forfeited the possibility of polyploid evolution. They have to rely exclusively upon tandem duplication of the already-existing genes as the means of generating new genes. Indeed, cloned mammalian DNA almost invariably reveals the presence of silent or defunct genes in the neighborhood of each still-functioning gene, thus attesting to constant experimentation with the mechanism of tandem gene duplication as well as to the extreme inefficacy of this mechanism. The above, then, is the cause of progressive desertification of the mammalian euchromatic region. It is well to remember that when applied to mammals, the genetic distance of one cross-over unit covers one million base pairs of DNA in which there may be no more than 30 or 40 still-functioning genes. It follows then that as far as mammals are concerned, there are no closely linked genes *sensu stricto*; the genetic close linkage is largely an illusion.

VII. SUMMARY

Evolution by natural selection is the mechanism which is content to rely upon variations of the same theme. Truly new innovations are seldom, if ever, encountered in the evolutionary process. In spite of apparently dazzling diversity in karyotypes of mammals, the underlying theme is the conservation of linkage relationships between genes. The extreme of this is the conservation *in toto* of the X-linkage group by all eutherian as well as marsupial mammals. When one encounters the apparently tight linkages of functionally sequential or even related genes, it is tempting to invoke natural selection for their maintenance. However, it is well to remember that the mammalian euchromatic region is a barren stretch of desert in which still functioning genes are scattered as though oases. *Sensu stricto*, there are no closely linked genes in mammals.

REFERENCES

1. **Ohno, S.,** *Evolution by Gene Duplication,* Springer Verlag, Heidelberg, 1970.
2. **Wurster, D. H. and Benirschke, K.,** Indian muntjak, *Muntijacus muntijak;* a deer with a low chromosome number, *Science,* 168, 1364, 1970.
3. **Gardner, H. L.,** Karyotypes of two rodents from Peru with a description of the highest diploid number recorded for a mammal, *Experientia,* 27, 1088, 1971.
4. **Leder, A., Swan, D., Ruddle, F., D'Eustachio, P., and Leder, P.,** Dispersion of α-like globin genes of the mouse to three different chromosomes, *Nature (London),* 293, 196, 1981.
5. **Arnasson, U.,** The role of chromosomal rearrangements in mammalian speciation with special reference to Cetacea and Pinnipedia, *Hereditas,* 70, 113, 1972.
6. **Liming, S., Yingying, Y., and Xinsheng, D.,** Comparative cytogenetic studies on the red muntjac, Chinese muntjac and their F₁ hybrids, *Cytogenet. Cell. Genet.,* 26, 22, 1980.
7. **Ohno, S., Becak, W., and Becak, M. L.,** X-autosome ratio and the behavior pattern of individual X-chromosomes in placental mammals, *Chromosoma,* 15, 14, 1964.
8. **Ohno, S.,** *Sex Chromosomes and Sex-Linked Genes,* Springer-Verlag, Berlin, 1967.

9. **Pathak, S. and Stock, A. D.,** Conserved G-banding pattern of mammalian X-chromosomes, *Genetics,* 78, 708, 1974.
10. **McKusick, V. A.,** *Mendelian Inheritance in Man,* 5th ed., The Johns Hopkins University Press, Baltimore, 1978.
11. **Shimizu, N., Shimizu, Y., Kondo, I., Woods, C., and Wenger, T.,** The bovine genes for phosphoglycerate kinase, Glucose-6-phosphate dehydrogenase, α-galactosidase and hypothanthine phosphoribosyl transferase are linked to the X-chromosome in cattle-mouse cell hybrids, *Cytogenet. Cell Genet.,* 29, 26, 1981.
12. **Cooper, D. W., VandeBerg, J. L., Sharman, G. B., and Poole, W. E.,** Phosphoglycerate kinase polymorphism in kangaroos provides further evidence for paternal X inactivation, *Nature New Biology,* 230, 155, 1971.
13. **Cooper, D. W.,** Directed gene change model for X chromosome in activation in Eutherian mammals, *Nature (London),* 230, 292, 1971.
14. **Ohno, S.,** *Major Sex Determining Genes,* Springer-Verlag, Berlin, 1979.
15. **Jost, A.,** Action de la testosterone sur l'embryon male castre de lapin, *C.R. Soc. Biol. Paris,* 141, 275, 1947.
16. **Wilson, J. D. and MacDonald, P. C.,** Male pseudohermaphroditism due to androgen resistance: testicular feminization and related syndromes, in *The Metabolic Basis of Inherited Disease,* Stanbury, J. B., Wyngaarden, J. B., and Fredrickson, D. S., Eds., McGraw-Hill, New York, 1978, chap. 42.
17. **Cattanach, B. M., Iddon, C. A., Charlton, H. M., Chiappa, S. A., and Fink, G.,** Gonadotropin-releasing hormone deficiency in a mutant mouse with hypogonadism, *Nature (London),* 269, 338, 1977.
18. **Wachtel, S. S., Ohno, S., Koo, G. C., and Boyse, E. A.,** Possible role of H-Y antigen in primary sex determination, *Nature (London),* 257, 235, 1975.
19. **Simpson, J. L., Blagowidow, N., and Martin, A. O.,** XY gonadal dysgenesis: Genetic heterogeneity based upon clinical observations, H-Y antigen status, and segregation analysis, *Human Genet.,* 58, 91, 1981.
20. **Ferguson-Smith, M. A.,** X-Y chromosomal interchange in the aetiology of true hermaphroditism and of XX Klinefelter's syndrome, *Lancet,* 2, 475, 1966.
21. **de la Chapelle, A.,** The etiology of maleness in XX men, *Human Genet.,* 58, 105, 1981.
22. **Gropp, A., Winking, H., Frank, F., Noack, G., and Fredga, K.,** Sex-chromosome aberrations in wood lemmings *(Myopus schisticolor), Cytogenet. Cell Genet.,* 17, 343, 1976.
23. **Lyon, M. F.,** Gene action in the X-chromosome of the mouse *(Mus musculus* L.), *Nature (London),* 190, 372, 1961.
24. **Cattanach, B. M. and Williams, C. E.,** Evidence of non-random X chromosome activity in the mouse, *Genet. Res.,* 19, 229, 1972.
25. **Therman, E., Sarto, G. E., Palmer, G. C., Kallio, H., and Denniston, C.,** Position of human X inactivation center on X_q, *Human Genet.,* 50, 59, 1979.
26. **Lyon, J. B.,** The X-chromosome and the enzymes controlling muscle glycogen: Phosphorylase kinase, *Biochem. Genet.,* 4, 169, 1970.
27. **Baverstock, P. R., Addams, M., Polkinghorne, R. W., and Gelder, M.,** A sex-linked enzyme in birds: Z-chromosome conservation and lack of dosage compensation, *Nature (London),* 296, 763, 1982.
28. **Ohno, S.,** Ancient linkage groups conserved in human chromosomes and the concept of frozen accidents, *Nature (London),* 244, 259, 1973.
29. **Lalley, P. A., Minna, J. D., and Francke, U.,** Conservation of autosomal gene syntheny groups in mouse and man, *Nature (London),* 274, 160, 1978.
30. **Lundin, L-G.,** Evolutionary conservation of large chromosomal segments reflected in mammalian gene maps, *Clin. Genet.,* 16, 72, 1979.
31. **Gill, T. J., III, Kunz, H. W., Schaid, D. J., Vandeberg, J. L., and Stolc, V.,** Orientation of loci in the major histocompatibility complex of the rat and its comparison to man and the mouse, *J. Immunogenet.,* (in press) 1982.
32. **Ohno, S.,** So much "junk" DNA in our genome, in *Evolution of Genetic Systems,* Smith, H. H., Ed., Brookhaven Symp. No. 26, Gordon and Breach, Inc., New York, 1972, 366.
33. **Miozzari, G. F. and Yanofski, C.,** Gene fusion during the evolution of the tryptophan operon in Enterobacteriaceae, *Nature (London),* 227, 486, 1979.
34. **Little, P. F. R., Flavell, R. A., Kooter, J. M., Annison, G., and Williamson, R.,** Structure of the human fetal globin gene locus, *Nature (London),* 278, 227, 1979.
35. **Honjo, T., Kataoka, T., Yaoita, Y., Shimizu, A., Takahashi, N., Yamawaki-Kataoka, Y., Nikaido, M., Nakai, S., Obata, M., Kawakami, T., and Nishida, Y.,** Organization and reorganization of immunoglobulin heavy chain genes, *Cold Spring Harbor Sym. Quant. Biol.,* 45, 913, 1980.
36. **Ferris, S. D. and Whitt, G. S.,** Loss of duplicate gene expression after polyploidization, *Nature (London),* 265, 258, 1977.

Chapter 2

PATTERNS AND MODES OF CHROMOSOMAL EVOLUTION IN REPTILES

John W. Bickham

TABLE OF CONTENTS

I. INTRODUCTION

A. General Comments

Living reptiles are derived from the once dominant group of land vertebrates. Although reptiles flourished during the entire Mesozoic, their demise during the end of that Era was sudden. They were replaced by the mammals as the dominant land vertebrates by the early Tertiary. Reptiles have never reattained their previous level of diversity. Presently, about 45 families of living reptiles are recognized and approximately 180 extinct families are known from the fossil record.[1] Most of the latter are from the Mesozoic.

Our knowledge of chromosomal evolution in reptiles necessarily is limited to extant forms. Although we may never be able to relate reptile chromosomes directly to bird or mammal chromosomes[2] or be able to construct the ancestral karyotypes for reptiles, birds, and mammals, the study of reptile cytogenetics has proven to be richly rewarding. Much has been learned about the nature and patterns of chromosomal evolution in reptiles and this has been compared to patterns found in mammals, birds, and other groups. Reptiles offer us the opportunity to study the early stages of sex chromosome evolution. Studies on *Anolis, Sceloporus*, and others have provided important information on the role of chromosomal rearrangement in the process of speciation.

Recent advances in technology have made it possible to band reptile chromosomes. The banding techniques allow for a more sophisticated application of chromosomal data to evolutionary studies. Homologous chromosomes can be identified among different species, genera, and even families of reptiles using the G-band procedure. With this, the precise nature of chromosomal rearrangements can be determined. Heterochromatin can be differentially stained with the C-band technique. In addition, there are procedures for identifying late replicating DNA, the nucleolar organizer regions, fast-reassociating DNA, A-T or G-C rich DNA, and many others. These techniques are extensively applied in human cytogenetics and certain areas of cell biology, but only a few have been applied to reptiles. The majority of chromosomal data available for reptiles are from studies of standard, or non-differentially-stained, karyotypes. This means that diploid number, size, and centromere position of chromosomes is all that is known of the karyotypes of most reptiles.

The purpose of this chapter is not to tabulate the diploid numbers and fundamental numbers (number of chromosomal arms) of all reptile species. Gorman[3] expertly reviewed the literature up to 1973 and an up-date of his work was recently published by Peccinini-Seale.[4] Rather, I would briefly review the state of knowledge of cytogenetics for each order of reptiles with emphasis on particular exemplary studies, together with polyploidy and sex determination and modes of chromosomal evolution. The major purposes of this chapter are to delineate patterns of karyotypic evolution in reptiles, to attempt to interpret these patterns in light of current theories of chromosomal evolution, and to suggest future areas of research that appear to be particularly fruitful.

B. Reptile Classification

Table 1 presents a classification of reptiles including orders, suborders, and families according to Goin et al.[5] For the most part, reptilian classification at the family level and above is stable. There is fluctuation in regard to level of classification of some families and subfamilies, which does not, however, affect an understanding of reptilian chromosomal evolution. Therefore, I present a rather conservative scheme of classification and consider it unproductive to debate the merits of various classifications except where karyology contributes to an understanding of taxonomic problems.

Table 1
THE CLASSIFICATION OF LIVING REPTILES[5]

Class Reptilia
  Order Testudines
    Suborder Cryptodira
      Family Cheloniidae
      Family Dermochelyidae
      Family Emydidae
      Family Testudinidae
      Family Platysternidae
      Family Staurotypidae
      Family Chelydridae
      Family Kinosternidae
      Family Dermatemydidae
      Family Carettochelyidae
      Family Trionychidae
    Suborder Pleurodira
      Family Pelomedusidae
      Family Chelidae
  Order Rhynocephalia
    Family Sphenodontidae
  Order Squamata
    Suborder Amphisbaenia
      Family Amphisbaenidae
      Family Trogonophidae
    Suborder Sauria
      Family Anniellidae
      Family Gekkonidae
      Family Pygopodidae
      Family Xantusiidae
      Family Agamidae
      Family Chamaeleonidae
      Family Iguanidae
      Family Anguinidae
      Family Xenosauridae
      Family Helodermatidae
      Family Lanthanotidae
      Family Varanidae
      Family Dibamidae
      Family Cordylidae
      Family Lacertidae
      Family Scincidae
      Family Teiidae
    Suborder Serpentes
      Family Anomalepidae
      Family Leptotyphlopidae
      Family Typhlopidae
      Family Acrochordidae
      Family Aniliidae
      Family Boidae
      Family Uropeltidae
      Family Xenopeltidae
      Family Colubridae
      Family Elapidae
      Family Viperidae
  Order Crocodilia
    Family Crocodilidae
    Family Gavialidae

II. REVIEW OF REPTILE KARYOLOGY

A. Order Crocodilia

Standard karyotypes for all 21 living species of crocodilians were reported by Cohen and Gans,[6] making this one of the most intensively studied orders of reptiles. Diploid numbers range from 30 to 42 and the number of chromosomal arms ranges from 56 to 62. Cohen and Gans presented a complex diagram indicating the possible paths by which karyotypes of the various species could have evolved.

The data presented by Cohen and Gans[6] have been reanalyzed here to present a much simplified chromosome phylogeny, on the basis of several assumptions. First, the chromosomes are divided into two groups. The five largest biarmed chromosome pairs of *Alligator* (and the presumed biarmed and uniarmed homologues in other genera) make up one group. The 11 smallest chromosome pairs of *Alligator* (and the presumed homologues in other genera) make up the second group. In my analysis I consider which of the first group of chromosomes (pairs 1 to 5) are present as a single biarmed pair or two acrocentric pairs. The second group of chromosomes are merely enumerated and no attempt is made to distinguish between them since their number and small size make it difficult to accurately determine centromere position. For this reason I feel that many of the reported karyotypic differences between certain species are unjustified.[6] For example, *Crocodilus acutus* and *C. johnsoni* supposedly differ from *C. niloticus* and *C. intermedius* in that one of the small chromosome pairs is submetacentric in the former pair of species and metacentric in the latter. However, the differences are so slight as to be highly questionable.

Figure 1 represents a suggested phylogeny of crocodilian karyotypes. The ancestral karyotype of the group was possibly similar to the karyotypes of *Alligator*, *Tomistoma*, and

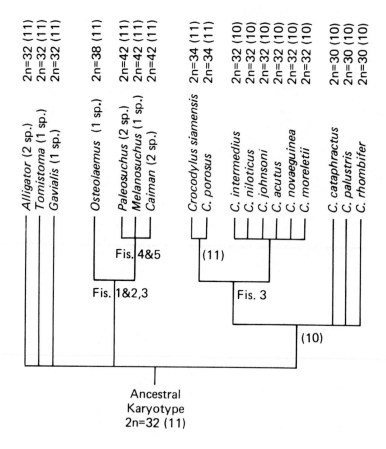

FIGURE 1.   Proposed phylogeny of karyotypes of crocodilians. Chromosome pairs
1 to 5 are macrochromosomes. The number of pairs of microchromosomes is given
in parentheses. A single reversal is the reattainment of 11 pairs of microchromosomes
in *Crocodylus siamensis* and *C. porosus*. The number of microchromosomes is altered
by fissions or fusions. Fis. is fission.

Gavialis (these three being indistinguishable). There are two reasons for considering this
the primitive karyotype: 1. These three genera are not closely related. In fact, Brazaitis[7]
included them in the families Alligatoridae, Crocodilidae, and Gavialidae, respectively. 2.
The karyotypes of all other species of crocodilians can be derived from the proposed ancestral
karyotype with a minimum of only 2 convergent events.

Two major clades can be identified based on the chromosomal data. The first includes
the South American caimans *(Paleosuchus, Melanosuchus,* and *Caiman)* plus the West
African *Osteolaemus*. The data indicate that all four genera share apparent fissions of pairs
1, 2, and 3. In addition, the three South American genera share apparent fissions of pairs
4 and 5. This interpretation of apparent chromosomal fission events suggests that *Osteolaemus*
is the sister group to the South American genera, a relationship alluded to[6] but not currently
recognized in taxonomy. According to Brazaitis,[7] *Osteolaemus* is in the Crocodilidae and
the South American caimans are in the Alligatoridae. The relationship suggested by karyology
is reasonable based on the many faunal similarities between Africa and South America, such
as the distribution of pelomedusid turtles.

The second major clade includes all species of *Crocodylus*. The karyotypes of *C. cataphractus, C. palustris,* and *C. rhombifer* can be derived from the ancestral crocodilian
karyotype by the centric fusion of two small acrocentric chromosome pairs. This fusion
results in a karyotype with $2n = 30$ and 10 pairs of small chromosomes that I propose is
ancestral for the genus *Crocodylus*.

A fission in pair 3 defines a clade that includes the rest of the species of *Crocodylus* divided into two separate groups. *Crocodylus intermedius, C. niloticus, C. johnsoni, C. acutus, C. novaeguinea* and *C. moreletii* all have $2n = 32$ with 10 pairs of small chromosomes. *Crocodylus siamensis* and *C. porosus* have increased the number of small pairs to 11, and the diploid number to 34, presumably by a centric fission event.

This reinterpretation of the data of Cohen and Gans[6] through simplifying assumptions and use of cladistic methodology presents a rather straight-forward and reasonable phylogeny. Though making assumptions about chromosome homology based on standard karyotypes is hazardous,[8] yet it should be possible to test this phylogeny through banding procedures.

It is disappointing that in the 12 years since the appearance of the paper by Cohen and Gans, their work has not been followed with comprehensive chromosome banding analysis. Singh and Ray-Chaudhuri,[9] using autoradiographic techniques, demonstrated the presence of large blocks of late-replicating DNA (heterochromatin) distributed on the five pairs of large biarmed chromosomes of *Crocodylus palustris*. Therefore, the application of the C-band technique may be useful in understanding the pattern of chromosomal variation in Crocodilia. A relatively large amount of heterochromatin may explain why crocodilians have the highest DNA content of all reptiles.[10]

B. Order Testudines

The living turtles are divided into two major groups. The suborder Pleurodira (side-necked turtles) contains two families (about 50 species), limited to the southern continents. Chelids occur in South America and Australia and pelomedusids in South America and Africa. The suborder Cryptodira (hidden-necked turtles) contains 11 families (about 180 species) and primarily is distributed on the northern continents. However, the marine turtles (Cheloniidae and Dermochelyidae) are virtually world-wide in distribution in the marine environment. It can be concluded that the two suborders represent northern (Laurasia) and southern (Gondwanaland) radiations. Some cryptodires are found on the southern continents but no pleurodires occur on the northern ones.

In 1973, Gorman[3] regarded the turtles as the group of reptiles most neglected by cytotaxonomists. This situation has changed in the past decade and turtles are now, in many respects, the best-studied group of reptiles and the only one for which an extensive body of chromosomal banding data exists.

Bull and Legler[11] presented karyotypes of 13 of the 14 genera and reviewed all the literature on pleurodiran karyotypes. The most striking feature of the pattern of karyotypic variation in the side-necked turtles is a dichotomy of diploid numbers between chelids ($2n = 50$-64) and pelomedusids ($2n = 26$-36) (see Figure 2). Within the pelomedusids, there appear to be distinctive radiations in South America and Africa. The South American species are all $2n = 26$-28. The African species are all $2n = 34$-36, except *Podocnemis (Erymnochelys) madagascarensis* from Madagascar, which is $2n = 28$.[12] This species is, however, more closely related to the South American forms than to the African species. It usually is considered congeneric with the South American *Podocnemis*, although Rhodin et al.[12] suggest it is distinct at the generic level.

Chelids also appear to represent two distinctive radiations, one in Australia and one in South America. The Australian one consists of a series of "short necked" genera, all with $2n = 50$, and a "long-necked" genus *(Chelodina)* with $2n = 54$. The South American radiation has several genera with diploid numbers ranging from 58 to 64 and at least one *(Chelus)* with $2n = 50$. Banding comparisons are needed to determine the relationship between *Chelus* and the "short-necked" genera of Australian chelids.

Chromosomal banding data are available for a few species of pleurodires.[11] G-bands for three genera of pelomedusid turtles, *Pelusios* (Africa), *Pelomedusa* (Africa), and *Podocnemis* (South America) show a high degree of homology among the macrochromosomes of the

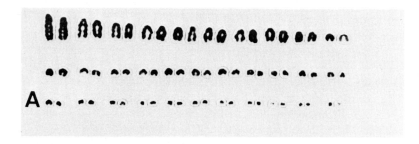

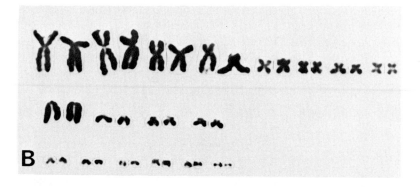

FIGURE 2.   (A) Karyotype of a chelid turtle, *Platemys platycephala,* $2n = 64$, the highest diploid number in the family with an all acrocentric complement; (B) karyotype of a pelomedusid turtle, *Pelomedusa subrufa,* $2n = 36$; pelomedusids have low diploid numbers and no overlap in the range of diploid numbers with chelids.

three genera. Each genus, however, possesses distinctive karyotypic features. C-bands of the same animals show both centromeric and intercalary bands of heterochromatin on many of the larger chromosomes. C-band data for several chelid genera and species show some variation in amount and distribution of heterochromatin.

A large body of chromosomal data is available for the suborder Cryptodira, including both standard and chromosome banding data. Recent reviews with details of chromosome data are by Bickham[13] and Bickham and Carr.[14] Over half of the 180 species of Cryptodira have been studied for standard karyotypes,[14] and chromosome banding data are available for about 28%.

The pattern of karyotypic variation in cryptodires is basically conservative. Generally, all or most of the species within a family or subfamily appear karyotypically identical. However, comparisons between families reveal appreciable amounts of variation.[13,15]

Unexpectedly, sea turtles (Cheloniidae) appear karyotypically identical to the Central American river turtles *(Dermatemys mawii),* possessing a karyotype with $2n = 56$. Both cheloniids and dermatemydids are among the most ancient of living turtle lineages and have fossil records extending back to the Cretaceous.[18] Although both families possess certain primitive characteristics, they are morphologically quite distinct from one another. All available evidence indicates an early evolutionary divergence between them. It is concluded that the karyotypic similarity between the two families is the result of their having retained a primitive karyotype. The $2n = 56$ karyotype (Figure 3) is considered to be primitive for the suborder Cryptodira[13-15,17] although a karyotype like that of trionychids could instead be the ancestral condition.

Members of the families Trionychidae and Carettochelyidae are the most karyotypically divergent cryptodiran turtles. Whereas all other taxa have diploid numbers in the 50 to 56 range, trionychids have $2n = 66$ and carettochelyids have $2n = 68$.[19] Chromosome banding comparison between *Trionyx* and *Chelonia* (which possesses the presumed primitive kary-

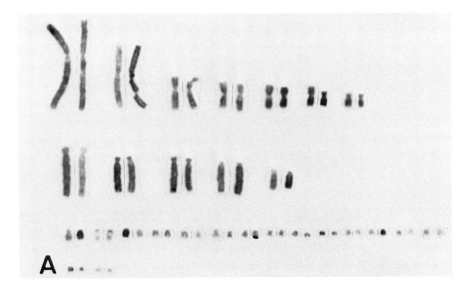

FIGURE 3. Karyotype of the green turtle, *Chelonia mydas*, $2n = 56$, the proposed primitive karyotype of cryptodiran turtles.

otype for all non-trionychoid cryptodires[14]) indicates relatively few homologous elements shared between the two genera. *Chelonia* has 12 pairs of macrochromosomes, only 4 of which are identical to those of *Trionyx*. This karyotypic difference is consistent with the taxonomic placement of trionychids and carettochelyids in the superfamily Trionychoidea, distinct from all other cryptodires.

Although species in different families usually differ in karyotype, those of some emydids and some testudinids are identical.[14,20] These two families are obviously closely related[18,21-22] and have been considered by some authors to be confamilial,[23,24] although most recent classifications consider them distinct at the family level. The emydids are divided into two subfamilies, Emydinae and Batagurinae, that are chromosomally distinct (with a few exceptions). The batagurines are considered the most primitive emydid group,[23] and the testudinids are thought to have evolved from a batagurine ancestor.[25,26] Karyological data support this conclusion.[14,20,27,28] Members of the testudinid genus *Geochelone* (and some other genera) have karyotypes which appear identical to species of several batagurine genera (such as *Chinemys, Sacalia,* and others). G-band comparisons revealed no discernable differences between the macrochromosomes of *Geochelone elongata* and *Chinemys reevesii*[20] and this karyotype ($2n = 52$) is considered to be the primitive one for both families.[20,28-30] Based on the geographic distribution of testudinids and emydids possessing the primitive karyotype, Dowler and Bickham[20] recognize that an Asian origin of the Testudinidae is consistent with the karyological data, although an American or European origin cannot be ruled out entirely.

The chromosomes of kinosternid and staurotypid turtles have been studied by several workers.[31-35] All three staurotypid species possess $2n = 54$.[34] Nearly all the species in the Kinosternidae (including only *Kinosternon* and *Sternotherus*) have been studied, giving $2n = 56$.[14] Sites et al.[35] found variation in amount and distribution of heterochromatin. *Kinosternon bauri, K. subrubrum,* and *Sternotherus minor* all possess heterochromatic blocks that stain positive by both the C-band and the G-band techniques. This heterochromatin was absent entirely in *K. scorpioides* and its distribution varied among the three species that possessed it. Heterochromatin was found on the short arms of three acrocentric macrochromosomes in *K. bauri,* but only on numerous microchromosomes in *K. subrubrum* and *S. minor.*

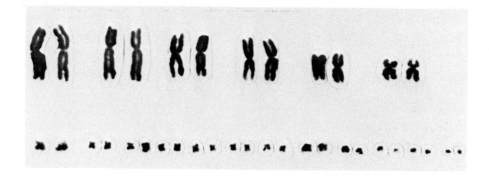

FIGURE 4.   Karyotype of an iguanid lizard, *Crotaphytus collaris*, 2n = 36, 12 macrochromosomes and 24 microchromosomes (12 + 24) proposed to be the primitive lizard karyotype.[3]

The nature of the variation of heterochromatin in kinosternids has both systematic and zoogeographic implications. Its absence in *K. scorpioides,* the only Neotropical species studied, is probably the primitive condition. G-band positive (A-T rich) heterochromatin is rare in cryptodiran turtles. Accordingly, the center of origin of the family could have been in Central America or Mexico. The presence of heterochromatin (the derived condition) in all three species from the southeastern U.S. suggests that they share a common ancestor. The similarity of the distribution of heterochromatin in *K. subrubrum* and *S. minor* implies a relationship between them and suggests that the genus *Sternotherus* may have evolved from a southeastern U.S. species of *Kinosternon*. However, until the primitive and derived conditions regarding the distribution of heterochromatin are known, no conclusions can be made regarding the relationships of the three southeastern U.S. species of kinosternids.

C. Order Rhynchocephalia

The only living member of this order, the tuatara of New Zealand, has a diploid number of 36.[3] Although this number is common in snakes and lizards, the chromosome morphology of the tuatara karyotype is quite distinct. All but three or four of the smallest pairs could be considered macrochromosomes. No sex chromosome dimorphism was found.

D. Order Squamata
1. Suborder Sauria

The lizards are the most speciose group of reptiles and have received the maximum attention from cytosystematists. The taxonomic arrangement of the group comprises 16 families. Some of them have been intensively studied but others are in need of considerable work. In spite of their abundance, ease of collection, and presence of karyotic diversity, there have been virtually no studies applying modern cytogenetic procedures, such as chromosome banding, to lizards.

The Iguanidae is one of the most diverse families of lizards and several patterns of karyotypic variation have been discovered within this group. The primitive karyotype (Figure 4) appears to be 2n = 36 with 12 biarmed macrochromosomes and 24 microchromosomes (the 2n = 36 12 + 24 karyotype[36]). This karyotype is known to occur in numerous iguanid genera including morphologically both primitive and derived taxa. Paull et al.[36] offer a cogent argument for accepting this karyotype as primitive for the family, with which I agree. An alternative hypothesis is that the biarmed macrochromosomes have originated through the process of Robertsonian fusion.[37,38]

Paull et al.[36] suggest there are two major patterns of karyotypic variation in iguanids. The first pattern is one of conservatism. Many iguanid genera share identical karyotypes or show a very low level of karyotypic diversity. In almost all cases these genera have relatively few species and there is little geographic overlap, as in *Basiliscus, Ctenosaura, Cyclura, Phrynosoma,* and many others.

The second pattern of variation is seen in the iguanid genera *Sceloporus, Anolis* and *Liolaemus.* These are all highly speciose genera in which many species differ in karyotype. Often, several congeneric species occur sympatrically without interbreeding. This pattern has been interpreted to be the result of chromosomal rearrangements being incorporated in contiguous populations and acting as reproductive isolating mechanisms.[3,36,39] The resulting chromosomal, or stasipatric, speciation provides a proliferation of biological species at a rate much faster than the usual geographic mode thought to characterize the karyotypically conservative genera and species mentioned above.

Two of the three karyotypically diverse iguanid genera *(Sceloporus* and *Anolis)* are well studied. There are approximately 200 species in *Anolis,* karyotypes of about 80 of which are known.[3,36] The diploid numbers range from $2n = 25$ to 48. Chromosomal variation is common within and among species groups. For example, *A. monticola* has a karyotype ($2n = 48$) which can be derived from closely related species with the $2n = 36$ primitive karyotype by a series of centric fissions.[40]

Such centric fissions, as have occurred in the evolution of the *A. monticola* karyotype, are the exception rather than the rule. Most *Anolis* species that differ from the primitive $2n = 36$ $12 + 24$ karyotype have lowered the diploid number by reducing the number of microchromosomes.[3] For example, the Beta anoles have a "typical" karyotype of $2n = 30$ with 14 biarmed macrochromosomes and only 16 pairs of microchromosomes. Therefore, the fusion of microchromosomes is an important mechanism in karyotypic evolution of *Anolis.*

An exemplary study on *Anolis* by Gorman and Atkins[41] illustrates the degree of karyotypic diversity in the genus. There are two major species groups of *Anolis* inhabiting the Lesser Antilles. The *roquet* group occupies the southern part of this island chain and is found from Grenada, the southernmost island, north to Martinique. The *bimaculatus* group is found from Dominica through the rest of the island chain and also on some of the Greater Antilles.

The two species groups differ in karyotype as well as osteology and electrophoretic mobility of lactic dehydrogenase (LDH). Two karyotypes are found in the *roquet* group; the primitive iguanid karyotype ($2n = 36$ $12 + 24$) and $2n = 34$ $12 + 22$. There is no male sex chromosome heteromorphism. The *bimaculatus* group species have derived karyotypes (reduced $2n$) and male heterogamety and can be divided into three subgroups based on chromosomes and LDH.[41] The *acutus* group is found west of the Lesser Antilles; the *bimaculatus* and *wattsi* groups are restricted to the Lesser Antilles. The *acutus* group may be karyotypically intermediate between the primitive *roquet* group and the more highly derived *bimaculatus* and *wattsi* groups.

Two invasions have been visualized of the Lesser Antilles by *Anolis.*[41] One invasion came from South America and proceeded north along the southern part of the island chain (*roquet* group). The other invasion was from the Greater Antilles south along the island chain (*bimaculatus* group). There is no geographic overlap between the species groups presumably because of competitive exclusion.

The genus *Sceloporus* has been intensively studied by several workers, notably C. J. Cole, W. P. Hall, and J. W. Sites, Jr. The most extensive discussion of the pattern of karyotypic variability was presented by Hall[39] and reviewed recently by Paull et al.[36] The family Iguanidae originated in the Cretaceous. However, the sceloporine genera are thought to be relatively young, no older than the Miocene, and probably evolved concomitant with the development of the North American deserts. The genus *Sceloporus* is probably the most recently derived of the sceloporine genera because it possesses certain derived morphological characters not found in any of the closely related genera. There seem to be two distinct

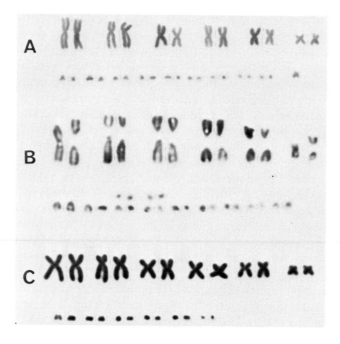

FIGURE 5.   Chromosomal variation in the genus *Sceloporus*. (A) The
$2n = 32$ cytotype of *S. grammicus* has a karyotype similar to that
proposed as primitive for the genus ($2n = 34$). Figured here is the
karyotype of a male with $2n = 31$, with an unpaired small metacentric
chromosome and an X_1X_2Y sex chromosome complement. (B) Evolution in *S. grammicus* has occurred through centric fission. The karyotype figured is from a ♀ representing the most highly derived race
($2n = 44$). All macrochromosomes (except two pairs heterozygous for
fissions) and many microchromosomes have apparently undergone centric fission relative to the $2n = 32$ cytotype. (C) Karyotype of *S.
goldmani*, of the *scalaris* group of small-scaled species $2n = 24$,
presumably evolved from a $2n = 34$ ancestor by the reduction (fusion)
of microchromosomes.

lineages within the genus, the more primitive small-scaled, small-sized species and the
highly derived large-scaled, large-sized species. The small-scaled lineage is composed mostly
of species with the presumed primitive *Sceloporus* karyotype ($2n = 34$, $12 + 22$). Among
this group the most karyotypically highly derived species groups (the *scalaris* (Figure 5C)
and *merriami* groups) also are highly differentiated ecologically. The large-scaled lineage
is composed of species that are karyotypically derived, with only *S. orcutti* known to have
the primitive $2n = 34$ karyotype. Within this genus, as within the family as a whole,
karyotypic evolution rather closely parallels morphological evolution.

The most variable species of *Sceloporus*, in terms of the karyotype, is *S. grammicus*
(Figure 5).[39,42] Diploid numbers vary from $2n = 32$ to $40+$.

Another iguanid genus, *Polychrus*, is presumably an ancient South American group of
arboreal lizards comprised of six species. Diploid numbers range from $2n = 20$ to 30. The
karyotypes of all species are highly divergent from the primitive iguanid karyotype as well
as from one another. Paull et al.[36] conclude that this genus might represent the few survivors
of a once diverse lineage, mostly replaced by the more progressive *Anolis* radiation.

Next to iguanids, the teiids are the karyologically best-known family of lizards. Teiidae
is restricted in distribution to North and South America with the center of diversity being
South America (a rich North American fossil record may indicate South America is not the

center of origin). Karyological studies are consistent with morphological studies that have divided the family into two major groups, the macroteiids and the microteiids.[43] Among the macroteiids, two subgroups are apparent. The *Dracaena* group has diploid numbers in the $2n = 34$ to 38 range. Karyotypes of this group[43] are easily derivable from the proposed primitive lizard karyotype ($2n = 36$ $12 + 24$). For example, *Crocodilurus lacertinus* has a $2n = 34$ $12 + 22$ karyotype that could have evolved through the loss of one pair of microchromosomes.

The *Ameiva* group contains species with diploid numbers in the $2n = 46$ to 56 range. Gorman[43] concludes that a karyotype of $2n = 50$ with 26 acrocentric macrochromosomes and 24 microchromosomes, like that of *Kentropyx striatus* is primitive for the group, which could have evolved from a *Dracaena* group ancestor primarily through centric fission.

The *Ameiva* group contains the diverse and chromosomally variable genus *Cnemidophorus*, studied intensively by C. H. Lowe and his colleagues. Variation within the genus is found primarily at the species group level. Lowe et al.[44] recognize five species groups: *tigris, deppei, sexlineatus, lemniscatus,* and *tessalatus.* Fritts[45] recognizes another species group, the *cozumela* complex, which is also karyotypically distinct. Species within each group usually have similar karyotypes, although they may differ by centromere placement in some chromosomes.[45] Differences among species groups are due mostly to centric fusions. Lowe et al.[44] conclude that the *deppei* group karyotype ($2n = 52$ with 28 acrocentric macrochromosomes and 28 microchromosomes) is primitive for the genus. Gorman[43] believes that the *lemniscatus* group karyotype is primitive because of its near identity to the karyotype of *Ameiva ameiva.* The question of the primitive karyotype is somewhat clouded by the karyotype of *C. lacertoides* which is more similar to that of *Ameiva* and *Kentropyx* than to any other *Cnemidophorus.*[47] Several species of *Cnemidophorus* are known to be parthenogenetic.

The Old World family Lacertidae is considered to be closely related to the Teiidae but the pattern of karyotypic variation is quite different. Most lacertid species have a $2n = 38$ karyotype composed of a graded series of acrocentric elements (including two microchromosomes). This karyotype is found in numerous species and genera.[3] The only known variation in the family is in *Lacerta* in which most species are $2n = 38$,[3,48,49] but some populations of *L. vivipara* have $2n = 36\female, 35\male$[49] and *L. parva* has $2n = 24$ with 14 biarmed and 10 acrocentric chromosomes.

The family Agamidae occurs in Australia, Africa, and Eurasia and is considered closely related to the iguanids. Considering the diversity of the group, it has been poorly studied. The primitive $2n = 36$ $12 + 24$ lizard karyotype is found in several genera and is presumed primitive for the family.[3,50] Karyotypes with $2n = 46$ or 48, in which the primitive biarmed macrochromosomes have apparently undergone centric fission, frequently are also found.[50-53] Species with diploid numbers of $2n = 30$,[50] 32,[54] and 34[55] also are known and apparently have evolved through the loss or fusion of microchromosomes.

The family Agamidae occurs in Australia, Africa, and Eurasia and is considered closely related to the iguanids. Considering the diversity of the group, it has been poorly studied. The primitive $2n = 36$ $12 + 24$ lizard karyotype is found in several genera and is presumed primitive for the family.[3,50] Karyotypes with $2n = 46$ or 48, in which the primitive biarmed macrochromosomes have apparently undergone centric fission, frequently are also found.[50-53] Species with diploid numbers of $2n = 30$,[50] 32,[54] and 34[55] also are known and apparently have evolved through the loss or fusion of microchromosomes.

The family Cordylidae includes two subfamilies, Cordylinae and Gerrhosaurinae (Gerrhosauridae of some classifications). A diploid number of $2n = 34$ with $12 + 22$ characterizes most species in both subfamilies for which karyotypes are known.[56] Matthey[57] reported $2n = 36$ $12 + 24$ for *Gerrhosaurus flavigularis.* If this report is accurate, then $2n = 36$ could be the primitive karyotype of the group. Most likely, however, it is inaccurate and $2n = 34$ is primitive as Olmo and Odierna[56] believe. This family may be closely related to the

Teiidae. The latter may also have had a $2n = 34$ karyotype as ancestral and is found today in *Crocodilurus*. A karyotype with $2n = 44$ characterizes *Cordylus giganteus* and *C. cataphractus* and seemingly results from the centric fission of 5 macrochromosomal pairs.

The family Xantusiidae, restricted to North and Central America and Cuba, is of uncertain taxonomic relationship. Karyotypes of 10 species, ranging in diploid number from 36 to 40, were reported by Bezy.[58] Intraspecific chromosomal variation was found among populations of *Xantusia vigilis* as well as *X. henshawi*. The primitive karyotype of the family was considered to be the alpha karyotype of *X. vigilis* ($2n = 40$ with 18 macrochromosomes and 22 microchromosomes). The karyotypic affinities of the family seem to be with the family Teiidae, particularly the microteiids.

The family Varanidae is characterized by a diploid number of 40. Karyotypic variation is restricted to variable centromere placement in three pairs of macrochromosomes (acrocentric, submetacentric, metacentric) and two pairs of microchromosomes,[59,60] occurring as a result of pericentric inversions.[59] Intraspecific variation (inversion polymorphism) was reported by Singh[55] for *Varanus bengalensis*. Karyotypic variation is useful in identifying species groups in *Varanus*. Six different groups were recognized based on karyotype,[59] suggesting that the subgenus *Varanus* may be a composite taxon because it includes species from three different groups as defined by karyotype. The genus *Varanus* is suggested to have originated in Asia with subsequent radiations in Australia and an invasion of Africa by way of the Middle East. This phylogeny is consistent with biochemical data.[60]

The family Gekkonidae is found in both the New and Old World and is the second largest family of lizards (approximately 650 species). The "typical" gecko karyotype has many acrocentric chromosomes and there is no break between macrochromosomes and microchromosomes. Relatively little is known about the karyotypes of members of this family except the Australian species.[61-66] A great deal of intraspecific chromosomal variation (in the form of chromosomal races) has been reported for Australian geckos (see King[67]). Species for which chromosomal races are known include *Diplodactylus tessellatus*, *D. vittatus*, *Phyllodactylus marmoratus*, *Gehyra variegata*, *G. punctata*, *G. australis*, and *Hemidactylus frenatus*.

Diploid number variation is marked in geckos, ranging from $2n = 32$ to 46. Robertsonian fusions/fissions as well as pericentric inversions have played a role in the karyotypic evolution. Because of the high degree of variability, this family holds a great deal of promise for cytogenetic research.

The relationship of the Gekkonidae to other lizard families is uncertain. The karyotypic characteristics of geckos (no break between microchromosomes and macrochromosomes and karyotypes rich in acrocentrics) differentiate them from nearly all other families. However, the family Pygopodidae of the Australia and New Guinea region is similar to geckos in these respects.[68] These two families are usually placed together in the same infraorder. Some lacertids also are similar in karyotype but are not considered closely related.

The family Scincidae is the largest family of lizards and one of the most poorly sampled for karyotypes. In general, the karyotypes of skinks are relatively conservative. Diploid numbers vary from $2n = 24$ to 32. Karyotypes are usually divisible into macrochromosomes and microchromosomes, but not clearly. The presence or absence of microchromosomes has been argued by some authors.[3,61] Karyotypes of closely related species usually appear identical.[61] Intrageneric variation is known in *Mabuya*[55] and intraspecific variation in *Scincella laterale*.[69] King[61] noted karyotypic stability in Australian skinks but the members of the subfamilies Lygosominae and Scincinae were characterized by $2n = 30$ and $2n = 32$, respectively.

The family Anniellidae consists of one genus and two species in western North America; the karyotypes of both species are known.[70] *Anniella geronimensis* has $2n = 36$ and *A. pulchra* two races with $2n = 22$ and $2n = 20$. Supernumerary microchromosomes occur in *A. p. pulchra*. The relationship of Anniellidae to other lizard families is uncertain[70] but they are usually considered closely related to the Anguidae. Karyotypes do not resolve this uncertainty because the karyotype of *A. geronimensis* is similar to that of certain anguids but that of *A. pulchra* is similar to that of varanids.[70]

Karyotypes of only ten of the approximately 67 species of Anguidae are known.[3,71] The diploid number varies from $2n = 30$ to 48. *Diploglossus costatus*, a member of the primitive diploglossines, has the primitive $2n = 36$ $12 + 24$ lizard karyotype.[3] Chromosomal variation in the family can be attributed to Robertsonian fissions of macrochromosomes as well as both reduction and increase in the numbers of microchromosomes.

The family Helodermatidae has 2 species, one of which has been karyotyped.[3] *Heloderma suspectum* has $2n = 38$ and differs from the primitive lizard karyotype ($2n = 36$ $12 + 24$) by a Robertsonian fission of one pair of macrochromosomes.

The Chamaeleonidae has a range of diploid numbers of $2n = 20$ to 36.[3,72] Several species (*Chamaeleo* and *Brookesia*) have the primitive $2n = 36$ $12 + 24$ lizard karyotype. Variation is accounted for by reduction of the number of microchromosomes with a corresponding increase in the number of small macrochromosomes.

2. Suborder Amphisbaenia

Karyotypes are known from species of both of the two families of amphisbaenians.[73-75] Both species studied of Trogonophidae have the typical lizard karyotype ($2n = 36$ $12 + 24$) as do some species of Amphisbaenidae. This is evidence for a close relationship between lizards and amphisbaenians.

Diploid number variation in Amphisbaenidae ranges from $2n = 30$ to 50,[3] some but not all of which can be accounted for by Robertsonian fissions. Pericentric inversions may also be involved in the chromosomal evolution of the group.[3]

MacGregor and Klosterman[75] observed $2n = 42$ and 44 in the karyotypes of two species of *Bipes (B. biporus* and *B. canaliculatus,* respectively). Details of the karyotype differ in centromere position of three pairs of macrochromosomes and in the number of macrochromosomal pairs. These authors also describe the lampbrush chromosomes found in the two species and report DNA content. The diploid number for *B. biporus* ($2n = 42$) differs from $2n = 40$ reported by Huang et al.[73]

3. Suborder Serpentes

Snakes have received much less attention from cytosystematists than have lizards. Nonetheless, the body of data from standard karyotypes is adequate to suggest general patterns of karyotypic variation in this important group. In addition, there are several studies using modern cytogenetic procedures that are very informative with reference to the accuracy of conclusions based on the standard karyotype data.

Over half of the snake species studied possess karyotypes of $2n = 36$ with 16 macrochromosomes and 20 microchromosomes. This arrangement has been proposed as the primitive karyotype of all snakes[3,76,77] and it has been found in many species of "lower snakes" including members of the families Leptotyphlopidae, Xenopeltidae, and Boidae. It is also found in many of the "higher snakes" including the families Colubridae, Acrochordidae, and Viperidae. Within the colubrids and viperids, the $2n = 36$ karyotype occurs within each subfamily for which specimens have been studied.[3] This karyotype has not been found in any species of the higher snake family Elapidae.[78] There are no karyotypic data for the families Anomalepidae, Aniliidae, and Uropeltidae.

Except for the Boidae, the lower snakes have not been well studied. Two species of *Typhlops* (Typlopidae) have $2n = 32$[3,79] and one each of *Leptotyphlops* (Leptotyphlopidae) and *Xenopeltis* (Xenopeltidae) have $2n = 36$. Within the Boidae, diploid numbers range from $2n = 34$ to 44 with $2n = 36$ the most common.[3,79,80]

Within the higher snakes, the extremely diverse Colubridae have been studied by numerous investigators.[3,4,77,79,81-83] Diploid numbers range from $2n = 24$ to 50, $2n = 36$ being the most commonly observed.

Karyotypes have proven useful in the taxonomy of colubrids. *Natrix,* formerly considered to comprise species from North America, Europe, Asia, and Africa, was partitioned into four genera by Rossman and Eberle[81] based on a study of karyotypes, blood proteins, scutellation, and cranial osteology. Two Asian species *(Sinonatrix)* have $2n = 42$, three European species *(Natrix)* have $2n = 34$, and the American species *(Nerodia)* have $2n = 36$. No African species *(Afronatrix)* were studied. The American *Nerodia,* along with the other New World natricine genera, are now placed in the tribe Thamnophiini.[81]

The family Viperidae includes two subfamilies (Crotalinae, pit vipers, and Viperinae, vipers). All crotalines thus far studied have $2n = 36$.[3,77,84,85] Viperines vary from $2n = 36$ to 42,[3,4,79] mainly due to Robertsonian fissions.[86]

The family Elapidae includes the sea snakes (Hydrophiinae, sometimes considered a family), coral snakes, cobras, and their relatives. None have yet been reported to possess the typical snake $2n = 36$ karyotype.[3,4] The diploid number varies from $2n = 32$ to 44. The number of microchromosomes varies from 16 to 22. The number of macrochromosomal arms varies from 20 to 24 but most have 20.[78] Centric fission plays a role in karyotypic variation in the sea snakes and both fissions and pericentric inversions are important in terrestrial elapids. The karyotypes of sea snakes and terrestrial elapids have many similarities.[78]

The proposal that the $2n = 36$ karyotype, found in the majority of snake species for which karyotypes are known, is the ancestral karyotype[3,76] assumes a high degree of chromosomal homology among snake species possessing this karyotype. Chromosomal banding data from five species of Boidae and three species of Colubridae[80] suggest this assumption may be unwarranted. There is little direct homology between boids with $2n = 36$ and a colubrid with $2n = 36$.[80] However, there is complete homology of chromosome 1 and the long arm of 2 indicating these elements have been conserved for a considerable period of time.[80] Significant chromosomal variation was also found within each family and even within genera suggesting that snake karyotypes are not as conservative as many workers have assumed based on studies of standard chromosome morphology.

Mengden and Stock[80] suggested that the primitive snake karyotype probably possessed a diploid number higher than $2n = 36$, a larger number of acrocentrics and no sharp distinction between macrochromosomes and microchromosomes. If this suggestion is true, then numerous snake lineages have independently evolved $2n = 36$ karyotypes that are only superficially similar. This finding is one of the most important advances in snake cytosystematics.

Mengden and Stock[80] identified a number of different types of chromosomal rearrangements which occurred during the evolution of the snake karyotypes. These included Robertsonian fusions and fissions, inversions, changes in amount and distribution of heterochromatin, and an ostensible case of centric transposition. They also demonstrated the conversion of microchromosomes to macrochromosomes by means of heterochromatic addition.

III. SEX DETERMINATION

There are two general modes of sex determination in reptiles. In *environmental sex determination* the sex of a juvenile is determined by a particular environmental parameter encountered during some critical developmental state. Certain species of reptiles possess temperature-dependent sex determination (TSD) including crocodilians, lizards, and turtles.[87] Under this system nest temperature determines the sex of the developing embryo.[87-93] *Genotypic sex determination* is a system in which the sex of an individual is (usually) irreversibly determined by some gene or genes inherited from either parent, at the moment of fertilization.

The adaptive significance of these two systems and their evolution in reptiles is discussed by Bull.[87] In this chapter, genotypic sex determination and specifically evolution of sex chromosomes are discussed.

Many, but not all, species of reptiles that possess genotypic sex determination have morphologically differentiated sex chromosomes. Bull and Vogt[88] demonstrated the absence of TSD in a trionychid turtle (*Trionyx spiniferus*). This species, and others in the genus *Trionyx,* do not have identifiable sex chromosomes based on either standard chromosome morphology or chromosomal banding comparisons.[19] All available data, as well as theoretical considerations, indicate that species with known TSD do not possess sex chromosomes and that species with sex chromosomes do not also show TSD.[87] However, under certain conditions the two systems may not be mutually exclusive.[87] In general, however, when sex chromosomes are found, genotypic sex determination can be assumed, but not vice versa.

There are two general types of sex chromosome systems. The male heterogametic type is termed XX♀/XY♂ and the female heterogametic type ZZ♂/ZW♀. Both are known in reptiles (for reviews see Bull[87] and Peccinini-Seale[4]). Female heterogamety is found in some lizards and most snakes. Male heterogamety is found in two genera of turtles and some lizards. Multiple sex chromosome systems, resulting from sex chromosome-autosome translocations, are known in both lizards and snakes.

The generalized model of sex chromosome evolution proposed by Muller[94] and Ohno[95,96] assumes that sex chromosomes have evolved from undifferentiated autosomes that carried the sex determining genes. A chromosomal rearrangement, such as a pericentric inversion that includes the sex determining genes, isolates part of the Y or W chromosome and prevents crossing over. Because the Y or W is never in the homozygous condition, it is free to accumulate deleterious mutants and gradually becomes heterochromatic as it loses function not associated with sex determination. Dosage compensation would regulate the expression of genes on the X or Z chromosomes in the homozygous state. Highly developed sex chromosome systems would manifest gross differences in size and banding patterns between the X and Y or Z and W. Sex chromosome systems in the early stages of development would exhibit an overall similarity, particularly in size, of the sex chromosomes.

Sex chromosomes in both early and late stages of evolution are found in reptiles. This, along with the presence of TSD in some, makes reptiles an ideal group in which to study the phenomenon of sex chromosome differentiation. In fact, both early and late stages of differentiation are found in snakes, and the degree of chromosomal heteromorphism increases with degree of morphological evolution.[95] The primitive snakes, such as boids, mostly lack dimorphic sex chromosomes. In the single known case of sex chromosome heteromorphism in a boid (*Acrantophis dumereli*), the Z and W differ only by a pericentric inversion.[80] Advanced snakes, such as viperids and elapids, show distinct differences supporting Muller's hypothesis of sex chromosome differentiation.

Similar support can be found among other reptiles. In lizards both female and male heterogamety is found, as well as numerous species with no dimorphic chromosomes. Male heterogamety is known in iguanids, lacertids, teiids, scincids, and pygopodids and female heterogamety in gekkonids, lacertids, and varanids. Sex chromosomes likely have evolved numerous times in lizards and probably are all of relatively recent origin. The only studies of lizard sex chromosomes using banding techniques[62,66,97] indicate they are in the early stages of differentiation.

Sex chromosomes are known in two genera of turtles. Both *Staurotypus*[34,98] and *Siebenrockiella,*[99] studied using banding procedures, seem to be in early stages of differentiation (Figure 6). This is expected according to Muller's model because the taxonomic distribution of sex chromosomes in turtles, as in lizards, suggests recent phylogenetic origin. However, the nature of the rearrangements in both turtle sex chromosome systems, is not precisely as predicted by the model. In *Staurotypus*, the X has differentiated and the Y remained un-

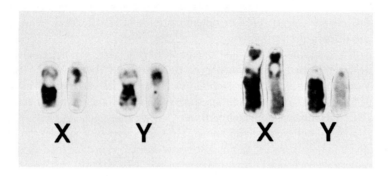

FIGURE 6.   Sex chromosomes of two species of turtles, *Siebenrockiella crassicolis*, the X and Y on the left, and *Staurotypus salvinii,* the X and Y on the right. Each chromosome is shown for G-bands (left member of each pair) and C-bands. The sex chromosomes of *Staurotypus* are not homologous with those of *Siebenrockiella* and different types of chromosomal rearrangements were involved in differentiation of the two sex chromosome systems.

changed.[98] In *Siebenrockiella*, the entire short arm of the X has become heterochromatic but the Y has not.[89] These data do not mean that Muller's model is necessarily incorrect, but suggest that some sex chromosome systems are evolving according to different constraints. Interestingly, the sex chromosomes of the two turtle genera are on different (non-homologous) chromosome pairs, indicating that genotypic sex determination has evolved independently involving different chromosomes and perhaps even different genetic systems.

The most interesting data on reptile sex chromosomes are the biochemical and chromosomal banding studies of snakes.[80,100-102] The W chromosomes of colubids and elapids are heterochromatic[80,102] and possess a W chromosome specific satellite DNA.[100,101] The W chromosomes of snakes (and birds), although entirely heterochromatic, possess a distinct banded appearance. This suggests a heterogeneous nature of the snake and bird W heterochromatin with some regions more highly repetitive than others.[102] This supports the proposal that heterochromatinization is an important mechanism in sex chromosome differentiation. It has even been suggested that sex chromosomes can differentiate merely by the heterochromatinization of euchromatin on the Y or W.[100,102] However, the evidence for this is not compelling because there are no examples of this in an early stage of differentiation.

To summarize, reptile sex chromosomes are in early stages of differentiation (with the exception of advanced snakes). This affords an opportunity to study the evolution of sex chromosome systems, particularly in species with male heterogamety. Such studies should include meiotic analyses (particularly pachytene pairing), chromosomal banding analyses, and biochemical studies of satellite DNA (where appropriate).

IV. PARTHENOGENESIS AND POLYPLOIDY

Parthenogenesis in reptiles recently was reviewed by Cole.[103] He listed 26 species of lizards and one snake, known or suspected of possessing all-female populations. The snake is a typhlopid and the lizards are representatives of the families Teiidae, Lacertidae, Xantusiidae, Gekkonidae, Agamidae, and Chameleonidae. An iguanid has subsequently been added to the list.[104]

Most of these species are entirely parthenogenic but a few consist of both bisexual and all-female populations. The latter include *Cnemidophorus lemniscatus,*[105,106] *Lepidophyma flavimaculatum,*[58] and *Basiliscus basiliscus.*[104] Many of the parthenogenetic species have evolved through interspecific hybridization,[44,46,103,107-113] some clearly have not,[58,104] and others need to be studied further.[103,113]

There is an extensive body of literature on the ecology, genetics, cytogenetics, distribution, and histocompatibility of parthenogenetic lizards. Two theories have emerged to account for the evolution of parthenogenesis. The first is evolution through hybridization.[103] Under this model, bisexual species hybridize in an ecotone between the habitats occupied by the two parental species. Hybrids are reproductively less fit but may enjoy phenotypic heterosis. Female hybrids that reproduce parthenogenetically enjoy both phenotypic heterosis and greater reproductive success than either bisexual hybrids or parentals. Thus, parthenogenetic clones are established in ecotonal habitats.[114] Hybridization between diploid parthenoforms and a bisexual species results in triploids that may also reproduce parthenogenetically. Both diploid and triploid parthenogenetic species are known in *Cnemidophorus* and one species *(C. tessalatus)* has both diploid and triploid populations.[115] *Cnemidophorus tessalatus* has evolved through the hybridization of *C. tigris* with *C. septemvittatus* to produce the diploid parthenoform. Subsequent hybridization between *C. tessalatus* and *C. sexlineatus* resulted in the triploid *C. tessalatus* populations.

Under the hybridization model, parthenogenetic reproduction results somehow from the act of hybridization. An alternative model is evolution of parthenogenesis without hybridization,[103,116] as the result of a fortuitous gene mutation or gene combination within a population of an ordinary, bisexually reproducing species. Under ecological conditions, such as low population density or situations that favor dispersers,[58] parthenogenetic females possess an advantage because of higher reproductive rate and the ability of a single individual to found a population. Cuellar[116] suggested that parthenoforms must evolve in isolation from bisexual species to protect them from competition and hybridization. Thus, parthenoforms only occur in disclimax habitats created by fire, glaciers, or flooding and devoid of the bisexual species. However, some are island forms that have taken advantage of the particular characteristics of parthenogenetic reproduction that allow them to be good colonizers (high intrinsic rate of population increase and the ability of one individual to found a population).

Cuellar's[116] model discounts the importance of hybridization in directly initiating parthenogenetic reproduction. He was criticized for this[106,107,113] because, among other things, the parthenogenetic species of *Cnemidophorus* in North America are all clearly interspecific hybrids. Cuellar[117] does not deny these forms are hybrids but merely questions whether hybridization in itself can result in parthenogenetic reproduction.

The reptile literature indicates that both the hybridization and nonhybridization models are correct.[107] Moreover, it is obvious that not only a number of ecological conditions, but also various types of genetic mechanisms might be involved in the evolution of parthenogenetic reproduction (see Peccinini-Seale[4]).

Parthenogenetic reproduction involves the production of unreduced eggs. This can be accomplished by a premeiotic doubling of the chromosomes[118] or some other mechanism.[119] When a diploid parthenoform mates with a male of a bisexual species, the result is a triploid offspring. A few triploid parthenogenetic species of lizards are known but there are no tetraploid species or populations. However, tetraploid individuals have been found that appear to be hybrids between a triploid parthenogenetic species and a bisexual species.[44,120] This conclusion appears to be supported by the production of a tetraploid hybrid resulting from a cross between *Cnemidophorus sonorae* (3n) and *C. tigris* (2n) in the laboratory.[108]

Although polyploidy has played an important role in the evolution of some groups of lower vertebrates, it is rare in reptiles. No polyploid snakes, turtles, or crocodilians have been reported. Except for the triploid parthenogenetic species and occasional tetraploid hybrids, there are only a few examples of polyploid lizards reported. Bezy[58] reported a diploid-triploid mosaic specimen of *Lepidophyma flavimaculatum* (Xantusiidae). All other individuals in the population were diploid parthenogenetic females. Witten[54] reported a triploid *Amphibolorus nobbi* (Agamidae). All other individuals examined by him were diploids. These examples probably represent spontaneously produced polyploids. The triploid male *Amphibolorus* had highly irregular meiosis and was undoubtedly sterile.[54]

One other report of triploidy is a $3n = 96$ for 2 individuals of *Platemys platecephala*, a South American chelid turtle.[11] The diploid number for this species is 64.[32] It is unclear if the triploids are the result of spontaneous mutation or are maintained in populations.

V. PATTERNS OF CHROMOSOMAL VARIATION

A. General Comments

Reptiles display a variety of patterns of karyotypic variation. There are some highly successful and diverse groups, such as emydine turtles, that are virtually monomorphic in terms of karyotype. The other end of the spectrum is represented by *Anolis* and *Sceloporus* in which closely related species, and even conspecific populations, are chromosomally well differentiated. Another pattern frequently observed is the predominance of one particular type of rearrangement being incorporated during the evolution of a lineage. This is termed karyotypic orthoselection by White.[121] The reasons for such disparity in pattern of variability has often been discussed. Many factors may promote chromosomal evolution. They include low vagility and social structuring of populations that promote small deme size and inbreeding.[122-127] Other factors, not related to population size, include mutation rate,[128,129] meiotic mechanisms that ensure proper segregation in a chromosomal rearrangement heterozygote,[130-132] and directional selection. The latter may, under certain conditions, result in the fixation of a new chromosomal rearrangement if it provides some selective advantage.[15,130-134] Additionally, historical factors, such as founder events and population bottlenecking, can play important roles.[129,135,136]

B. Elucidation of Ancestral Karyotypes

An understanding of the mechanisms of chromosomal evolution, as well as the application of karyotypic data to taxonomy and phylogeny reconstruction, necessitates the determination of primitive and derived character states (character polarity). Various methods for determining primitive karyotypes have been proposed (see Paull et al.[36] and King[38]). The single best method is to use G-band information for the determination of homologies, and to utilize suitable out-group taxa for comparisons. For example, the primitive karyotype of the Emydidae is identical to that found in certain batagurine turtles ($2n = 52$) rather than like the $2n = 50$ karyotype found in emydines and some batagurines. Comparisons to out-group taxa such as the Testudinidae and Platysternidae revealed that some species of testidinids have karyotypes identical to the $2n = 52$ batagurines and *Platysternon* is more similar to the batagurine than the emydine karyotype.[28] It strengthens the argument that the batagurines are thought to be primitive based on osteology.[23] However, because a group is primitive or derived morphologically does not necessarily mean the same is true of the karyotype.

We may never know what the ancestral karyotype of reptiles was because banding analyses between classes or even among the orders of reptiles may reveal no homologies. However, it may be possible to determine the ancestral karyotypes of the orders or suborders. Banding data from turtles indicate remarkable conservation of G-band homology.[13] If the same pattern is found in the other reptilian orders, we may expect to determine primitive and derived states.

Speculation about the ancestral karyotypes of groups such as lizards, for which no G-band data are available, is highly tenuous. The primitive karyotype of lizards has been proposed to be $2n = 36$, similar to that found so frequently in a number of families of lizards as well as amphisbaenians.[3,36] Gorman and Paull et al. argue that broad taxonomic distribution implies primitiveness. An alternative hypothesis is that a higher diploid number with all acrocentric elements is primitive.[37,38] It assumes that the predominant mode of karyotypic evolution is by centric fusion and that fission is rare or never occurs. If the biarmed macrochromosomes of the $2n = 36$ karyotypes are found to be homologous across many of the higher categories, it would argue strongly for their primitive nature. The alternative hypothesis, that identical centric fusions occurred independently in many different families of lizards and in amphisbaenians, hardly would be believeable.

The primitive karyotype of snakes may not be the common $2n = 36$ karyotype proposed as ancestral by Gorman.[3,78] Mengden and Stock[80] found little homology between boids and colubrids with $2n = 36$, but the data are based on too few taxa for this conclusion to be considered unequivocal. However, this should serve as a caveat to workers attempting to reconstruct ancestral karyotypes by assuming broad homology among superficially similar, nondifferentially stained karyotypes.

C. Role of Chromosomal Rearrangement in Speciation

The classical hypothesis of karyotype evolution by White[126,137-139] was further elaborated by numerous workers.[39,124,127,140] White proposed in his original model of stasipatric speciation that chromosomal rearrangements become fixed in small populations. He assumed that the heterozygote for a chromosomal rearrangement exhibits partial sterility caused by the production of some percentage of unbalanced gametes due to meiotic malassortment. Because stochastic processes are important in determining the genetic (or chromosomal) makeup of small populations,[129] occasionally a rearrangement will reach fixation due to genetic drift or inbreeding. The new homozygous population will represent an incipient species due to partial reproductive isolation from the parental population. Large sized populations cannot undergo this process because natural selection plays a more decisive role than stochastic processes in the determination of the genetic makeup of such populations.

Given that scenario for chromosomal evolution, animals with small population sizes should exhibit greater karyotypic diversity and more rapid rates of karyotypic evolution than animals with large population sizes.[124,127] Accordingly, biological characteristics of species which promote population subdivision should be correlated with karyotypic variability. Such characteristics include low vagility, territoriality, and social structure.[124,126-128] Whether or not such a model of chromosomal evolution is applicable to any reptilian group will be considered in this review.

Several groups of lizards possess patterns of karyotypic variation that ostensibly support stasipatric speciation[38] or some other form of chromosomal speciation such as parapatric[140] or cascading chromosomal speciation.[39] These lizards exhibit extensive populational variation in karyotype. Chromosomally differentiated populations are distributed parapatrically with narrow zones of hybridization. However, before concluding chromosomal speciation to be the only explanation of intraspecific chromosomal polytypy, several other parameters must be investigated, such as meiosis of hybrids and patterns of protein variability. Under allopatric speciation, rearrangements are fixed in an allopatric population due to stochastic or selective processes but do not result in reproductive isolation.[141] King[38] has referred to this as secondary chromosomal allopatry. His primary chromosomal allopatry model, although different in detail from stasipatric speciation, is a form of chromosomal speciation.

Table 2 details the chromosomal, distributional, and genetic characteristics of several modes of speciation, including the chromosomal modes. Although King[38] proposes that several species of lizards exemplify chromosomal modes, only *Sceloporus grammicus* has been studied in enough detail to test this hypothesis. Several other lizard species, however, (but no turtle, crocodilian, or snake species) also show populational variation and are thus potential tests of the models.[38]

Chromosomal variation in *S. grammicus* was studied by Hall[39,142] and by Sites.[42,143] This species has extensive geographic variation in karyotype and where hybrid zones are known

Table 2
**CHROMOSOMAL, DISTRIBUTIONAL, AND GENETIC CORRELATES OF
3 MODES OF SPECIATION**

| | Mode | | |
Correlates	Stasipatric[125,137-139]	Primary chromosomal allopatry[38]	Secondary chromosomal allopatry[38]
Pattern of distribution of primitive and derived cytotypes	Mosaic	Derived cytotypes at periphery	Derived cytotypes at periphery
Gene flow through contact zones	Always restricted	May or may not be restricted	May or may not be restricted
Hybrid semi-sterility	Due to chromosome structural rearrangement	May be due to genotypic imbalance or structural rearrangement	Due to genotypic imbalance
Population subdivision	Highly subdivided	May or may not be subdivided	May or may not be subdivided
Exhibit karyotypic orthoselection	No	No	May or may not
Balanced polymorphisms of the type that distinguish cytotypes	No	No	Yes

they are narrow.[142] Hall[39] believed that it has low vagility, which promotes chromosomal rearrangement fixation in small demes. The evolution of the cytotypes appeared to be due to a linear progression of fixation of centric fissions from $2n = 32 \rightarrow 34 \rightarrow 36 \rightarrow 40 \rightarrow 44$. Hall[39] proposed that the rearrangements act as a reproductive isolating mechanism and that these chromosome races might be good biological species. A more detailed study suggests that this is not the case, at least in part of the species range. Sites found chromosomal polymorphisms, including inversions and centric fissions and fusions, seemingly common and widespread in some cytotypes.[42] Widespread polymorphisms are not expected in animals with small deme size. That a particular kind of rearrangement can act as a reproductive isolating mechanism and at the same time be present as a frequently observed polymorphism can hardly be believed.

Electrophoretic studies[143] indicate high levels of heterozygosity and relatively low F_{ST} values (a measure of population subdivision), both consistent with large population sizes. In addition, the pattern of geographic variation of alleles suggests that the different cytotypes studied ($2n = 32, 34,$ and 36) are not reproductively isolated. The data (chromosomal and biochemical) all are consistent with the alternative hypothesis that chromosomal rearrangements become fixed in allopatric populations and that they play no direct role in speciation. This model is equivalent to King's[38] secondary chromosomal allopatry.

To conclude, *S. grammicus* does not fit any of the chromosomal modes of speciation. Further studies of this nature are required to determine if chromosomal speciation is a viable explanation for chromosomal polytypy in other lizard species.

D. Role of Natural Selection in Chromosomal Evolution

Chromosomal speciation might occur in reptiles, but alternative hypotheses might better explain the observed patterns of karyotypic variation. These models are based on the assumption that chromosomal rearrangements can be fixed, even in large populations, if they confer increased fitness to the individual heterozygous *or* homozygous for the new rearrangement.[15]

Assuming natural selection plays the major role in promoting rearrangement fixation, small deme size and the biological characteristics that promote population subdivision may

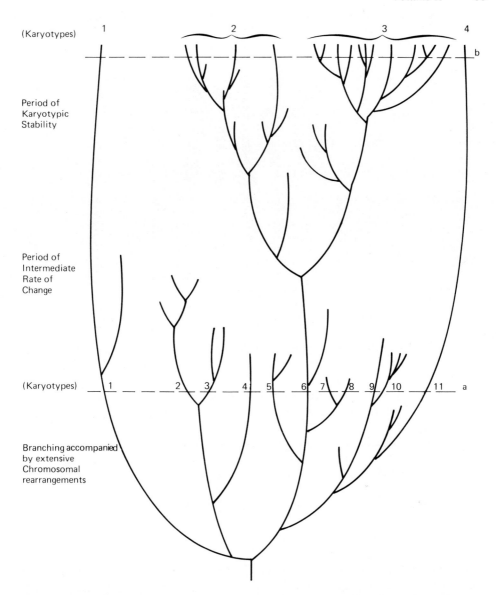

FIGURE 7.   The canalization model of chromosomal evolution hypothesizes that extensive karyotypic rearrangement occurs during the initial radiation of an order. The lineage evolves through a period of intermediate rate of change and into a period of karyotypic stability. Karyotypic comparisons of taxa that diverged during the period of stability reveals karyotypic identity. Comparisons between taxa that diverged near the base of the tree reveal considerable karyotypic divergence. In this figure, 11 different karyotypes are present near the base of the tree but only 4 are at the top.

not necessarily be associated with chromosomal change. Bickham and Baker[15] noted that the age of a lineage seemed to be negatively correlated with karyotypic variability and they formulated the canalization model as an alternative explanation of karyotypic evolution (Figure 7). Under this model, lineages experience rapid chromosomal evolution when they first break into a new, broad, adaptive zone, for example, during the origin of a new order. Chromosomal rearrangement incorporation is, then, implicated in the process of adaptive diversification. As time progresses and the adaptive zone becomes filled, the rate of rearrangement incorporation decelerates because the accumulation of many adaptive gene se-

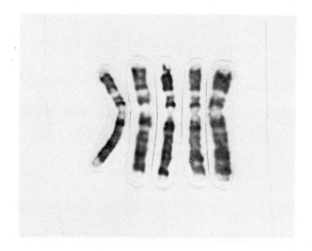

FIGURE 8.   G-bands of chromosome no. 1 from 5 families of cryptodiran turtles (left to right: Testudinidae, Trionychidae, Cheloniidae, Emydidae, Kinosternidae). These taxa may have diverged as long ago as 200 million years.[13]

quences (karyotypic arrangement) reduces the likelihood of a new arrangement being beneficial. This process continues until an "optimal karyotype" evolves, at which time chromosomal rearrangement virtually ceases.

Studies of turtles, bats, and rodents indicate that younger lineages are more variable than older lineages. They have chromosomal variation at lower taxonomic levels (species and conspecific populations in rodents) than older lineages (genera of bats and families of turtles). Furthermore, in turtles the rate of karyotypic evolution was more than twice as fast during the diversification of the families than is characteristic of modern species.[13] These facts do not support a model requiring stochastic processes in small demes.

VI. CONCLUDING REMARKS

A. Further Research in Reptile Cytosystematics

The surface of reptile cytosystematics barely has been scratched. Standard karyotypic data are lacking for many lizard and snake groups and chromosome banding data are available for less than one percent of all reptiles. Chromosome banding studies are needed on a wide variety of reptiles as are detailed population cytogenetic studies of particular groups. One of the most urgent issues to be addressed is the determination of the primitive karyotype of lizards. This information could clarify the taxonomic and phylogenetic relationships among certain families of lizards as well as the evolutionary relationships within certain families, such as iguanids.

Reptiles also offer the opportunity to study the evolution of sex chromosomes. Modern staining and biochemical studies are needed to fully understand the sequence of events involved in sex chromosome differentiation.

I have emphasized the need for the application of modern cytogenetic techniques in reptile cytosystematics. Standard karyotype analyses are certainly needed for many groups and they often result in important findings. Many workers do not have the necessary facilities to do banding analyses, but by employing the classical techniques they can still contribute to the progress of this field. I believe that in the immediate future we will see a proliferation of interest in reptile chromosomes and an increased sophistication in this area of science.

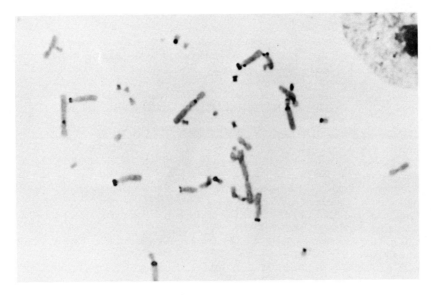

A

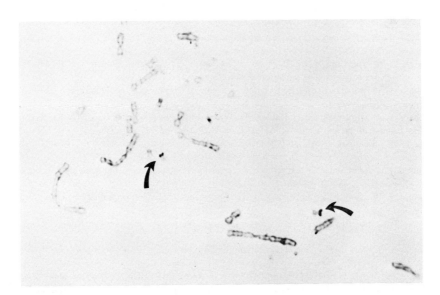

B

FIGURE 9. (A) C-bands of *Geochelone elongata*. Constitutive heterochromatin is present as both centromeric bands and short arms. (B) Silver staining reveals the location of the nucleolar organizer region (arrows) in *G. elongata*.

B. Acknowledgments

My studies of reptile chromosomes have been funded by the National Science Foundation (DEB-7713467, DEB-7921519). I gratefully acknowledge M. D. Engstrom, K. McBee, J. L. Carr, J. J. Bull, and J. W. Sites, Jr. for commenting on an early draft of this manuscript. I am grateful to the following for providing aid in constructing the figures: K. McBee (Figure 2), M. Engstrom (Figure 4), J. Sites and M. Haiduk (Figure 5), R. Baker (Figure 7), and R. Dowler (Figure 9).

REFERENCES

1. **Romer, A. S.,** *Vertebrate Paleontology,* 3rd ed., University of Chicago Press, Chicago, 1966.
2. **Stock, A. D. and Mengden, G. A.,** Chromosome banding pattern conservatism in birds and nonhomology of chromosome banding patterns between birds, turtles, snakes, and amphibians, *Chromosoma,* 50, 69, 1975.
3. **Gorman, G. C.,** The chromosomes of the Reptilia, a cytotaxonomic interpretation, in *Cytotaxonomy and Vertebrate Evolution,* Chiarelli, A. B. and Capanna, E., Eds., Academic Press, New York, 1973, chapter 12.
4. **Peccinini-Seale, D.,** New developments in vertebrate cytotaxonomy. IV. Cytogenetic studies in reptiles, *Genetica,* 56, 123, 1981.
5. **Goin, G. C., Goin, O. B., and Zug, G. R.,** *Introduction to Herpetology,* 3rd ed., W. H. Freeman and Co., San Francisco, 1978.
6. **Cohen, M. M. and Gans, C.,** The chromosomes of the order Crocodilia, *Cytogenetics,* 9, 81, 1970.
7. **Brazaitis, P.,** The identification of living crocodilians, *Zoologica,* 58, 59, 1973.
8. **Bickham, J. W. and Baker, R. J.,** Implications of chromosomal variation in *Rhogeessa* (Chiroptera: Vespertilionidae), *J. Mammal.,* 58, 448, 1977.
9. **Singh, L. and Ray-Chaudhuri, S. P.,** DNA replication pattern in the chromosomes of *Crocodylus palustris* (Lesson), *The Nucleus,* 16, 33, 1973.
10. **Atkin, N. B., Mattinson, G., Becak, W., and Ohno, S.,** The comparative DNA content of 19 species of placental mammals, reptiles and birds, *Chromosoma,* 17, 1, 1965.
11. **Bull, J. J. and Legler, J. M.,** Karyotypes of side-necked turtles (Testudines: Pleurodira), *Can. J. Zool.,* 58, 828, 1980.
12. **Rhodin, A. G. J., Mittermeier, R. A., Gardner, A. L., and Medem, F.,** Karyotypic analysis of the *Podocnemis* turtles, *Copeia,* 1978, 723, 1978.
13. **Bickham, J. W.,** Two-hundred-million-year-old-chromosomes: deceleration of the rate of karyotypic evolution in turtles, *Science,* 212, 1291, 1981.
14. **Bickham, J. W. and Carr, J. L.,** Taxonomy and phylogeny of the higher categories of cryptodiran turtles based on a cladistic analysis of chromosomal data, *Copeia,* 1983, 918, 1983.
15. **Bickham, J. W. and Baker, R. J.,** Canalization model of chromosomal evolution, *Bull. Carnegie Mus. Nat. Hist.,* 13, 70, 1979.
16. **Bickham, J. W., Bjorndal, K. A., Haiduk, M. W., and Rainey, W. E.,** The karyotype and chromosomal banding patterns of the green turtle *(Chelonia mydas), Copeia,* 1980, 540, 1980.
17. **Carr, J. W., Bickham, J. W., and Dean, R. H.,** The karyotype and chromosomal banding patterns of the Central American river turtle *Dermatemys mawii, Herpetologica,* 37, 92, 1981.
18. **Mlynarski, M.,** Testudines, in *Encyclopedia of Paleoherpetology,* Kuhn, O., Ed., Fischer, New York, 1976, chap. 7.
19. **Bickham, J. W., Bull, J. J., and Legler, J. M.,** Karyotypes and evolutionary relationships of trionychoid turtles, *Cytologia,* 48, 177, 1983.
20. **Dowler, R. C. and Bickham, J. W.,** Chromosomal relationships of the tortoises (Family Testudinidae), *Genetica,* 58, 189, 1982.
21. **Gaffney, E. S.,** A phylogeny and classification of the higher categories of turtles, *Bull. Amer. Mus. Nat. Hist.,* 155, 387, 1975.
22. **Chkhkvadze, V. M.,** Subclass classification of the testudinates [in Russian], Tezisy Dokladov, 16 Nachnaia Sessiia, Inst. Paleobiol., Posviaskennaia 50, Letiiu Sovetskoi Gruzii i Kommunist. Partii Gruzii, 10-11 Dekobria, p. 7, 1970.
23. **McDowell, S. B.,** Partition of the genus *Clemmys* and related problems in the taxonomy of the aquatic Testudinidae, *Proc. Zool. Soc. Lond.,* 143, 239, 1964.
24. **Frair, W.,** Taxonomic relations among sea turtles elucidated by serological tests, *Herpetologica,* 35, 239, 1979.
25. **Loveridge, A. and Williams, E. E.,** Revision of the African tortoises and turtles of the suborder Cryptodira, *Bull. Mus. Comp. Zool.,* 115, 163, 1957.
26. **Auffenberg, W.,** Checklist of fossil land tortoises (Testudinidae), *Bull. Fla. St. Mus. Biol. Sci.,* 18, 121, 1974.
27. **Bickham, J. W. and Baker, R. J.,** Chromosome homology and evolution of emydid turtles, *Chromosoma,* 54, 201, 1976.
28. **Haiduk, M. W. and Bickham, J. W.,** Chromosomal homologies and evolution of testudinoid turtles with emphasis on the systematic placement of *Platysternon, Copeia,* 1982, 60, 1982.
29. **Bickham, J. W.,** A cytosystematic study of turtles in the genera *Clemmys, Mauremys,* and *Sacalia, Herpetologica,* 31, 198, 1975.

30. **Carr, J. L.** Phylogenetic Implications of Chromosomal Variation in the Batagurinae (Testudines: Emy-didae),M.S. Thesis, Texas A & M University, and College Station, 1981.

31. **Barros, R. M., Ayres, M., Sampaio, M. M., Cunha, O., and Assis, F.,** Karyotypes of two subspecies of turtles from the Amazon region of Brazil, *Caryologia,* 25, 463, 1972.

32. **Barros, R. M., Sampaio, M. M., Assis, M. F., Ayres, M., and Cunha, O. R.,** General considerations of the karyotypic evolution of Chelonia from the Amazon region of Brazil, *Cytologia,* 41, 559, 1976.

33. **Bickham, J. W. and Baker, R. J.,** Karyotypes of some Neotropical turtles, *Copeia,* 1976, 703, 1976.

34. **Bull, J. J., Moon, R. G., and Legler, J. M.,** Male heterogamety in kinosternid turtles (genus *Staurotypus),* *Cytogenet. Cell Genet.,* 13, 419, 1974.

35. **Sites, J. W., Jr., Bickham, J. W., Haiduk, M. W., and Iverson, J. B.,** Banded karyotypes of six taxa of kinosternid turtles, *Copeia,* 1979, 692, 1979.

36. **Paull, D., Williams, E. E., and Hall, W. P.,** Lizard karyotypes from the Galapagos Islands: chromosomes in phylogeny and evolution, *Breviora,* 441, 1, 1976.

37. **Cole, C. J.,** Karyotypes and evolution of the *spinosus* group of lizards in the genus *Sceloporus, Amer. Mus. Novitates,* 2431, 1, 1970.

38. **King, M.,** Chromosome change and speciation in lizards, in *Evolution and Speciation, Essays in Honor of M. J. D. White,* Atchley, W. R. and Woodruff, D. S., Eds., Cambridge University Press, London, 1981, chap. 13.

39. **Hall, W. P.,** Comparative population cytogenetics, speciation, and evolution of the iguanid lizard genus *Sceloporus,* Ph.D. thesis, Harvard University, Cambridge, 1973.

40. **Webster, T. P., Hall, W. P., and Williams, E. E.,** Fission in the evolution of a lizard karyotype, *Science,* 177, 611, 1972.

41. **Gorman, G. C. and Atkins, L.,** The zoogeography of Lesser Antillean *Anolis* lizards — an analysis based upon chromosomes and lactic dehydrogenases, *Bull. Mus. Comp. Zool.,* 138, 53, 1969.

42. **Sites, Jr., J. W.,** Chromosome evolution in the iguanid lizard *Sceloporus grammicus.* I. Chromosome polymorphisms, *Evolution,* 33, 38, 1983.

43. **Gorman, G. C.,** Chromosomes and the systematics of the family Teiidae (Sauria, Reptilia), *Copeia,* 1970, 230, 1970.

44. **Lowe, C. H., Wright, J. W., Cole, C. J., and Bezy, R. L.,** Chromosomes and evolution of the species groups of *Cnemidophorus* (Reptilia: Teiidae), *Syst. Zool.,* 19, 128, 1970.

45. **Fritts, T. H.,** The systematics of the parthenogenetic lizards of the *Cnemidophorus cozumela* complex, *Copeia,* 1969, 519, 1969.

46. **Bickham, J. W., McKinney, C. O., and Mathews, M. F.,** Karyotypes of the parthenogenetic whiptail lizard *Cnemidophorus laredoensis* and its presumed parental species (Sauria: Teiidae), *Herpetologica,* 32, 395, 1976.

47. **Cole, C. J., McCoy, C. J., and Achaval, F.,** Karyotype of a South American teiid lizard, *Cnemidophorus lacertoides, Am. Mus. Novitates,* 2671, 1, 1979.

48. **Chevalier, M., Dufaure, J. P., and Lecher, P.,** Cytogenetic study of several species of *Lacerta* (Lacertidae, Reptilia) with particular reference to sex chromosomes, *Genetica,* 50, 11, 1979.

49. **Darevsky, I. S., Kupryanova, L. A., and Bakradze, M. A.,** Residual bisexuality in parthenogenetic species of lizards of the genus *Lacerta. Zh. Obshch. Biol.,* 38, 772, 1977.

50. **Moody, S. and Hutterer, R.,** Karyotype of the agamid lizard *Lyriocephalus scutatus* (L., 1758), with a brief review of the chromosomes of the lizard family Agamidae, *Bonn. Zool. Beitr.,* 29, 165, 1978.

51. **Sokolovsky, V. V.,** A comparative karyological study of the lizards of the family Agamidae. I. Chromosome complements of eight species of the genus *Phrynocephalus, Tsitiologiya,* 16, 920, 1974.

52. **Sokolovsky, V. V.,** A comparative karyological study of the lizards of the family Agamidae. II. Karyotypes of five species of the genus *Agama, Tsitologiya,* 17, 91, 1975.

53. **Bhatnagar, A. N. and Yoniss, Y.,** The chromosome cytology of two lizards, *Agama ruderata* and *Mabuya aurata septemtaeniata, Caryologia,* 30, 399, 1977.

54. **Witten, G. J.,** A triploid male individual *Amphibolorus nobbi nobbi* (Witten) (Lacertilia: Agamidae), *Aust. Zool.,* 19, 305, 1978.

55. **Singh, L.,** Study of mitotic and meiotic chromosomes in seven species of lizards, *Proc. Zool. Soc., Calcutta,* 27, 57, 1974.

56. **Olmo, E. and Odierna, G.,** Chromosomal evolution and DNA of cordylid lizards, *Herpetologica,* 36, 311, 1980.

57. **Matthey, R.,** Nouvelle contribution a l'etude des chromosomes chez Sauriens, *Rev. Suisse Zool.,* 40, 281, 1933.

58. **Bezy, R. L.,** Karyotypic variation and evolution of the lizards in the family Xantusiidae, *Contrib. Sci.,* 227, 1, 1972.

59. **King, M. and King, D.,** Chromosomal evolution in the lizard genus *Varanus* (Reptilia), *Aust. J. Biol. Sci.,* 28, 89, 1975.

60. **Holms, R. S., King, M., and King, D.,** Phenetic relationships among varanid lizards based upon comparative electrophoretic data and karyotypic analyses, *Biochem. Syst. Ecol.,* 3, 257, 1975.

61. **King, M.,** Chromosomes of two Australian lizards of the families Scincidae and Gekkonidae, *Cytologia,* 38, 205, 1973.

62. **King, M.,** Chromosomal and morphometric variation in the gekko *Diplodactylus vittatus* (Gray), *Aust. J. Zool.,* 25, 43, 1977.

63. **King, M.,** Karyotypic evolution in *Gehyra* (Gekkonidae: Reptilia). I. The *Gehyra variegata-punctata* complex, *Aust. J. Zool.,* 27, 373, 1979.

64. **King, M. and Hayman, D.,** Seasonal variation of chiasma frequency in *Phyllodactylus marmoratus* (Gray) (Gekkonidae-Reptilia), *Chromosoma,* 69, 131, 1978.

65. **King, M. and King, D.,** An additional chromosome race of *Phyllodactylus marmoratus* (Gray) (Reptilia: Gekkonidae) and its phylogenetic implications, *Aust. J. Zool.,* 25, 667, 1977.

66. **King, M. and Rofe, R.,** Karyotypic variation in the Australian gekko *Phyllodactylus marmoratus* (Gray) (Gekkonidae: Reptilia), *Chromosoma,* 54, 75, 1976.

67. **King, M.,** Karyotypic studies of some Australian Scincidae (Reptilia), *Aust. J. Zool.,* 21, 21, 1973.

68. **Gorman, G. C. and Gress, F.,** Sex chromosomes of a pygopodid lizard, *Lialis burtonis, Experientia,* 26, 206, 1970.

69. **Wright, J. W.,** Evolution of the X_1X_2Y sex chromosome mechanism in the scincid lizard *Scincella laterale* (Say), *Chromosoma,* 43, 101, 1973.

70. **Bezy, R. L., Gorman, G. C., Kim, Y. J., and Wright, J. W.,** Chromosomal and genetic divergence in the fossorial lizards of the family Anniellidae, *Syst. Zool.,* 26, 57, 1977.

71. **Stamm, B. and Gorman, G. C.,** Notes on the chromosomes of *Anolis agassizi* (Sauria: Iguanidae), *Smithson. Contr. Zool.,* 176, 52, 1975.

72. **Robinson, M. D.,** Karyology, phylogeny and biogeography of the namaqua chamaeleon, *Chamaeleo namaquensis* Smith, 1831 (Chamaeleonidae, Reptilia), *Beaufortia,* 28, 153, 1979.

73. **Huang, C. C., Clark, H. F., and Gans, C.,** Karyological studies on fifteen forms of amphisbaenians (Amphisbaenia-Reptilia), *Chromosoma,* 22, 1, 1967.

74. **Huang, C. C. and Gans, C.,** The chromosomes of 14 species of amphisbaenians (Amphisbaenia, Reptilia), *Cytogenetics,* 10, 10, 1971.

75. **Macgregor, H. and Klosterman, L.,** Observations on the cytology of *Bipes* (Amphisbaenia) with special reference to its lampbrush chromosomes, *Chromosoma,* 72, 67, 1979.

76. **Becak, W. and Becak, M. L.,** Cytotaxonomy and chromosomal evolution in Serpentes, *Cytogenetics,* 8, 247, 1969.

77. **Baker, R. J., Mengden, G. A., and Bull, J. J.,** Karyotypic studies of thirty-eight species of North American snakes, *Copeia,* 1972, 257, 1972.

78. **Gorman, G. C.,** The chromosomes of *Laticauda* and a review of karyotypic evolution in the Elapidae, *J. Herpetol.,* 15, 225, 1981.

79. **DeSmet, W. H. O.,** The chromosomes of 23 species of snakes, *Acta Zool. Path. Antverpiensia,* 70, 85, 1978.

80. **Mengden, G. A. and Stock, A. D.,** Chromosomal evolution in Serpentes; A comparison of G and C chromosome banding patterns of some colubrid and boid genera, *Chromosoma,* 79, 53, 1980.

81. **Rossman, D. A. and Eberle, W. G.,** Partition of the genus *Natrix,* with preliminary observations on evolutionary trends in natricine snakes, *Herpetologica,* 33, 34, 1977.

82. **Sharma, G. P. and Nakhasi, U.,** Chromosomal polymorphism in three species of Indian snakes, *Cytobios,* 24, 127, 1979.

83. **Sharma, G. P. and Nakhasi, U.,** Karyological studies on six specimens of Indian snakes (Colubridae: Reptilia), *Cytobios,* 27, 177, 1980.

84. **Zimmerman, E. G. and Kilpatrick, C. W.,** Karyology of North American crotaline snakes (Family Viperidae) of the genera *Agkistrodon, Sistrurus,* and *Crotalus, Can. J. Genet. Cytol.,* 15, 389, 1973.

85. **Gutierrez, J. M., Taylor, R. T., and Bolanos, R.,** Cariotipos de diez especies de serpientes costarricenses de la familia Viperidae, *Rev. Biol. Trop.,* 27, 309, 1979.

86. **Saint-Girons, H.,** Caryotypes et evolution des viperes europeennes (Reptilia, Viperidae), *Bull. Soc. Zool. Fr.,* 102, 39, 1977.

87. **Bull, J. J.,** Sex determination in reptiles, *Quart. Rev. Biol.,* 55, 3, 1980.

88. **Bull, J. J. and Vogt, R. C.,** Temperature-dependent sex determination in turtles, *Science,* 206, 1186, 1979.

89. **Mrosovsky, N.,** Thermal biology of sea turtles, *Amer. Zool.,* 20, 531, 1980.

90. **Mrosovsky, N. and Yntema, C. L.,** Temperature dependence of sexual differentiation in sea turtles: implications for conservation practices, *Biol. Conserv.,* 18, 271, 1980.

91. **Pieau, C.,** Différenciation du sexe chez les embryons d'*Emys obicularis* L. (Chélonien) ussus d'oeufs incubés dans le sol, *Bull. Soc. Zool. France,* 100, 648, 1975.

92. **Yntema, C. L.,** Effects of incubation temperatures on sexual differentiation in the turtle, *Chelydra serpentina, J. Morphol.,* 150, 453, 1976.

93. **Yntema, C. L.,** Temperature levels and periods of sex determination during incubation of eggs of *Chelydra serpentina, J. Morphol.,* 159, 17, 1979.

94. **Muller, H. J.,** A gene for the fourth chromosome of *Drosophila, J. Exp. Zool.,* 17, 325, 1914.

95. **Ohno, S.,** *Sex Chromosomes and Sex-Linked Genes,* Springer-Verlag, Berlin, 1967.

96. **Ohno, S.,** *Major Sex-Determining Genes,* Springer-Verlag, Berlin, 1979.

97. **Bull, J. J.,** Sex chromosome differentiation: an intermediate stage in a lizard, *Can. J. Genet. Cytol.,* 20, 205, 1978.

98. **Sites, J. W., Jr., Bickham, J. W., and Haiduk, M. W.,** Derived X chromosome in the turtle genus *Staurotypus, Science,* 206, 1410, 1979.

99. **Carr, J. L. and Bickham, J. W.,** Sex chromosomes of the Asian black pond turtle, *Siebenrockiella crassicollis* (Testudines: Emydidae), *Cytogenet. Cell Genet.,* 31, 178, 1981.

100. **Singh, L., Purdom, I. F., and Jones, K. W.,** Satellite DNA and evolution of sex chromosomes, *Chromosoma,* 59, 43, 1976.

101. **Singh, L., Purdom, I. F., and Jones, K. W.,** Behaviour of sex chromosome associated satellite DNAs in somatic and germ cells in snakes, *Chromosoma,* 71, 167, 1979.

102. **Mengden, G. A.,** Linear differentiation of the C-band pattern of the W chromosome in snakes and birds, *Chromosoma,* 83, 275, 1981.

103. **Cole, C. J.,** Evolution of parthenogenetic species of reptiles, in *Intersexuality in the Animal Kingdom,* Reinboth, R., Ed., Springer-Verlag, Berlin, 1975, 340.

104. **Böhme, W.,** Indizien für natürliche Parthenogenese beim Helmbasilisken, *Basiliscus basiliscus* (Linnaeus 1758), *Salamandra,* 11, 77, 1975.

105. **Peccinini-Seale, D. and Frota-Pessoa, O.,** Structural heterozygosity in parthenogenetic populations of *Cnemidophorus lemniscatus* (Sauria, Teiidae) from the Amazonas Valley,*Chromosoma,* 47, 439, 1974.

106. **Vanzolini, P. E.,** Parthenogenetic lizards (technical comment), *Science,* 201, 1152, 1978.

107. **Cole, C. J.,** Parthenogenetic lizards (technical comment), *Science,* 201, 1154, 1978.

108. **Cole, C. J.,** Chromosome inheritance in parthenogenetic lizards and evolution of allopolyploidy in reptiles, *J. Hered.,* 70, 95, 1979.

109. **Brown, W. M. and Wright, J. W.,** Mitochondrial DNA analyses and the origin and relative age of parthenogenetic lizards *(genus Cnemidophorus), Science,*203, 1247, 1979.

110. **Neaves, W. B.,** Adenosine deaminase phenotypes among sexual and parthenogenetic lizards in the genus *Cnemidophorus* (Teiidae), *J. Exp. Zool.,* 171, 175, 1969.

111. **McKinney, C. O., Kay, F. R., and Anderson, R. A.,** A new all female species of the genus *Cnemidophorus, Herpetologica,* 29, 361, 1973.

112. **Maslin, T. P.,** Conclusive evidence of parthenogenesis in three species of *Cnemidophorus* (Teiidae), *Copeia,* 1971, 156, 1971.

113. **Wright, J. W.,** Parthenogenetic lizards (technical comment), *Science,* 201, 1152, 1978.

114. **Wright, J. W. and Lowe, C. H.,** Weeds, polyploids, parthenogenesis, and the geographical and ecological distribution of all-female species of *Cnemidophorus, Copeia,* 1968, 128, 1968.

115. **Wright, J. W. and Lowe, C. H.,** Evolution of the alloploid parthenospecies *Cnemidophorus tesselatus* (Say), *Mamm. Chromo. Newslett.,* 8, 95, 1967.

116. **Cuellar, O.,** Animal parthenogenesis, *Science,* 197, 837, 1977.

117. **Cuellar, O.,** Parthenogenetic lizards (technical comment), *Science,* 201, 1155, 1978.

118. **Cuellar, O.,** Reproduction and the mechanism of meiotic restitution in the parthenogenetic lizard *Cnemidophorus uniparens, J. Morph.,* 133, 139, 1971.

119. **Peccinini-Seale, D.,** Estudo da meiose em machos de *Cnemidophorus lemniscatus* (Sauria, Teiidae) e quantidade de DNA na espermatogenese, *Cienc. Cult. (suppl.),* 24, 161, 1972.

120. **Neaves, W. B.,** Tetraploidy in a hybrid lizard of the genus *Cnemidophorus* (Teiidae), *Breviora,* 381, 1, 1971.

121. **White, M. J. D.,** Chromosomal repatterning-regularities and restrictions, *Genetics,* 79, 63, 1975.

122. **Lewis, H.,** Speciation in flowering plants, *Science,* 152, 167, 1966.

123. **Fredga, K.,** Chromosomal changes in vertebrate evolution, *Proc. Roy. Soc. London,* 199, 377, 1977.

124. **Bush, G. L., Case, S. M., Wilson, A. C., and Patton, J. L.,** Rapid speciation and chromosomal evolution in mammals, *Proc. Nat. Acad. Sci. U.S.A.,* 74, 3942, 1977.

125. **White, M. J. D.,** *Modes of Speciation,* W. H. Freeman, San Francisco, 1978.

126. **Arnason, U.,** The role of chromosomal rearrangement in mammalian speciation with special reference to the Cetacea and Pinnipedia, *Hereditas,* 70, 113, 1972.

127. **Wilson, A. C., Bush, G. L., Case, S. M., and King, M. C.,** Social structuring of mammalian populations and rate of chromosomal evolution, *Proc. Nat. Acad. Sci. U.S.A.,* 72, 5061, 1975.

128. **McClintock, B.,** Mechanisms that rapidly reorganize the genome, *Stadler Symp., Univ. of Missouri,* 10, 25, 1978.

129. **Lande, R.,** Effective deme sizes during long term evolution estimated from rates of chromosomal rearrangement, *Evolution,* 33, 234, 1979.

130. **Bickham, J. W. and Baker, R. J.,** Reassessment of the nature of chromosomal evolution in *Mus musculus, Syst. Zool.,* 29, 159, 1980.

131. **Baker, R. J. and Bickham, J. W.,** Karyotypic evolution in bats: evidence of extensive and conservative chromosomal evolution in closely related taxa, *Syst. Zool.,* 29, 239, 1980.

132. **Elder, F. F. B. and Pathak, S.,** Light microscopic observations on the behavior of silver-stained trivalents in pachytene cells of *Sigmodon fulviventer* (Rodentia, Muridae) heterozygous for centric fusion, *Cytogenet. Cell Genet.,* 27, 31, 1980.

133. **Bengtsson, B. and Bodmer, W. F.,** On the increase of chromosome mutations under random mating, *Theoret. Pop. Genet.,* 9, 260, 1976.

134. **Baker, R. J.,** Chromosome flow between chromosomally characterized taxa of a volant mammal, *Uroderma bilobatum* (Chiroptera: Phyllostomatidae), *Evolution,* 35, 296, 1981.

135. **Templeton, A. R.,** Modes of speciation and inferences based on genetic distances, *Evolution,* 34, 719, 1980.

136. **Patton, J. L. and Yang, S. Y.,** Genetic variation in *Thomomys bottae* pocket gophers: macrogeographic patterns, *Evolution,* 31, 697, 1977.

137. **White, M. J. D.,** Models of speciation, *Science,* 159, 1065, 1968.

138. **White, M. J. D.,** Speciation in the Australian morabine grasshoppers: the cytogenetic evidence, in *Genetic Mechanisms of Speciation in Insects,* White, M. J. D., Ed., Australia and New Zealand Book Co., Sydney, 1974, 57.

139. **White, M. J. D.,** Chain processes in chromosomal speciation, *Syst. Zool.,* 27, 285, 1978.

140. **Bush, G. L.,** Modes of animal speciation, *Annu. Rev. Ecol. Syst.,* 6, 339, 1975.

141. **Moran, C.,** Spermatogenesis in natural and experimental hybrids between chromosomally differentiated taxa of *Caledia captiva, Chromosoma,* 81, 579, 1981.

142. **Hall, W. P. and Selander, R. K.,** Hybridization of karyotypically differentiated populations in the *Sceloporus grammicus* complex (Iguanidae), *Evolution,* 27, 226, 1973.

143. **Sites, J. W. and Greenbaum, I. F.,** Chromosome evolution in the iguanid lizard *Sceloporus grammicus.* II. Allozyme variation, *Evolution,* 37, 54, 1983.

Chapter 3

CHROMOSOMES IN EVOLUTION OF COLEOPTERA

Niilo Virkki

TABLE OF CONTENTS

I. INTRODUCTION

In the largest of all animal orders, Coleoptera, an extensive array of chromosomal characteristics can be expected, and is indeed found. Still the main impression is that of a certain conservatism. Monocentry is the rule, and metacentry prevails. One chiasma per bivalent is the rule, and two, in different arms, is the cytological maximum, although genetic data[1,2] suggest higher frequencies as well.

Causes for a low chiasma frequency are interference, pairing and crossing over adjustments, or structural details (e.g., heterochromatin).[3] Discovery of a special type of constitutive heterochromatin in *Chilocorus* chromosomes[4] offered a structure-based possibility of explanation for the invariably chiasmaless arms so often seen in the beetles. We call such heterochromatin ''cryptic,''[3] because it is isocyclic at metaphases and thus indistinguishable from euchromatin. Under treatment with colchicine, the euchromatic arms condense more rapidly than the heterochromatic ones. Originally, the heterochromatic character of the arms was defined by a genetic criterion: they were dispensable. Smith's colchicine test still is the only way to identify the cryptic heterochromatin morphologically. C-banding does not mark it satisfactorily.[5,6]

In Smith's terminology, chromosomes with two arms of a similar capacity of condensation under the colchicine test are called *monophasic*. They can be either euchromatic or heterochromatic. Chromosomes having one arm euchromatic and the other arm heterochromatic are *diphasic*.

The phenology of diphasism has since grown more complex. Smith's test revealed a block of cryptic heterochromatin **within an arm** of *Coeloxis bicornis* × chromosome (Figure 1). ''Naturally occurring diphasism'' was reported within the arms of *Hermaeophaga cubana*,[3] and between the arms in *Bolbites ornitis* chromosomes,[8] without any treatment. Although ''induced'' and ''natural'' diphasism might be closely related phenomena with similar evolutionary implications, the latter is already covered by the customary terminology of heterochromasy and differential condensation. Thus the terms monophasic and diphasic should be limited to manifestation of cryptic heterochromatin through inductive treatments. Once the structural basis for these differences is understood, there might be a re-evaluation of the terminology.

Although the significance of Smith's discovery is related to that of C-banding techniques, the test has not been widely explored. Whole-arm diphasism was induced in various *Anthonomus grandis* autosomes, combining a short colchicine treatment with hypotony,[9] the latter being considered essential. In *Cyrsylus volkameriae*, the sex chromosomes and the four largest autosomes are heterochromatic monophasics, the rest of the autosomes, euchromatic monophasics.[10]

These experiments also have a bearing on the attempts at a linear estimation of euchromatin vs. heterochromatin in the nuclei. Hypotony and colchicine treatments, both standard techniques of chromosome preparation, might induce changes in the measures, critical study of which has not yet been made.

II. THE ANCESTRAL COLEOPTERAN KARYOTYPE

A group called Oligoneoptera (or Endopterygotida[11]) emerged at the beginning of the Carboniferous era. The Panorpoid complex (leading to the Mecoptera, Siphonaptera, Diptera, and Lepidoptera) was separated from it in the late Carboniferous.[12] The other three contemporary orders all show peculiar sex determination mechanisms. Neuroptera shows asynaptic, precessional sex chromosomes, and Hymenoptera haplodiploidy. Exceptions are so few that these standard mechanisms must date back to the origin of the two orders. The Coleoptera, diversified to become the largest of all animal orders, show an array of sex determination

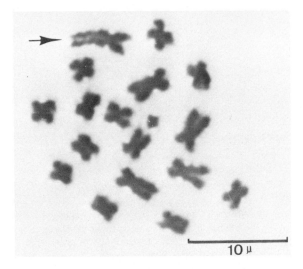

FIGURE 1.   *Coeloxis bicornis* (Scarabaeidae: Dynastinae); $2n = 18$. Smith's colchicine test shows diphasism within an arm of the X chromosome (arrow). (From Martins, V. G.- Citogenética de dezenove espécies de Scarabaeoidea — Coleoptera, Master's thesis, Universidade Estadual Paulista, Rio Claro, Brazil, 1982, 29. With permission.)

mechanisms including neuropteroid and hymenopteroid ones, but the sex parachute (Xy_p) prevails.

Such circumstances may indicate that differentiation of sex chromosomes of the late oligoneopterans must have reached a point where their primitive mode of association, by chiasmata or ectopic pairing, had started to fail. The modern mouse might be just arriving at such a phase: the presence of a synaptinemal complex in its sex bivalent[13] suggests a recent loss of chiasmate pairing. The crisis of pairing, orientation, and segregation of sex chromosomes was mastered in the three non-panorpoid orders in different ways.

Moreover, the late oligoneopteran karyotypes may have shared further architectural and functional characteristics that became delegated to the incipient off-branching new orders. Thus, present-day Coleoptera shares most of its diverse $X + Y$ sex chromosome systems with the Neuroptera and Diptera, its Xy_p with the Megaloptera,[14] and its arrhenotoky with Hymenoptera. It seems impossible to defend a monophyletic origin of all these post-oligoneopteran oddities, even though each, especially arrhenotoky and Xy_p, must be an extremely rare event.

Because the karyotype formula $9 + Xy_p$ is basic for Coleoptera Polyphaga,[3] because the same autosomal number occurs in the primitive Adephagan family Cicindelidae, and because the same sex chromosome system has been reported in some other Adephagans,[15-19] should this formula be proposed as basic for the entire Coleoptera?

The autosomal number offers no serious barrier to the assumption[20] that the 18 pairs ancestral for Adephaga[21] are the result of a complete series of centric fissions of 9 original metacentrics. The 18 pairs of Adephaga tend to consist of numerous metacentrics, but this is a problem encountered repeatedly in coleopteran karyotypes supposedly derived through fissions. The secondary metacentry is due either to accretion of heterochromatic arms → diphasism or to pericentric inversions, if not to centromeric shifts, which follow the fissions. Weber,[22,23] Mossakowski and Weber,[24-26] Wilken,[27] Nettmann,[28-29] and Serrano[30-32] have indeed described diphasic autosomes in carabids treated with hypotony, with or without colchicine, but these involve a minor portion of the karyotype.

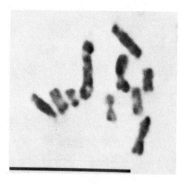

FIGURE 2.   *Itu zeus* (Myxophaga: Torridincolidae). M
I, 9 + XY; bar: 10 μm. From Mesa, A., and Fontanetti,
C. S., 33ª Reun. An. Soc. Brasil. Prog. Ci., Salvador
de Bahia, Brazil,1981. With permission.

It is more difficult to relate the adephagan Xy with the polyphagan Xy_p. In addition to the reports on Xy_p occurring in the Adephaga, more accurate ultrastructural documentation is required. After analysis of the pairing mode of the sex multivalent of *Cicinedela* by Giers[33] and Hammersen,[34] it seems possible that the association of the caraboid Xy depends on a heterochromatic and/or telomeric adhesion, which might be even more ancient than the nucleolar Xy_p of Polyphaga. At the same time, recent work has shown parachutes without a nucleolus (Section V.A).

The cytology of the remaining two suborders is little known, but is in agreement with the alleged ancient $2n = 20$. *Micromalthus debilis* ($2n = 20$ in the female) represents the Archostemata somewhat questionably, because some systematicists prefer to retain it in Cantharoidea,[35] where its arrhenotoky is even less plausible than in the primitive Archostemata. Mesa and Fontanetti[36] gave the first report concerning Crowson's Myxophaga: 9 + XY in *Itu zeus* Reichardt (Torridincolidae), which could very well belong to Polyphaga (Figure 2). There are no serious objections to the hypothesis that the Coleoptera emerged with 9 pairs of autosomes (probably monophasic euchromatic metacentrics), but it is still an open question whether Xy_p is the original sex chromosome system for all beetles.

III. RECENT KARYOLOGICAL ADVANCES IN COLEOPTERA

Since the publication of our monograph,[3] the study of beetle chromosomes has progressed in all permanently inhabited continents, but only slightly in Africa, Australia, and North America.

Caraboids have been studied by Weber and his associates[24-29,33,34] in Germany, by Serrano[21,30-32,37-39] in Spain, and by Yadav and Karamjeet[40,41] in India. Meloids,[42,43] hydrophilids,[42,44,45] cantharoids,[46] elaterids,[47] buprestids,[42,48] bostrychids,[49] cerambycids,[43,48,50-52] and lucanids[53] have attracted little attention, most data coming from Asia and tropical America. The quantitative DNA studies in *Dermestes* (Dermestidae) by Fox and his associates (Scotland)[54-58] have been extended to the borderland between organismal and molecular evolution. The Coccinellidae,[42,59,60] Tenebrionidae,[61-65] Passalidae,[66-68] and Scarabaeidae[69-79] have been studied to some extent by Indian, Argentinian, and Brazilian authors.

Chrysomelidae has drawn more attention.[43,80-108] Now the modal formula for the primitive subfamily Criocerinae is $7 + Xy_p$, and Petitpierre[88] suggests that it might be ancestral as well. For the Chrysomelinae, $11 + Xy_p$ continues to be modal, but the primitive *Timarcha* usually shows $9 + Xy_p$, implying that this formula is ancestral here. The very closely related, mostly allopatric, European Xy_p-species of *Timarcha* (called the *goettingensis* complex[86,90]) may differ only in minor chromosomal rearrangements, but species occupying mountain niches and North Africa tend to possess increased chromosome numbers. While most Old World species contain a variable number of metacentric autosomes, the North American *T. (Americanotimarcha) intrigata* has $2n = 44$, $21 + Xy_p$, and most chromosomes acrocentric. This species obviously has a long evolutionary history apart from its Old World relatives. Raising this taxon to the status of a new genus *Americanotimarcha* is karyologically substantiated. On the basis of exo- and endophenotypic traits, including comparison of chromosome measurements, *Timarcha* is now divided into 11 subgroups.[86]

New information on Bruchidae is limited to one paper[109] which establishes the karyotype for a French strain of *Acanthoscelides obtectus*, a standard to be used in future comparisons with karyotypes of other provenances.

Takenouchi[110-129] has continued his extensive studies on Japanese Curculionoidea. Other data come from Finland[130-135] and India.[136-142] Sharma and Pal[142] reported $7 + Xy_p$ in *Echinocnemus* sp. as the lowest for Curculionidae, but the record is $6 + X + 1 - 4$ *ss* in *Gelus californicus*.[143] A recent study of the karyotype of the boll weevil, *Anthonomus grandis*, by North et al.,[9] disqualifies the earlier reports by Lue et al.[144,145] Lanier, continuing his studies on the North American Scolytidae, has published a cytotaxonomic synopsis of the genus *Dendroctonus*.[146]

Since the publication of the global chromosome list at the end of our monograph,[3] several authors have published lists limited by systematics and/or region. The important ones are those of Caraboidea,[21] Japanese Curculionoidea,[123] and Indian Curculionoidea.[138] The global lists of Scarabaeidae[79] and Tenebrionidae[65] are practically identical with the corresponding parts of our 1978 list; only a few references have been added.

IV. TRENDS IN AUTOSOMAL EVOLUTION

Fox's quantitative studies in *Dermestes* spp. have been continued by Rees[54] and Rees et al.[55] Earlier results on heterochromatic rank order of the species were confirmed, with DNA renaturation, melting temperature, densitometry, centrifugation to CsCl equilibrium, and banding as criteria. Very high quantities (up to 65%) of satellite DNA were found. The conclusion is that visible blocks of constitutive heterochromatin are principally responsible for the ranks and contain a fast renaturing fraction of DNA and satellite DNA. These blocks are marked by C-banding, but Q- and G-bands are so unique for species that they do not help in studying phyletic affinities. Consequently, not only quantitative changes in constitutive heterochromatin but also frequent changes in euchromatic sequences distinguish different *Dermestes* species from one another.

Degree of homology of unique DNA sequences as determined by thermal stability of the hybrid molecules (*D. ater* or *D. maculatus cum* bulk DNA of each other species) supports the taxonomist's viewpoint (Figure 3). Distances between the species determined this way do not necessarily reflect the age of separation of the species, because the rate of the diverging DNA changes remains unknown.

Smith's *Chilocorus* studies and the B chromosomes of Coleoptera indicate that the constitutive heterochromatin in itself has no epigenetic significance in the beetles.[3] The same conclusion emerges from further quantitative DNA studies on *Dermestes*.[56,57] Data obtained from Carabidae are at variance with this view.

Mossakowski and Weber[24-26] found both positive and negative correlations between exophenotypic characteristics and size of the variable heterochromatic arms of the A autosomes of *Carabus*.[22] Increase in heterochromatin can increase or decrease the expression of most diverse characteristics but its possible role in formation of races could not be determined. The authors suggest that the arms are reserves of rDNA called to activity where an increased

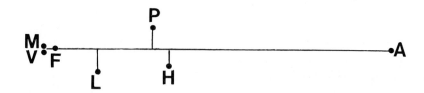

FIGURE 3.   Phenetic relationships of 7 *Dermestes* karyotypes as estimated by differences in thermal denaturation (1 cm corresponds to 1 C°$\triangle$Tm). Subgenus *Dermestinus: maculatus maculatus* (M), *maculatus vulpinus* (V),*frischii* (F); subgenus *Dermestes* s. str.: *lardarius* (L), *peruvianus* (P), *haemorrhoidalis* (H), *ater* (A). (From McLean,[58] based on Rees data.[54] With permission of R. W. Rees.)

rRNA production is required. A reduced capacity to modify major gene effects seems to be the price that novel karyotypes, arisen through heterochromatin-dispensing rearrangements, have to pay.

Serrano[31] has described distal collochores in the meiotic chromosomes of bembidines. This condition must arise polyphyletically, because earlier reports have come from Lampyridae, Elateridae: Pyrophorini, the flea beetles Ocedionychina,[3] Chrysomelidae: Megalopodinae,[147] and Stylopidae.[148] Only Lampyridae and Pyrophorini might feasibly share a common origin of this phenomenon.

A. Robertsonian Translocations

Whole-arm or Robertsonian translocations play an important role in the karyotype evolution of the beetles. Because they are more easily recognized than other rearrangements, they overwhelmingly lead the records on evolutionary changes of beetle karyotypes. Genetically gentle rearrangements,[149] they are attractive for speculating major evolutionary sequences like the origin of the adephagan karyotypes through a series of centric fissions, or formation of *Chilocorus* and *Blaps* karyotypes through fissions and subsequent fusions, as well as for explaining differences by one or more chromosome pairs within a species. The latter type of difference has been reported, especially by Takenouchi for the Curculionidae,[111,112,120,123] and one case, in *Catapionus gracilicornis*, was substantiated by a closer comparison of the differing karyotypes. Where intraspecific polymorphism does not accompany such discrepancies, sibling species might be involved.

Metacentric chromosomes are far more common in the beetles than are telo- or acrocentrics. They also prevail among the supposed products of centric fissions, as in the karyotypes of Passalidae:Proculini,[150] in the polymorphous galerucine *Acalymma trivittatum*,[151] in *Chrysolina carnifex* and *haemoptera*,[89] and in *Americanotimarcha intrigata*.[90] It is possible that fission telo- or acrocentrics soon turn to metacentrics, either through pericentric inversions, centromeric shifts, or accretion of new, heterochromatic arms. This Coleopteran trend is opposite to that of Acridoidea.[152] Such "postfission metacentrics" should be capable of undergoing fissions again, in order to explain such extreme discrepancies between closely related karyotypes as reported by Sharma and Sood[93] for *Oides*.

1. Chilocorus *Karyoclines*

This is the classical story of the role of diphasics in the evolution of beetle karyotypes.

In 1955, Smith encountered 3 male specimens of *Chilocorus stigma* (Coccinellidae), each one chromosomally different. One had 11 bivalents, 1 trivalent, and an unpaired chromosome. The other had 9 bivalents, 2 trivalents, and the unpaired one; and the third, 7 bivalents, 3 trivalents, and the unpaired chromosome.[153] The unpaired chromosome was originally interpreted as the sex chromosome, but this was later amended.[154] The trivalent in all three specimens was actually an X_1X_2Y, the unpaired chromosome being a supernumerary. The

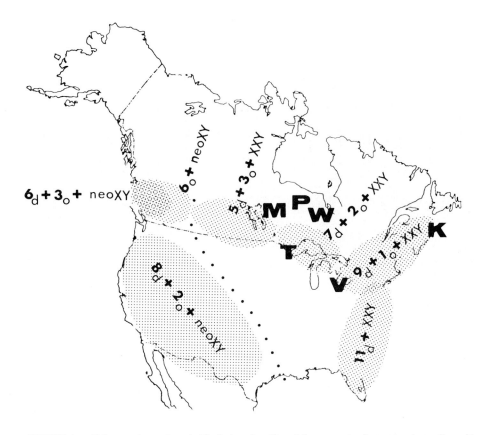

FIGURE 4. *Chilocorus* karyoclines in North America. Dotted line separates the western karyocline with its fixed species ($8_d + 2_o$ + neoXY in *orbus, fraternus, cacti;* $6_d + 3_o$ + neoXY in *tricyclus,* and 6_o + neoXY in *hexacyclus*) from the eastern *stigma* karyocline. Karyotypes given for the latter (from 11_d + XXY to $5_d + 3_o$ + XXY) are maximum fusion homozygotes encountered. (According to data in Smith,[154] and Smith and Virkki.[3])

remaining two trivalents were morphologically distinguishable from one another and from the X_1X_2Y. The bivalents were unichiasmate with free arms of a conspicuous size. Although large heteropycnotic masses were encountered in pachytene, no allocycly was seen in mitotic or meiotic metaphases. The diphasic nature of the autosomes was later confirmed by colchicine test.[4]

The modal formula for Coccinellidae is $9 + Xy_p$. Only the tribe Chilocorini shows consistently higher numbers.[3] Smith postulated two Robertsonian trends in the evolution of *Chilocorus* karyotypes, one, more ancient, of centric fissions followed by formation of heterochromatic second arms in the fission products. This might have raised the autosomal number to 13 pairs, although it is possible that one of these pairs had already joined the X chromosome to form a neoXY system prior to the fission series. The second trend engaged exceptional Robertsonian fusions, where two diphasic metacentrics gave rise to two monophasic metacentrics, one euchromatic and the other heterochromatic. The latter was a genetically insignificant supernumerary destined to erosion and vanishing, after having floated in the population.

Figure 4 summarizes the main results of the survey of the North American *Chilocorus* karyotypes. From the polymorphous *Ch. stigma,* only fusion homozygotes have been marked on the map. *Ch. stigma* occupies a vast territory east of the Rocky Mountains, but only the transsection studied by Smith is marked.

Florida has only unfused, diphasic (d) bivalents, plus the sex trivalent: $11_d + X_1X_2Y$.

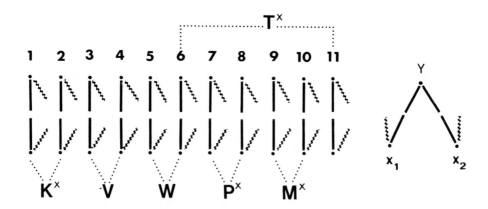

FIGURE 5.   Schematic presentation of the initial *Chilocorus stigma* karyotype in Florida: $11_d + XXY$. K to M: spatially successive fusions as they occur along the *stigma* karyocline. T: Thunder Bay fusion. Morphologically (including banding) recognizable fusions marked with x. (According to data in Smith,[154] and Smith and Virkki.[3])

This karyotype extends far up to the North and may include up to 5 supernumeraries, apparently resulting from accumulation of the heterochromatic monophasics arisen through incorporation of two pairs of diphasics into the sex chromosome system. In Nova Scotia, the K fusion (K from the locality where collected) appears, producing, when homozygous, the karyotype $9_d + 1_o + X_1X_2Y$ (o = a ring bivalent formed by two euchromatic monophasic metacentrics). Towards Ontario, three additional fusions, V, W, and P, occur, and homozygotes up to $7_d + 2_o \times XXY$ were found. One more fusion, M, occurs from Manitoba to Alberta, and homozygotes up to $5_d + 3_o + X_1X_2Y$ were found.

Each of these fusions can occur only between two given autosomes. Thus, fusion chains are not formed. Only once, in a Thunder Bay population, the 6th diphasic that elsewhere in Ontario participates in the W fusion has fused with the 11th diphasic (Figure 5). A chain-of-four occurs where T and W fusions meet.

The story differs beyond Saskatchewan. The sex chromosome system encountered in Alberta and British Columbia is neoXY, that is, one fusion step earlier than X_1X_2Y. The formula of this karyotype is $6_d + 3_o + $ neoXY, that of Alberta, $6_o + $ neoXY. Polymorphism was not found. Because intercrossing between these populations resulted in lowering of fertility, Smith separated them as two species independent from *stigma: Ch. tricyclus* Smith and *Ch. hexacyclus* Smith.[155] They evolved from a hypothetical ancestor having $12_d + $ neoXY. Intermediate karyotypes $8_d + 2_o + $ neoXY still survive in three species of the Western U.S.: *Ch. orbus, fraternus,* and *cacti.* These five form the so-called "Cyclus" group. Their common ancestor has obviously been a polymorphous species as *stigma* is today. Isolation provided by the mountains has aided speciation in the West.

If we track both the Cyclus and *stigma* karyoclines to their putative common origin, $13_d + Xy_p$, arisen from the modal $9 + Xy_p$ through 4 centric fissions and subsequent accretion of 8 heterochromatic arms, this putative karyotype would have had n.f. = 56, $2n = 28$. Heterochromatic arms must also have been present in the remaining, nonfissioned autosomes. A neoXY formation would reduce these numbers to 51 (52 where the neoY is metacentric) and 26, respectively. The ultimate fusion product, $6_o + $ neoXY of *hexacyclus,* has n.f. = 28, $2n = 14$, e.g., both arm and chromosome number reduced to half from the putative original numbers, with about 50% loss of the total chromatin. Still the genetic differences, especially between the wholly intercrossable *stigma* karyoforms, must be negligible.

Chilocorus spp., exotic to North America, tend to have lower chromosome and arm numbers.[3] *Ch. kuwanae* from Japan has one fusion ring, all other chromosomes, excluding

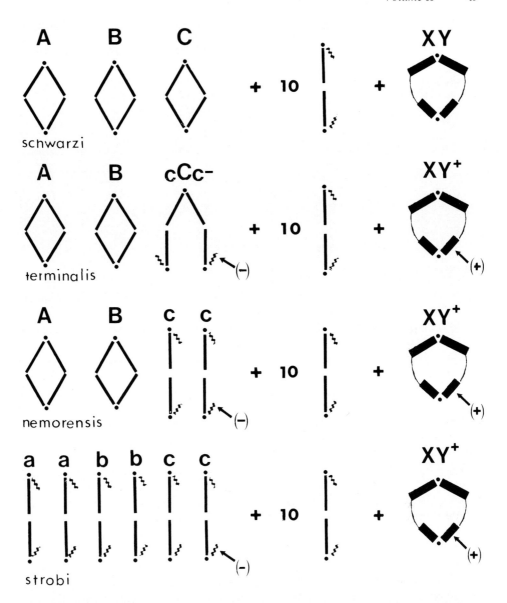

FIGURE 6. Examples of *Pissodes* karyotypes: one non-SIS "fusion" karyotype (*schwarzi*), and three fissioned SIS karyotypes (*terminalis, nemorensis,* and *strobi*). (According to data in Smith,[162] and Smith and Virkki.[3])

the neoY, being unichiasmate metacentrics (n.f. = 39, 2n = 20). *Ch. similis* (introduced to the U.S. and rediscovered by Smith[156]) and *renipustulatus,* also with one ring bivalent, have all unfused autosomes and the neoY telocentric (n.f. = 23, 2n = 20). The close exophenotypic relation of these species to *kuwanae* is' thus contrasted with lack of 16 apparently heterochromatic arms in the former. Thus the dispensable arms of diphasics have been deleted without simultaneous fusions. Pathak et al.[157] have reported a similar case in the rodent genus *Peromyscus* (n.f. = 96 to 56, 2n = 48).

Chilocorus and *Peromyscus* provide stunning examples of how much dispensable chromatin there can be in the karyotypes. They also explain the frequent discrepancies between chromosome data and exophenotypic systematics. The eukaryotic chromosomes may contain much inert chromatin in such strategic positions that dramatic rearrangements are possible without significant genetic effects.

Table 1
**RECIPROCAL CROSSES BETWEEN *PISSODES TERMINALIS* AND
*P. SCHWARZI***

| A. *terminalis* × *schwarzi* | | | | B. *schwarzi* × *terminalis* | | | | |
| XC/XC | | XC/YC | | XC/XC | | XC/Y$^+$cc$^-$ | | |

	XC	YC			XC	Xcc^{--}	Y$^+$C	Y$^+$cc$^-$
XC	XC/XC	XC/YC		XC	XC/XC	XC/Xcc$^-$	XC/Y$^+$C	XC/Y$^+$cc$^-$

Note: Recombinations containing ($+$) or ($-$) alone do not materialize.

2. Semi-incompatibility System of Pissodes

Diphasics and euchromatic monophasics occur in the karyotypes of *Pissodes* (Curculionidae). Since the colchicine test does not reveal the diphasism satsfactorily in *Pissodes,* the heterochromatic character of the second arms has been inferred from their dispensability and, in some instances, from an obvious accretion.[158]

A karyotype formula basic for *Pissodes,* $14 + XY_p$, occurs, among others, in *P. nemorensis* (Figure 6). There are twelve diphasic rod bivalents, two large monophasic ring bivalents, and an XY_p, where the Y_p is oversized. An increase of this number can take place through centric fission of the monophasics, culminating in *P. strobi* $(16 + XY_p)$. The principal candidates are the largest autosomes, labeled in their unfused, diphasic condition (as in *strobi*) as aa, bb, and cc pairs. Corresponding euchromatic fusion metacentrics were labeled A, B, and C pairs.

A key for understanding some genetic consequences of Robertsonian translocations was found in the male of *P. terminalis,*[159] which shows a permanent heterozygosity for the c-fusion (cCc), the female being the homozygote, CC. A complex sex chromosome system was ruled out because the trivalent cCc segregates independently from XY_p. Why was, then, the trivalent cCc excluded from the female, and the ring bivalent CC from the male?

A balanced lethal system became unlikely since no significant embryonal mortality was encountered.[160,161] Instead, an incompatibility system was assumed to operate at fertilization, a system with one factor ($+$) located in the unusually large Y_p chromosome, and another ($-$) in one of the unfused autosomes of the trivalent. Gametes bearing only one of these factors are incapable of fertilization.

The hypothesis was found to be valid in numerous hybridization experiments. Thus reciprocal crosses between *terminalis* and *schwarzi*, an ABC homozygote, produced karyotypes true to the parents, and new recombinations expected from the cross *schwarzi* × *terminalis* did not materialize (Table 1).These results indicate that the ($-$) factor is located near the centromere of a c chromosome and becomes eliminated by the centric fusion cc$^-$ → C.

Further hybridization experiments have shown that this SIS (semi-incompatibility system) is widely distributed in *Pissodes* although not detectable in karyotypes without marker chromosomes. The non-SIS, homozygous fusion species *schwarzi* is useful as a source of the marker C. Its reciprocal crosses to an aabbcc homozygote, like *strobi*, produced cytologically identical F_1 karyotypes, differing in the presence vs. absence of the ($+$) and ($-$) factors in the male (Table 2). Backcrosses to the female *schwarzi* proved the correctness of this assumption. The Y$^+$cc$^-$ male produced only two of the possible recombinations, whereas all expected karyotypes materialized in the progeny of the YC male (Table 3).

Table 2

RECIPROCAL CROSSES BETWEEN *PISSODES STROBI* AND *P. SCHWARZI*

A. *schwarzi* × *strobi*
XC/XC   Xcc/Y + cc −

	Xcc	Xcc −	Y⁺cc	Y⁺cc −
XC	XC/Xcc	XC/Xcc⁻	XC/Y⁺cc	XC/Y⁺cc⁻

B. *strobi* × *schwarzi*
Xcc/Xcc   XC/YC

	XC	YC
Xcc	Xcc/XC	Xcc/YC

Note: Recombinations containing (+) or (−) alone do not materialize.

Table 3

BACKCROSSES TO *SCHWARZI* OF THE MALES OBTAINED FROM RECIPROCAL CROSSES BETWEEN *PISSODES STROBI* AND *P. SCHWARZI*

A. *schwarzi* × *hybrid A*
XC/XC   XC/Y⁺cc⁻

	XC	Xcc⁻	Y⁺C	Y⁺cc⁻
XC	XC/XC	XC/Xcc⁻	XC/Y⁺C	XC/Y⁺cc⁻

B. *schwarzi* × *hybrid B*
XC/XC   Xcc/YC

	XC	Xcc	Ycc	YC
XC	XC/XC	XC/Xcc	XC/Ycc	XC/YC

Note: Recombinations containing (+) or (−) alone do not materialize.

The B: bb polymorphism of *Pissodes* is complicated by isomery (Section IV. B.2).

B. Inversions and Other Autosomal Rearrangements

1. General

Low chiasma frequency and limitation of chiasmata to the distal ends of the chromosomes are common coleopteran characteristics facilitating establishment of inversions. The opposite view can be held as well: heterosequentiality excludes chiasmata from the main part of the chromosomes.[162] Unique G-banding patterns in *Dermestes* spp.[55] can be indeed interpreted in terms of evolution by increasing heterosequentiality.

Under such chiasma distribution patterns, inversion bridges cannot be much expected.[71,163,164] Lanier[165] has reported them in certain scolytid hybrids, and Smith[3] interpreted persistent terminal adhesions leading to A I — M II bridges in *Chilocorus* hybrids as due to chiasma formation between duplicated ends, homozygous in the parents but relatively inverted in the hybrids.

Based on comparison of chromosomal structure, pericentric inversions have been reported in autosomes of the Cicindelidae,[39] Hydrophilidae,[45] Scarabaeidae,[8,71] and Chrysomelidae,[73,86,91,107] as well as in the sex chromosomes of *Cicindela*,[39] *Carabus*,[25,30] and Trechus.[31] To these must be added the numerous cases where a masking of a series of fissions by pericentric inversions is suspected: *Chirida* sp.,[108] *Chrysolina haemoptera*,[89] and *Oides* ssp.,[93] to mention only the most striking examples.

Deletions were suspected in the heterochromatic arms of *Carabus* spp.,[22,166,167] of some Bembidines,[32] and of the flea beetle *Alagoasa januaria*,[147] Hsiao and Hsiao[91] reported an

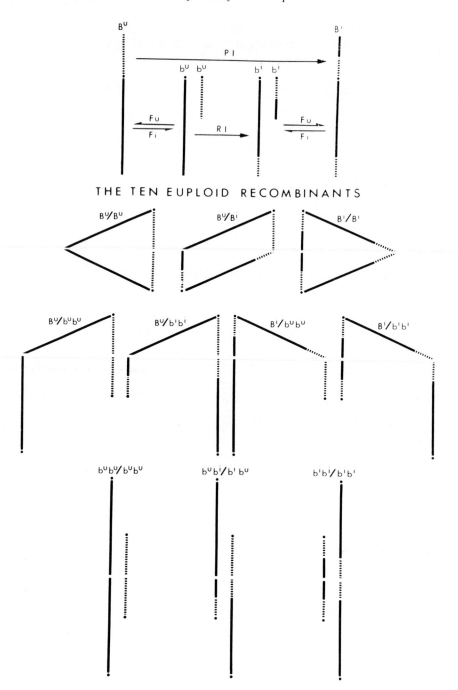

FIGURE 7.   Diagram illustrating alternative pathways in the origin of the B/bb isomery. PI = pericentric inversion; RI = reciprocal interchange; Fu = fusion; Fi = fission; superscripts u and i = unchanged and inverted or interchanged chromosomes. (From Smith, S. G., *Can. J. Genet. Cytol.*, 12, 526, 1970. With permission of the Genetics Society of Canada.)

interautosomal translocation in *Leptinotarsa defecta,* and Martins[68] found a chain-of-four at M I of one specimen of *Passalus mancus,* an obvious result of interchanges.

2. *Isomery in* Pissodes

Paradoxically, the best studied coleopteran inversions are possibly no inversions at all,

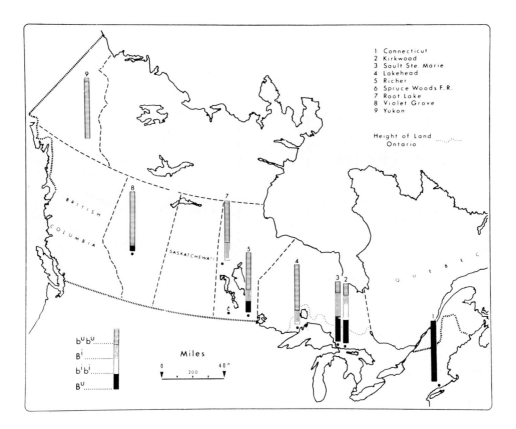

FIGURE 8. Karyocline of B^u, B^i, b^{u}b^u, and b^{i}b^i chromosomes in population samples of *Pissodes approximatus* and *P. canadensis*. (From Smith, S. G., *Can. J. Genet. Cytol.*, 12, 536, 1970. With permission of the Genetics Society of Canada.)

but reciprocal translocations followed by centric fusions (Figure 7). Structurally and functionally, the B^i autosomes thus restructured do not differ from pericentrically inverted ones, and are taken as such. All four forms: untranslocated acrocentrics (b^u) with corresponding fusion metacentrics (B^u), and translocated acrocentrics (b^i) with corresponding fusion ("inversion") metacentrics (B^i), occur in the *Pissodes approximatus-canadensis* complex and serve as markers in certain recombinations. In this karyocline, B^u prevails in the East, yielding to b^{u}b^u in the West (Figure 8).

Pericentric inversions involving breaks equidistant from chromosome ends produce structural isomers not detectable by the classical preparations. Meiotic configurations, however, are partially identifiable (Figure 7). Where B^u and B^i meet, a lopsided ring forms. B^u/b^{i}b^i and B^i/b^{u}b^u trivalents are asymmetrical, e.g., the associated acrocentrics are of notably different length. A combination of b^i and b^u chromosomes leads to two heteromorphous bivalents. A prerequisite for these configurations, chiasmata formed distally from the rearrangement, is always met.

Hybridization of species monomorphic (b^{i}b^i/b^{i}b^i and B^u/B^u) for the B:bb constitution produces only three types of meiotic association in F$_i$: the parental types, symmetrical rings or a pair of rod bivalents of notably different size, and the novel, asymmetrical trivalents (Figure 9). Figure 10 shows how the B:bb constitution of the *approximatus-canadensis* complex was checked by crossing to monomorphic species.

In view of the presence of interspecific B/b isomery, Smith held it possible that the isomery karyocline within the *approximatus-canadensis* complex is a result from introgressive hybridization among four species: b^{i}b^i/b^{i}b^i and B^u/B^u species of Eastern, and b^{u}b^u/b^{u}b^u and B^i/B^i species of Western distribution.

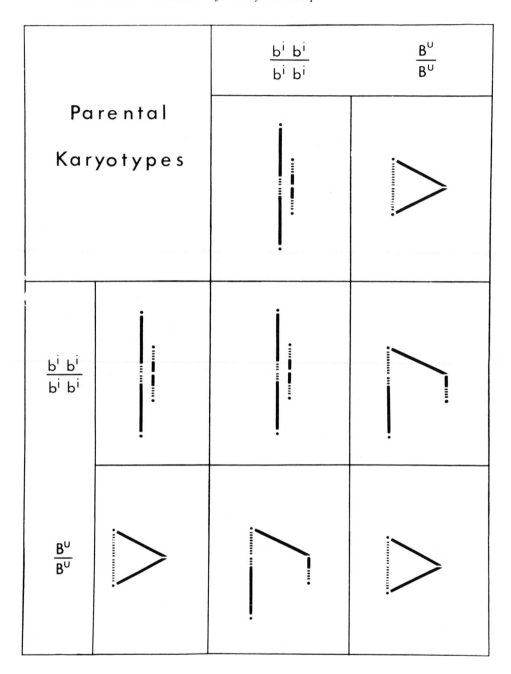

FIGURE 9.   Diagram of the B:bb karyotypes of F₁ males resulting from crossing monomorphic taxa. (From Smith, S. G., *Can. J. Genet. Cytol.*, 12, 509, 1970. With permission of the Genetics Society of Canada.)

Cytogenetic studies of *Pissodes* by Smith call for a drastic revision of the North American *Pissodes* taxonomy, which has been largely based on an alleged species specificity of food plant associations.[168] Twenty-two taxa accepted by Hopkins have now been reduced to 13 by the criteria of marker chromosomes, SIS, isomery, and hybrid viability. On the other hand, frequent viable hybrid karyoforms encountered are potential material for new processes of speciation when aided by adequate isolation mechanisms.

Although karyological evidence shows that Hopkins was overconfident in trusting in host

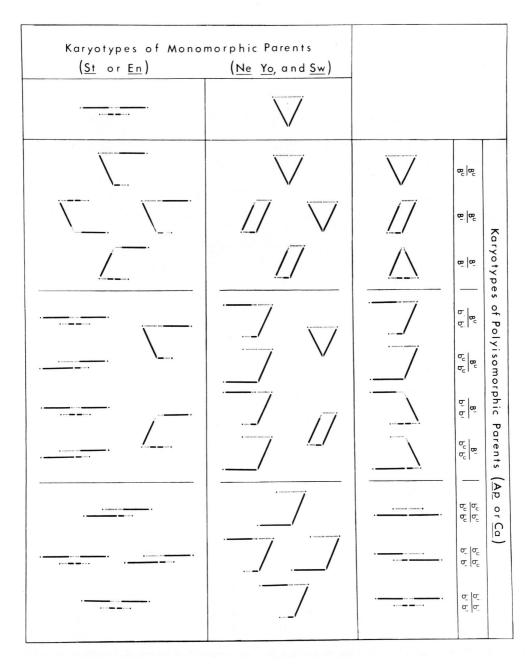

FIGURE 10.   Diagram illustrating the use of marker chromosomes to determine the B:bb constitution of unknown parents by M I complex(es) in F_1 hybrid males. In the progeny, the gametic contribution made by the monomorphic parent is placed above that of the polymorphic parent. (From Smith, S. G., *Can. J. Genet. Cytol.*, 12, 510, 1970. With permission of the Genetics Society of Canada.)

plants as indicators of species, *Chrysolina*[89] karyotypes are in agreement with a suspected beetle/plant coevolution. The evolution seems to have been conducted from an oligophagous group (of which one $10 + Xy_p$ species is known) via preferrers of Labiatae ($11 + Xy_p$ and $11 + X$) to preferrers of entirely new families (these last beetles have the highest chromosome

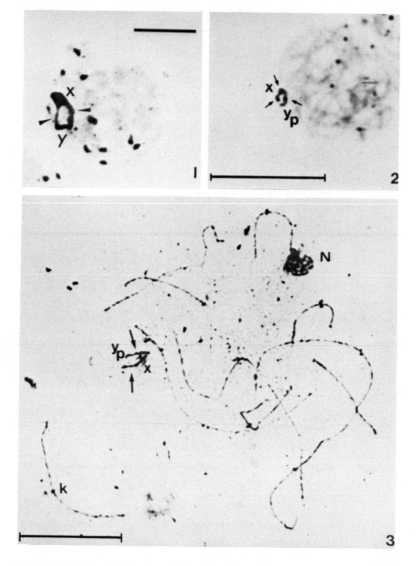

FIGURE 11.   Silver staining of prophasic nuclei of spermatogenesis in *Compsus maricao* (Curculionidae); and (2) and (3), *Chelymorpha variabilis* (Chrysomelidae); (1) Autosomal centromeric regions and Xy_p totally marked. Terminal association (arrowheads) of X_p and y_p; (2) as in 1; the arrows show the centromere of X_p and telomeric association zones; (3) X_p, y_p, and 10 autosomal synaptinemal complexes in a spreading preparation. N = nucleolus; k = autosomal centromere; arrows: centromere of X_p and y_p. Bars: 10 μm. [1] from Virkki unpubl.; [2] and [3] from Protiglioni, A. and Brum-Zorrilla.[248a]

numbers). Hsiao and Hsiao,[91] however, could not find a correlation between chromosomal data and host plants or geographical distribution, in the chrysomeline genera *Leptinotarsa* and *Labidomera*.

V. TRENDS IN SEX CHROMOSOME EVOLUTION

The coleopteran sex chromosomes are classified according to their mode of meiotic association in the male. This includes complicated multivalents where quite possibly many of the chromosomes, although markers of the sexual dichotomy, are not involved in the determination of sex. Therefore I prefer a genetically less compromising expression "sex chromosome system" to the much used "sex determination mechanism".

Karyotypic formulae, essentially written statements of the M I situation in the male, are equally exact as, but more informative than, the plain $2n$ numbers. The both logical and flexible rules reiterated by Smith and Virkki,[169] are not observed by all workers, and some confusion is the result.

Since publication of our monograph,[3] two new sex chromosome systems have been found for Coleoptera. Multiple sex chromosomes have been recorded for the first time in Caraboidea.[37,38] A naturally occurring, chiasmate XY_1Y_2,[37] as well as the scolytidean $neoX_pneoYy_p$ predicted by Smith and Virkki[3] is novel for the whole Coleoptera.

A. Xy_p, the Parachute Bivalent

Roughly half of the Polyphaga show the parachute association of sex chromosomes in the male meiosis. Since John and Lewis[170] interpreted it as to be nucleolar in nature, most beetle cytologists have agreed. Smith and Virkki[3] suggested that the sex nucleolus might be composed of two parts. The central part behaves as a prophasic nucleolus. Another superficial and late deposited layer (blister) lasts until A I and thus assures regularity of segregation of the sex chromosomes.

The John-Lewis hypothesis is now challenged by Uruguayan cytologists, whose preliminary microscopical and ultrastructural studies have not revealed NORs (nuclear organizing regions) in the sex chromosomes, nor nucleolus in Xy_p.[95-99] They suggest that the association is by telomeric adhesion. A Jerusalem team[171] similarly failed to see any nucleolus in the sex multivalent of *Blaps cribrosa*, where the orientation and segregation of the sex chromosomes are controlled by a "segregation body", probably composed of RNA/RNP. It might represent a colossally boosted "blister" of normal Xy_ps.[3]

The silver staining for NORs marks centromeric regions of autosomes and the sex chromosomes in their totality in the early spermatogenesis of both the chrysomelids studied in Montevideo and in the curculionids *Compsus maricao* and *Peridinetus signatus* now under study by me (Figure 11). This resembles closely what Quack and Noel[172] and Noel et al.[173] found in mammals. Centromeric markings could indicate the presence of nucleolar material as found in *Chilocorus*.[174] *Chilocorus* spp. seem to have no other NORs than the centromeric ones, in contrast to the cassidine *Chelymorpha variabilis*, with a brachial autosomal NOR[96] (Figure 11 [3]). In accordance with the slowness of the meiotic RNA turnover rate,[175] the total marking of Xy_p could mean RNA lagging *in situ*. Ultrastructural studies by the Montevideo team have not revealed such RNA, however. Silver marking is possibly therefore chromosomal condensation rather than any molecular structure.

In other beetles (Figure 12), the presence of a sex nucleolus seems unquestionable. It seems thus possible that there are both nucleolar and nonnucleolar parachutes. This would emphasize the importance of a persistent chromosome-to-chromosome contact in the maintenance of Xy_p just as in the adephagan sex chromosome associations.[33]

Although the parachute association modus is ancient, the structure of its component chromosomes need not be so. Variation of X_p:y_p size relationships in the *Monochamus scutellaris-oregonensis* complex[177] is the classical example of this. The y_p is often tiny, but can approximate the size of X_p, for instance in *Pissodes* weevils possessing the SIS system, which supposedly originated in an interchange between y_p and an autosome.[161] Exophenotypic evidence compels Lanier[146] to suggest *de novo* formation of Xy_p in a series of very closely related *Dendroctonus* species: $6 + Xy_p \rightarrow 5 + neoX_pneoYy_p \rightarrow 7 + Xy_p$. But even so, the parachute remains the same old continuum; only the associating chromosomes change.

1. Numerical Variation of X_p and y_p

The parachute has a limited capacity for controlling extra chromosomes. Since Kacker et

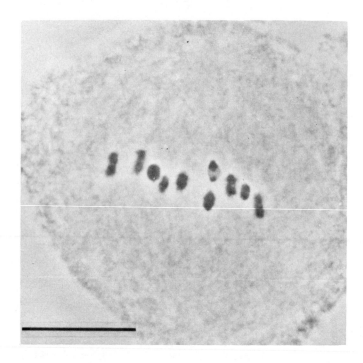

FIGURE 12.   *Pleocoma crinita* (Scarabaeidae); M I, $9 + Xy_p$.The lumen of the
parachute is filled by an opaque material. Acetocarmine plus phase contrast; bar:
20 μm.

al.[82] amended their interpretation of the sex multiple of *Aulacophora intermedia* from XXy_p^3
to a chiasmate XXY, only the cerambycids *Monochamus* are known to possess XXy_ps.[177]
Although the Xy_p variants of *Monochamus* show a correct association, one of the X chromosomes of the occasionally encountered XXy_p specimens may remain unassociated. Because the added length of $X_1 + X_2$ sums up to the length of the normal X, the two Xs are
obviously fission products. Thus the centric fission of X might have produced a "somato-
grammic mutation"; e.g., the allegedly predetermined mutual position of the interphasic to
early prophasic chromosomes[178-180] might have suffered a change that reduces the chances
of the extra X to be included in the parachute.

More data exist on extra y_ps. John and Shaw[181] found up to $Xyyy_p$ in some inbred *Dermestes
maculatus* and *frischii* strains. Four generations of selection, always favoring the multiple
condition,[182] produced even $Xyyyy_p$ animals. As a segregation device, the parachute with
multiple loads remained safe only up to Xyy_p. The third and the fourth y_p were often
outcrowded from the parachute but, remarkably, the univalent y_ps then tended to cosegregate
with the parachute y_ps. This hints at a possible orientative cooperation between the sex
chromosome centromeres[3] and contributes to the accumulation of the y_ps. As errors of y_p
segregation nevertheless occurred both in mitosis and meiosis, y_ps also are expected in
females, but this never occurred in *Dermestes*. Because the sex ratio was slightly but
significantly shifted to favor males, it is likely that y_ps are lethal in the female.

These *Dermestes* studies have provoked interest in extra y_ps, and many workers are now
reporting them in diverse polyphagans.[45,100,108,123,137,138,183]

The extra y_ps can apparently arise in all Xy_p species, either by accumulation of y_ps, or
by their centric fission. The segregational competence of the Xyy_p condition helps in their
perpetuation. There are no species with an established Xny_p system, however.

A   B

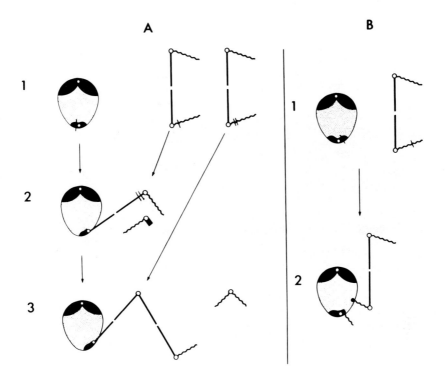

FIGURE 13. Incorporation of autosomes into sex parachute by means of centric fusion with y_p and neoX (A), or of segmental interchange with y_p (B). (A), 1 = initial stage; 2 = first incorporation: $X_p$neoXneoY_p as in *Pityogenes fossifrons,* plus a totally heterochromatic translocate; before this disappears, it may associate with the parachute by virtue of its y_p remnant; 3 = second incorporation: $X_p$neoXneo$Y_p$neoY^2 as in *Heikertingerella brevitarsis,* plus a totally heterochromatic reciprocal translocate. (B) 1 = initial stage; 2 = the segment derived from y_p incorporates the autosomal bivalent into the parachute, as, for instance, in *Pleocoma crinita.*[176] The orientation of the bivalent in relation to Xy_p becomes significant only if the segment derived from y_p contains genes that must be confined to the male. A similar association would probably result from segmental interchange with X_p. (From Smith, S. G. and Virkki, N., in *Animal Cytogenetics: Coleoptera,* John, B., Ed., Borntraeger, Berlin, 1978, 126. With permission.)

2. Incorporation of Autosomes in Xy_p

The Tenebrionid genus *Blaps* shows a series of sex multivalents from tri- to 18-valents. The hypothesis of Lewis and John[184] explains the origin of these multivalents up to the level 16 + XXXY, in *B. mucronata,* through incorporation of autosomes to the Xy_p system by repeated translocations with X_p. The initial condition was supposed to be a tetraploidized hybrid karyotype, which would tolerate loss of autosomes, and avoid autosomal multivalents at meiosis.

This hypothesis does not apply to all *Blaps* karyotypes and proposes availability of several monophasic metacentrics capable of forming one chiasma per arm. In addition, there is no convincing evidence for polyploidy in bisexual beetles, there is total loss of y_p and the theory does not explain why the X_p segment translocated to the autosome does not associate with the parachute. It has to consider position effects caused by rearrangement of euchromatic arms.

These limitations can be overcome if **diphasics** are incorporated into the parachute by repeated translocations to the y_p.[3,185,186] A translocation fusing a centromere-containing fragment of y_p with the heterochromatic arm of a diphasic autosome would result in a new viable chromosome consisting of the euchromatic arm and centromere of the initial diphasic, joined to the acentric fragment of y_p. The reciprocal translocate would be a totally heterochromatic

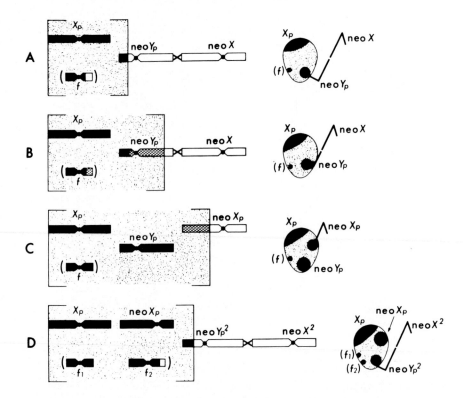

FIGURE 14.   Evolution of *Blaps mucronata*-like parachutes by repeated translocation of autosomes to Y chromosome. The reciprocal centric fragment (f) may or may not survive. Left column: heterochromatinization marked by a progressively darkening color, extension of parachute association by the stippled area; Right column: the presumed appearance of the parachute at M I. (A) Interchange between an autosome and one of the arms (plus centromere) of y_p has produced this unichiasmate mixed association, $X_p$neoXneoY$_p$. This is a factual parachute found in *Pityogenes fossifrons*. (B) Heterochromatinization has reduced the pairing segments of neoX and neoY$_p$ close to a minimum. (C) Further heterochromatinization has substituted the chiasma between neoX and neoY$_p$ by nucleolar association. (D) Second autosomal incorporation by a translocation between an autosome and neoY$_p$ has produced the mixed association, $X_p$neo-X$_p$neoX^2/neoY^2_p, which has been found (without fs) in *Blaps mucronata*, *mortisaga*, and *waltli*. One f (which could be marked as neoY^f_p) survives in the parachute of *tenuicollis*: $X_p$neoX$_p$neoX^2_pneoX^{3}neo^f_pneoY^3_p. (From Smith, S. G., and Virkki, N., in *Animal Cytogenetics: Coleoptera*, John, B., Ed., Borntraeger, Berlin, 1978, 129. With permission.)

chromosome consisting of the heterochromatic arm of the diphasic and the centric fragment of y_p (Figure 13A). Possible important genes of y_p have a chance to survive, because neochromosome can associate with the parachute by virtue of its y_p segment.

Further diphasics can be incorporated in the parachute through *Chilocorus*-type centric fusions resulting in a loss of the heterochromatic arms (Figure 13A3). Where shorter pieces of a diphasic and y_p are interchanged, the nucleolar association of the diphasic bivalent may remain ephemeral, and the rearranged y_p retains its status as sex chromosome (Figure 13B).This condition is sometimes seen in the nature (e.g., *Pleocoma crinita*.)[3,176]

The first proof for this hypothesis came from Syracuse (New York), where the situation in Figure 13A2 was recorded in a scolytid, *Pityogenes fossifrons*, by Lanier.[219] A similar trivalent was reported by de Vaio and Postiglioni[187] in a cassidine, *Botanochara angulata*. It, however, lacks the sex nucleolus. The northern race of *Blaps judaeorum*, of Israel, seems to have a similar condition.[171] In addition, the situation of Figure 13A3 is materialized in a Brazilian flea beetle, *Heikertingerella brevitarsis*.[188]

This idea was then applied to further species of *Blaps* (Figure 14). Figure 14A repeats the situation illustrated in Figure 13A2, and introduces the nomenclature. The X_p remains unchanged, the neochromosome (= the translocate diphasic/y_p) is called neoY_p, and the unchanged diphasic, neoX. The structural formula of the trivalent is thus $X_p$neoXneoY_p. The reciprocal heterochromatic neochromosome (f) may or may not survive, depending on its capability to associate with the parachute. B to D show the progressive heterochromatinization and substitution of the chiasmate association neoX/neoY_p by a nucleolar one. Translocation between a diphasic autosome and neoY_p results in the quadrivalent of *Blaps mucronata*, $X_p$neo$X_p$neoX^2neoY^2_p, with a chance of survival for the two totally heterochromatic reciprocal translocates (f^1 and f^2)

This scheme explains all but one of the known *Blaps* and *Caenoblaps* karyotypes; utilizes diphasics, resulting in a minimum of genetic punishment; explains the initial parachute association of neoY; leaves the original X_p untouched; and offers a survival chance to the possible important genes of y_p.

The only *Blaps* species which does not fit into this scheme is *cribrosa* (9 + 6X_p6neoX_p6neoY?). Even this karyotype can be derived through translocations from simpler ones,[3,185] but the complications are too numerous. A sextuplicate multiplication of the sex chromosomes suggested by Wahrman et al.[171] could have occurred in a parachute stage similar to Figure 13B2, from which the y_p has disappeared and the autosomal "bivalent" orientates inversely.

It is difficult to see how these courses of evolution could have succeeded in associations lacking a nucleolus as an adhesive agent, as has occurred at least in *Botanochara*[98,99] and *Cicindela*.[33] Studies including electron microscopy have shown that the nucleolus plays no role in their sex multivalents; instead, ectopic and/or telomeric adhesions do. The *Cicindela* affair cannot be, as was presumed earlier,[186] parallel to that of *Blaps*.

Although there is some evidence of increasing the autosomal number by fission in *Cicindela*, nothing supports incorporation of autosomes into the sex chromosome system; actually, the lowest autosomal number accompanies the simplest sex chromosome system: 8 + XY in *C. germanica*.[33]

Because the Xy_p has survived more speciation processes than any other heterosomal pairing modus, it can be justly viewed as the most competent sex bivalent. Studies on its structure and ultrastructure, together with improved methods of chromosome analysis, should determine how much of the theoretical "Glasperlenspiel" played with it and its modifications will survive.

B. Inherently Asynaptic Sex Chromosomes

Sex chromosomes starting to lose their ancestral, competent association at male meiosis may turn into errant univalents. The closest attempt at rescue would be shifting their centromeric cycle towards a mitotic "maturity" and thus amphiorientation, which would assure congression of the chromosomes, but not their reduction. The latter can be controlled by adhesion of chromatidal telomeres ("distal collochores"). An adhesion of moderate duration leads to a belated prereduction of the *Altica* type also found in the Diptera (Tipulidae), and interpreted in terms of the one-component hypothesis[189] of spindle function.[190] A more persistent adhesion leads to a rarer postreduction of the *Hyperaspis* type.[3] Within the same coccinellid genus, the segregation of synoriented asynaptics has been solved: the X and y of *H. billoti* move precociously to opposite poles at M I.[3] This "neuropteroid" system has polyphyletically emerged in the families Tenebrionidae *(Zopherus haldemani)* and Chrysomelidae *(Stenophyma* sp.). The primitive chrysomelids Megalopodinae and some of their more advanced kin in the subfamily Clytrinae show asynaptic, post-reductional sex chromosomes.[106,147] Yadav's[106] illustrations show similarity of A II of *Diapromorpha turcica* to the A I of *Altica*. It seems that the interchromatidal telomeric adhesion is switched from spermatocyte I to spermatocyte II in such clytrids.

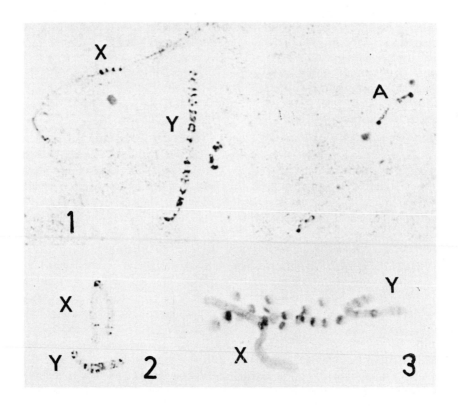

FIGURE 15.   *Omophoita octoguttata* (Chrysomelidae: Oedionychina); Ag-banding for NORs (1 and 2), and C-banding (3). (1), Diffuse diplotene; numerous Ag-bands in the sex chromosomes, especially the Y. One autosomal bivalent (A) showing procentric Ag-bands and adjacent nucleoli. (2) M I; numerous Ag-bands still present in the sex chromosomes. (3), Procentric C-bands in the autosomes and X, intercalar ones in the Y. 1300×.

The most elaborate mechanism for a prereductional segregation of asynaptic, synoriented sex chromosomes is found in the modern flea beetles Oedionychina. Their giant-size sex chromosomes do not move from the prometaphasic clump area, where a separate spindle portion is organized. Their "distance bivalent" is dissolved simultaneously with autosomal bivalents.

It is one of the puzzles of the karyokinesis theory how a correct segregation can be ascertained under such conditions. Virkki[191] suggested that the centromeres might be quantitatively different, the centromere of X requiring as much microtubule precursor as the y centromeres together. Provided that both hemispindles of the "sex spindle" have a limited pool of precursor available, then the only satisfactory final orientation at M I would be X/ny.

A further complication of the multiple distance pairing system is an interchange between X and an autosome, which leads to systems neoXY + y_1 . . . +y_n in some *Phenrica* and *Disonycha* spp.[3]

One is tempted to suggest that the Oedionychina system is a derivative of the megalopodine system through accretion of the sex chromosomes. In the Hepaticae, the mosses, the only group having large sex chromosomes of a similar meiotic behavior, gigantism develops in the autosomes, and is translocated on originally small sex chromosomes.[192] Unfortunately, the spermatocytes of the Megalopodinae are too small (512 sperm cells per bundle!) for a convenient comparison with Oedionychina. Disonychina, the chromosomally and taxonomically complex sister subtribe may be more useful in understanding evolution of the Oedionychina system.

Silver staining marks numerous sites (Ag-bands) of constant location in the diplotenic sex chromosomes, especially the Y, of the male Oedionychina (Figure 15). These sites are apparently responsible for production of large masses of "nuage" (RNP?) ultrastructurally similar to so-called chromatoid bodies.[105] This material passes to cytoplasm through the nuclear pores aggregated to "sieves" in the nuclear envelope of the spermatocyte I of *Alagoasa* and allied genera. This condition is derived from the random pore distribution in *Omophoita* spp. Because silver markings are found also at the cytoplasmic side of the nuclear sieve complexes,[147] gene amplification might be involved in production of the "nuage".

Though by C-banding criteria[104] (Figure 15 [3]), the large sex chromosomes are predominantly euchromatic, yet more probably they, like the heterochromatic arms of the A autosomes of *Carabus* spp.,[22,166] contain large rDNA reserves destined to production of rRNA. Why especially the Y chromosomes are involved in this affair in Oedionychina is still unknown.

Moderate doses (250r) of acute gamma radiation produce profuse rearrangements between the sex chromosomes, and between them and the autosomes.[103] Surprisingly, similar rearrangements have never been found in natural populations of Oedionychina, not even in samples collected from Morro-do-Ferro, a site of strong natural radiation in Poços de Caldas, Brazil.[147]

VI. EXTREME KARYOTYPES

In Coleopteran families and lower taxa, orthoselective fusion and fission series, combined with other rearrangements, have often led to altered chromosome numbers. Since Ferreira et al.[47] found the chromosomes of a Brazilian pyrophorine, *Chalcolepidius zonatus* (Figure 16), and Serrano[21] those of a carabid, *Ditomus capito obscuroides,* the total range of coleopteran chromosome numbers now extends from $2n = 4$, $1 + XY$, to $2n = 69$, $34 + X$.

The two acrocentric and two metacentric chromosomes of *Ch. zonatus* are of similar euchromatic structure. The luciferous Pyrophorini tend to lowered numbers, $8 + X$ being common. *Hemirrhipus lineatus,*[163] also from Brazil, has $4 + $ neoXY. This might have arisen from $8 + X$ through centric fusion of autosomes, and incorporation of the sex determining segment into one of the autosomes (Figure 17). A full series of pericentric inversions and subsequent centric fusions would lead to $1 + XY$; the acrocentry of one of the pairs in *Ch. zonatus* requires, however, one more pericentric inversion or centromeric shift. All constitutive heterochromatin that might have facilitated these rearrangements is dispersed at this stage. The regular association of the metacentrics in a ring bivalent suggests homosequentiality at least at the distal, chiasma-frequented parts. This slightly increases the alternative that the sex determining segment might be incorporated in one of the monochiasmatic acrocentrics.

The former low record for Coleoptera, $3 + XY$ (or Xy) in the West Indian flea beetle, *Homoschema nigriventre,*[193] necessarily has a similar fusion history accompanied by other rearrangements. Its origin is obscure because only one closely related karyotype, $4 + X_1X_2Y$ in *H. obesum,* is known. Wahrman[194] observed $3 + XY$ in a carabid, *Graphipterus serratus* and suggested it as the basic number for Caraboidea. He was contradicted because the evidence for polyploidy in evolution of bisexual beetles is not convincing, and for other valid reasons.[3,21] Exceptionally high chromosome numbers in the beetles are best regarded as products of fissions rather than of polyploidy, even where a multiple sex chromosome system is present, as in *Aulacophora (Rhaphidopalpa) femoralis*[195] and *A. foveicollis*[17] (both $28 + XXY$), and *A. intermedia*[82] ($27 + XXY$), species with close relatives having only half of that number, and Xy systems. The present high record, $34 + X$ in *Ditomus capito obscuroides,* indeed has a simple sex chromosome system, probably arisen through an incomplete (or complicated) fission-*cum*-inversion series from the ancestral caraboidean karyotype, $18 + X$.

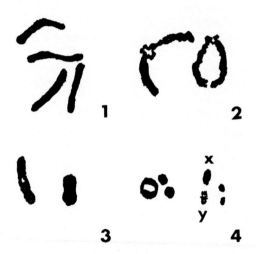

FIGURE 16.   *Chalcolepidius zonatus* (Elateridae: Pyrophorini), (1) to (3),and *Brasixenos* near *occidentalis* (Stylopoidea), (4). (1) $2n = 4$. (2), Diakinesis, rod and ring bivalent. (3), M I;3 + Xy. The slightly heteromorphic sex bivalent is marked x-y. (Courtesy of Ferreira, A., Instituto de Biociências, Universidade Paulista, Rio Claro, São Paulo, Brazil. With permission.)

An extreme karyotype has also been found in a putative beetle ally, *Brasixenos* near *occidentalis,* a stylopoid from Brazil.[148] The position of Stylopoidea (Strepsiptera) is uncertain. Some relate it with Hymenoptera,[196] most with Coleoptera.[11,35,197,198] According to Crowson,[198] its roots are to be sought among lymexyloids, not far from Mordellidae and Rhipiphoridae. For a long time, only one stylopoid karyotype was known: $2n = 16$, 7 + XY:XX, in *Achroschismus wheeleri.*[199] The new Brazilian karyotype comprises 3 + Xy. In the former species, all chromosomes look like similar small dots, but the latter has a slightly more asymmetrical karyotype (Figure 16[4]). Its sex bivalent could be a neoXY with a very small residual X segment. Such a karyotype could very well be one more of the polyphyletic coleopteran 3 + XY cases. As haplodiploidy is not involved in these two stylopoids, we can conclude that stylopoids are coleopterous rather than hymenoptenous insects.

VII. POLYPLOIDY AND PARTHENOGENESIS

In the past decade, notable progress has been achieved in this field by Finnish and Japanese beetle cytogeneticists. Three kinds of parthenogenesis have been encountered: arrhenotoky (sex determination by haplodiploidy), deuterotoky (both sexes produced parthenogenetically), and thelytoky (including pseudogamy).

Because switching from a normal bisexuality to arrhenotoky involves rebuilding of the sex determination mechanism, it does not occur readily. In Coleoptera, the change has taken place at both extremes of the phylogenetic system.

The classical coleopteran case of arrhenotoky is that of *Micromalthus debilis.*[200-203] Here the haplodiploidy is more a sex determination mechanism than a mode of reproduction, because the latter includes arrheno-, deutero-, and thelytoky, all in the larval stage (paedogenesis). Another hymenopteroid characteristic in male beetles is a monopolar M I — A I. These bizarre affairs are the more intriguing because *Micromalthus* is the sole chromosomally known representative of the primitive suborder Archostemata.

Further cases of arrhenotoky occur at the level of the highest derivation, in the tribe Xyleborini of Scolytidae (probably also in *Coccotrupes tanganus*[204]). Here females are mated

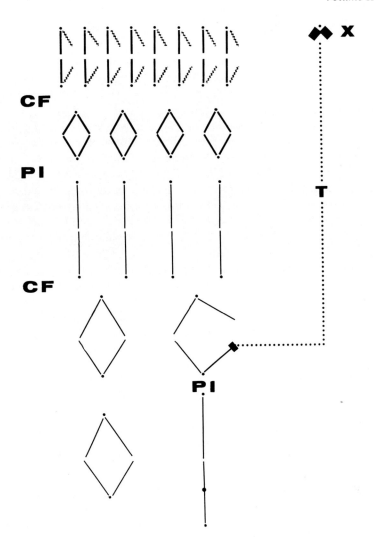

FIGURE 17. A hypothetic derivation of *Chalcolepidius zonatus* karyotype, 1 + XY, from the modal pyrophorinian 8 + X, through centric fissions (CF), pericentric inversions (PI), and incorporation to an autosome of the X segment containing sex determinators (T). Presence of diphasic autosomes in the initial karyotype is purely hypothetic; some kind of common architecture facilitating such serial rearrangements must be anticipated.

by their brothers soon after ecclosion, and give rise to both sexes. This is probably mechanically controlled by intermittent release of sperm from spermatheca. Females that remain unfertilized produce males only.[205-209] The haploid chromosome number of the male is conserved by two equational divisions in the meiosis. The same pattern is also found in occasional diploid males showing asynapsis of all chromosomes.

Thelytoky has been found in four families of beetles: Cisidae, Chrysomelidae, Ptinidae, and Scolytidae, and suspected in many others.[3] Chromosome pairing and the reduction division have been abolished, the maternal karyotype being inherited by the daughters without any genetic recombination. Only four thelytokous species are diploid or have diploid races: *Cis fuscipes*,[210] *Polydrosus mollis*,[211,212] *Scepticus insularis*,[213] and *Catapionus gracilicornis*.[122] All others have been polyploidized up to hexaploidy. Often the diploid bisexual race exists, either sympatrically with the parthenotes,[122,213,214] or geographically isolated from them (*Adoxus obscurus*,[215,216] and various Curculionid cases). The bisexuals tend to have a much more ancient and restricted distribution (but perhaps not so in the leaf beetles *Calligrapha*[217]). The pattern of geographic isolation of the parthenotes from their bisexual races can be traced back to the extension of ice of the last glaciation in Europe, but not so in Japan. Refuges that remained free of ice in Europe conserved the diploid bisexuals, whereas regions that emerged when the ice melted were colonized by polyploid parthenotes.[218] In Japan, such glacial time refuges are now also populated by polyploids.[126]

The ptinid *Ptinus clavipes mobilis* is pseudogametic (gynogenetic) and triploid, and so are some scolytids of the genus *Ips*.[219] In the former, the sperm is provided by males of the bisexual *P. clavipes*.[220] Sanderson and Jacob[221] held this to indicate a recent origin of thelytoky in *mobilis*. As long as *mobilis* depends on its putative parent taxon *clavipes*, it is confined to sympatry with it (which has a limited geographic meaning in these grain pests).

The thelytokous polyploids are autopolyploids. The hypothesis that thelytoky appears first, and then fertilization of an unreduced egg by a haploid spermatozoon produces a triploid female,[212,222,223] has been generally accepted. The latter part of this hypothesis is now challenged by Takenouchi.[124] Embryos developing in eggs of $2n$, $4n$, and $5n$ females of *Catapionus gracilicornis* (Curculionidae), fertilized by diploid bisexual males, did not show increased chromosome numbers. Cold treatment of eggs (0—7°C)[121,129] partially altered the ploidy level in three ways: haploid eggs of a bisexual race produced triploid and tetraploid embryos; original chromosome number of diploid and triploid parthenotes was doubled; ploidy level of a parthenote was reduced by one or more haploid sets.

In the first case, the shock was given during the meiotic divisions of fertilized eggs, which proceeded until the three polar bodies (→ triploidy) and also the sperm nucleus (→ tetraploidy) coalesced. In the second case, the shock was so much earlier that endopolyploidization of the egg chromosomes took place. The third case was due to enhanced "gonomery",[116,212,214,222,224] the tendency of one or more whole sets of chromosomes to become separated from the rest at the metaphase of the egg.

These experiments by Takenouchi stress that polyploidy in curculionids arises in females without participation of males and is controlled by naturally occurring low temperatures. In accordance with such a simple origin, it is comprehensible why polyploid parthenotes show a lower degree of allelic polymorphism than corresponding diploid bisexuals, a conclusion emerging from extensive starch gel electrophoresis studies by Suomalainen and his associates.[83,130-135,225-227] As the allele frequencies found in parthenote populations were not in a Hardy-Weinberg equilibrium, these authors concluded that evolution by mutation and selection still continues. The same was inferred already from morphometric discrepancies between different isoploid populations of *Otiorrhyncus scaber*,[228] a conclusion which was received with scepticism at that time.[153,229]

Although genetic recombination is abolished from apomictically thelytokous organisms, they still possess a notable adaptability to changing environments, because of different cloned genotypes, arisen through mutations subjected to selection.[223] Polyploidy increases heterosis by allowing more than two allelic enzymes per locus per cell. The effect is expressed most efficiently at the triploid level. Therefore, apparently, triploidy is the most common ploidy level in the apomictic parthenotes. An increasing heterosis initially improves the fitness of apomictic parthenotes, but accumulation of many recessive lethals finally reduces it.[226,227] Therefore, contemporary thelytoky tends to be of recent origin, contrary to arrhenotoky, which has arisen just a few times in the insects, and now tends to involve major taxonomic groups.

ADDENDUM

Some last-moment information received after the above review was closed and typed deserves to be mentioned.

A. Caraboidea

Serrano[230] will shortly report new chromosome data on Caraboids belonging to eleven tribes. By liaison of Dr. Serrano, I became aware of recent Polish studies on Carabidae.[231-233]

B. Cantharidae

Gambardella and de Vaio[234] have published a monograph on the morphology and biology of *Chauliognathus scriptus,* including chromosome studies. The species has $2n = 12 + X$; $6 + X$. All chromosomes are telocentric. One pair is quite large and probably arose through a centric fusion followed by a pericentric inversion. The X chromosome is postreductional.

C. Coccinellidae

Srivastava[235] reported a possible case of a balanced lethal system in *Epilachna* spp. in the surroundings of Allahabad: the largest autosome pair is invariably heteromorphous. As Professor Srivastava is now retired, the case is left to attention of a younger generation of Indian cytologists.

D. Scarabaeidae and Tenebrionidae

Vidal[236] observed late individualization of chromatids in procentric (C-bandable) heterochromites of a scarab, *Gromphas lacordairei* ($2n = 20$), and a Tenebrionid, *Scotobius tristis* ($2n = 18$). This appears to be a naturally occurring phenomenon parallel to the colchicine effect (apposition of heterochromatic chromatids) in *Chilocorus*.

E. Chrysomelidae

Petitpierre and his associates report further Chrysomelid studies on *Chrysolina* (Chrysomelinae),[237] European and Tasmanian Chrysomelinae,[238] *Cryptocephalus* (Cryptocephalinae),[239] and several European Alticinae.[240] Association of primitive ($n = 12$) *Chrysolina* with Labiatae is again encountered; karyotypes of this species group differ principally in arm ratios (due to pericentric inversions?) and presence or absence of y_p. Relation of host plant selection to karyotype evolution in Chrysomelidae has been discussed more thoroughly by Jolivet and Petitpierre.[241,242]

The Montevideo team of cytologists has been successful in discovering several cases of parachute complications in this and the following family. In *Botanochara* sp., externally very close to *B. angulata*, Panzera[243] found $20 + X^1_pX^2_p$neoXneoY$_p$, with both X_ps and most part of the autosomes telocentric. The two X_ps are presumably fission products of an originally metacentric X_p. The same formula was found also in *B. 12-verrugata*,[244] and in *B. bonariensis*,[245] a species resembling externally one of the color phases of *Stolas lacordairei*, which has most autosomes metacentric, one pair of them affected by a pericentric inversion.[246] These cassidine affairs are partly summarized in an abstract.[247]

F. Curculionidae

The Curculionid tribe Hylobiini, known for serial pericentric inversions, shows modified parachutes in Uruguay.[248] *Heilipus tremolerasi* has two karyoformulas, $14 + Xy_p$ and $13 + $ neoX$_p$neoYy$_p$, the latter paralleling the case just reported by Lanier[146] in *Dendroctonus* (see Section III). The parachute of *Heiliopodus scaber* is more complicated, the formula being, presumably, $13 \times X_p$neoXneoYy$_p$. This is not far from the case of *Heikertingerella brevitarsis* (Figure 13A3) the difference being that in *Heiliopodus*, the first autosome pair seems to be incorporated through an interchange with X_p, instead of y_p.

* * *

Additional recent reports (as of June 1983) include the following:

Wise et al.[249] have reported $19 + XXY$ for the curculionid *Anthonomus grandis*. (Compare with North et al.[9])

An abstract concerning the new low record of the coleopteran chromosome numbers[47] has now been widened to a brief, illustrated article.[250]

Mesa and Fontanetti[251] communicated a new coleopteran record of chiasmate sex multivalents: $X_1X_2X_3Y_1Y_2Y_3$ and $X_1X_2X_3Y_1Y_2$ in a Brazilian buprestid, *Euchroma gigantea*. These karyotypes also contain large swarms of fragments. The species has a wide distribution from Mexico to Argentina, with several subspecies determined.[252] Thus a future discovery of most interesting karyoclines seems possible.

Allele studies by electrophoresis have now been extended to the Japanese curculionid parthenotes. The first results of collaboration between Finnish and Japanese cytologists are already in press.[253]

Angus[254] revised the status of the European hydrophilids, *Helophorus aquaticus* L. and *H. aequalis* Thomson, relying much on chromosome banding, karyometry, and silver staining of NORs. The controversial G-banding[55] was thoroughly evaluated, and seems to be of use here.

Kacker's article of 1976[255] and most later papers published in Indian journals by this author and his associates have not been available for review.

REFERENCES

1. **Dawson, P. S.,** Linkage group IV of *Tribolium castaneum, Can. J. Genet. Cytol.,* 14, 675, 1972.
2. **Dawson, P. S.,** Sex and crossing over in linkage group IV of *Tribolium castaneum, Genetics,* 72, 525, 1972.
3. **Smith, S. G. and Virkki, N.,** *Animal Cytogenetics: Coleoptera,* John, B., Ed., Borntraeger, Berlin, 1978.
4. **Smith, S. G.,** Heterochromatin, colchicine, and karyotype, *Chromosoma,* 16, 162, 1965.
5. **Ennis, T. J.,** Chromosome structure in *Chilocorus* (Coleoptera: Coccinellidae). I.Fluorescent and Giemsa banding patterns, *Can. J. Genet. Cytol.,* 16, 651, 1974.
6. **Ennis, T. J.,** Studies on heterochromatin and chromosome banding in *Chilocorus* (Coleoptera: Coccinellidae), Ph.D. thesis, University of Toronto, Toronto, Canada, 1974.
7. **Martins, V. G.,** Citogenética de dezenove espécies de Scarabaeoidea (treze Passalidae e seis Scarabaeidae) — Coleoptera, Master's thesis, Universidade Estadual Paulista, Rio Claro — São Paulo, Brazil, 1982, 29.
8. **Vidal, O. R.,** Algunos aspectos de la meiosis en Coleópteros, *Actas IV Congr. Latinoam. Genét.,* 2, 153, 1980.
9. **North, D. T., Leopold, R. A., and Childress, D.,** Meiotic and mitotic chromosomes of the cotton boll weevil (Coleoptera: Curculionidae), *Can. J. Genet. Cytol.,* 23, 443, 1981.
10. **Virkki, N.,** A chiasmate sex quadrivalent in the male of an Alticid beetle, *Cyrsylus volkameriae* (F.), *Can. J. Genet. Cytol.,* 10, 898, 1968.
11. **Boudreaux, H. B.,** *Arthropod Phylogeny with Special Reference to Insects,* John Wiley & Sons, New York, 1979.
12. **Smart, J. and Hughes, N. F.,** The insect and the plant: Progressive palaeocological integration, in *Insect Plant Relationships,* van Emden, H. F., Ed., Royal Entomological Society of London, 1973, 143.
13. **Solari, J. A.,** The spatial relationships of the X and Y chromosomes during meiotic prophase in mouse spermatocytes, *Chromosoma,* 29, 217, 1970.
14. **Hughes-Schrader, S.,** Segregational mechanisms of sex chromosomes in Megaloptera (Neuropteroidea), *Chromosoma,* 81, 307, 1980.
15. **Brandmayr, P. and Kiauta, B.,** Cytotaxonomic notes on the endemic groundbeetles (Coleoptera, Adephaga, Carabidae) of Southeastern Alps and the Dinaric Karst. II. *Carabus (Platycarabus) creutzeri rinaldoi* Meggioclaro, 1958, and *C. (Megodontus) croaticus* Dejean, 1826, *Genen Phaenen,* 16, 73, 1973.
16. **Lahiri, M. and Manna, G. K.,** Chromosome complement and meiosis in nine species of Coleoptera, *Proc. 56th Indian Sci. Congr.,* 3, 448, 1969.
17. **Manna, G. K. and Lahiri, M.,** Chromosome complement and meiosis in forty-six species of Coleoptera, *Chrom. Inf. Serv.,* 13, 9, 1972.
18. **Sanyal, M.,**Meiosis in Xy_p males of two species of Carabidae, *Proc. 1st All India Congr. Cytol. Genet.,* Chandigarh, 1971, 57.
19. **Suortti, M.,** Spermatogenesis of some species of Dytiscidae (Coleoptera), *Ann. Zool. Fenn.,* 8, 390, 1971.

20. **Smith, S. G.**, The cytology of *Sitophilus (Calandra) oryzae* (L.), *S. granarius* (L.), and some other Rhynchophora (Coleoptera), *Cytologia*, 17, 50, 1952.
21. **Serrano, J.**, Chromosome numbers and karyotypic evolution of Caraboidea, *Genetics*, 55, 51, 1981.
22. **Weber, F.**, Die interspezifische Variabilität des heterochromatischen Armes eines Chromosoms bei der Gattung *Carabus* L. (Coleoptera), *Chromosoma*, 23, 288,1968.
23. **Weber, F.**, Korrelierte Formveränderungen von Nukleolus und nukleolusassoziiertem Heterochromatin bei der Gattung *Carabus* (Coleoptera), *Chromosoma*, 34, 261, 1971.
24. **Mossakowski, D. and Weber, F.**, Korrelation zwischen Heterochromatingehalt und morphologischen Merkmalen bei *Carabus auronitens* (Coleoptera), *Z. Zool. System. Evolutionforsch.*, 10, 291, 1972.
25. **Mossakowski, D. and Weber, F.**, Chromosomale und morphometrische Divergenzen bei *Carabus lineatus* und *C. splendens* (Carabidae). I. Ein Vergleich sympatrischer und allopatrischer Populationen, *Z. Zool. System. Evolutionsforsch.*, 14, 280, 1976.
26. **Mossakowski, D. and Weber, F.**, Chromosomale und morphometrische Merkmaldivergenzen bei *Carabus splendens-* und *lineatus-* Populationen, *Mitt. Dtsch. Ent. Ges.*, 35, 109, 1976.
27. **Wilken, H.**, Karyotyp-Analysen bei Carabiden, Staatsarbeit, University of Münster, Germany, 1973.
28. **Nettmann, H.-K.**, Karyotyp-Analyse bei Käfern der Gattungen *Argonum* und *Pterostichus* (Col. Carabidae), Dipl. Arbeit, University of Münster, Germany, 1975.
29. **Nettmann, H.-K.**, Karyotyp-Analysen bei Carabiden, *Mitt. Dtsch. Entom. Ges.*, 35, 113, 1976.
30. **Serrano, J.**, Cytotaxonomic studies of the tribe Carabini (Col., Caraboidea), *Genet. Iber.*, 32-33, 25-41, 1980.
31. **Serrano, J.**, Male achiasmatic meiosis in Caraboidea (Coleoptera, Adephaga), *Genetica*, 57, 131, 1981.
32. **Serrano, J.**, A chromosome study of Spanish Bembidiidae and other Caraboidea (Coleoptera, Adephaga), *Genetica*, 57, 119, 1981.
33. **Giers, E.**, Die Nicht-Homologen-Assoziation multipler Geschlechtschromosomen in der Spermatogenese von Cicindela hybrida (Coleoptera), Ph.D. thesis, University of Münster, Münster, Germany, 1977.
34. **Hammersen, M.**, Die Beeinflussung des Meioseablaufes durch Antibiotika, Staatexamensarbeit, University of Münster, Münster, Germany, 1979.
35. **Arnett, R. H.**, *The Beetles of the United States*, American Entomological Institute, Ann Arbor, Mich., 1968, 553.
36. **Mesa, A. and Fontanetti, C. S.**, Informações prelimineras sobre a citología de *Itu zeus* (Coleoptera, Myxophaga, Torridincolidae), *33ª Reun. An.Soc. Brasil. Progr. Ci.*, Salvador de Bahia, Brazil, 1981.
37. **Serrano, J.**, *Scarites buparius*, a caraboid beetle with an X_1X_2Y sex chromosome system, *Experientia*, 36, 1042, 1980.
38. **Serrano, J.**, XY_1Y_2, a new sex-chromosome system among the caraboid beetles, *Experientia*, 37, 693, 1981.
39. **Serrano, J.**, Diferencias cariotípicas entre *Cicindela maroccana pseudomaroccana*, y *C. campestris* (Col. Cicindelidae), *Bol. Asoc. Esp. Entomol.*, 4, 65, 1981.
40. **Yadav, J. S. and Karamjeet,** Chromosome number and sex-determining mechanism in thirty species of Caraboidea (Adephaga: Coleoptera), *Cordulia*, 6, 20, 1980.
41. **Yadav. J. S. and Karamjeet,** Chromosomal studies on three species of Cicindelidae (Adephaga: Coleoptera) from Haryana, *Zool. Anz.*, 206, 121, 1981.
42. **Dasgupta, J.**, Comparative cytology of seven families of Coleoptera, *Nucleus*, 20, 294, 1977.
43. **Ferreira, A. and Mesa, A.**, Estudos citológicos em tres espécies brasileiras de Coleópteros (Chrysomelidae, Cerambycidae e Meloidae), *Rev. Brasil. Biol.*, 37, 61, 1977.
44. **Vilardi, J. C.**, Estudios cromosómicos en seis especies del género *Tropisternus* (Coleoptera-Hydrophilidae), *Mendeliana*, 5, 9, 1981.
45. **Bidau, C. J.**, Estudios cromosómicos en cuatro especies del género *Berosus* (Coleoptera, Hydrophilidae), *Mendeliana*, 5, 19, 1981.
46. **Virkki, N.**, Pre-reduction of the X chromosome in Lycidae (Coleoptera: Cantharoidea), *Genetica*, 49, 229, 1978.
47. **Ferreira, A., Cella, D., and Ramos Tardivo, J.**, $2n = 4$: O menor número cromossômico dentre os Coleoptera, *Ciência e Cultura*, 34 (in press).
48. **Baragaño, R.**, Nuevo método para el estudio de cromosomas en Coleoptera a partir de hemocitos de estados larvarios, *Bol. Serv. Def. c. Plagas Hisp. Fitopatol.*, 4, 23, 1978.
49. **Virkki, N.**, Brief notes on the cytology of neotropical Coleoptera. II. *Apate monacha* F. (Bostrychidae), *J. Agric. Univ. P.R.*, 61, 519, 1977.
50. **Takenouchi, Y.**, A chromosome study on a cerambycid beetle, *Plectrura metallica* Bates (Cerambycidae: Coleoptera), *La Kromosomo*, 2, 10, 290, 1978.
51. **Takenouchi, Y.**, A chromosome study of two cerambycid beetles (Coleoptera: Cerambycidae), *J. Hokkaido Univ. Educ. Sect. IIB30*, 53, 1980.
52. **Kudoh, K., Abe, A., and Saitoh, H.**, Chromosome studies on beetles. Some chromosomal aspects of eleven species of Lamiinae (Cerambycidae), *Sci. Rep. Hirosaki Univ.*, 21, 53, 1974.

53. **Abe, A., Kudoh, K., and Saitoh, K.,** Chromosome studies on beetles. VIII. A revised and supplemental chromosome study in the genera *Prosopocoilus, Nipponodorcus* and *Macrodorcus* of the Lucanidae, *Sci. Rep. Hirosaki Univ.,* 23, 50, 1976.
54. **Rees, R. W.,** Molecular studies on the DNA of *Dermestes* beetle species, Ph.D. thesis, University of Aberdeen, Aberdeen, Scotland, 1976.
55. **Rees, R. W., Fox, D. P., and Maher, E. P.,** DNA content, reiteration and satellites in *Dermestes,* in *Current Chromosome Research,* Jones, K. and Brandham, P. E., Ed., North Holland, Amsterdam, The Netherlands, 1976, 33.
56. **Al-Taweel, A. A. and Fox, D.,** Germ cell differentiation and kinetics in the testis of *Dermestes* (Coleoptera), *Cytologia* (in press).
57. **McIntosh, R. V.,** DNA contents and mitotic cycles in *Dermestes,* Honours thesis, University of Aberdeen, Aberdeen, Scotland, 1976.
58. **McLean, R.,** Variation in ribosomal and satellite DNAs in the genus *Dermestes,* Ph.D. thesis, University of Aberdeen, Aberdeen, Scotland, 1982.
59. **Takenouchi, Y.,** A chromosome study on a lady-bird beetle, *Propylea japonica* (Thunberg, 1781) (Coccinellidae: Coleoptera), *J. Hokkaido Univ. Educ. IIB,* 30, 1, 1979.
60. **Yadav, J. S. and Pillai, R. K.,** Chromosome studies of six species of Coccinellidae (Coleoptera: Insecta) from Haryana (India), *Nucleus,* 22, 104, 1979.
61. **Dasgupta, J.,** Chromosomes of some Indian Tenebrionidae (Insecta: Coleoptera), *Proc. Indian Acad. Sci.,* 81B, 1-6, 1975.
62. **Sharma, G. P., Handa, S. M., and Sharma, S.,** Karyometrical analysis of three species of Tenebrionid beetles (Coleoptera), *Res. Bull. Panjab Univ.,* 28 III-IV, 197-199, 1977.
63. **Yadav, J. S. and Pillai, R. K.,** Karyological studies on Tenebrionidae (Coleoptera). III, *Zool. Anz.,* 193, 323, 1974.
64. **Yadav, J. S. and Pillai, R. K.,** Evolution of karyotype in Tenebrionidae (Coleoptera: Insecta), *Proc. Dunn Dobzhansky Symp. Genet.,* p. 280, 1976.
65. **Yadav, J. S., Pillai, R. K., and Karamjeet,** Chromosome numbers of Tenebrionidae (Polyphaga: Coleoptera), *Biologia,* 26, 31, 1980.
66. **Mesa, A., Ferreira, A., and Martins, V. G.,** Estudo citológico em três espécies de Passalidae Brasileiros, *Ciência e Cultura,* 29, 594, 1977.
67. **Mesa, A., Ferreira, A., and Martins, V. G.,** The chromosomes of an Australian passalid, *Aulacocyclus edentulus* Macl. (Coleoptera, Passalidae, Aulacocyclinae), *J. Austr. Entomol. Soc.,* 17, 385, 1978.
68. **Martins, V. G.,** Citogenética de dezenove espécies de Scarabaeoidea (treze Passalidae e seis Scarabaeidae) — Coleoptera, Master's thesis, Universidade Estadual Paulista, Rio Claro, São Paulo, Brazil, 1982.
69. **Vidal, O. R., Riva, R., and Giacomozzi, R. O.,** Números cromosómicos de Coleóptera de la Argentina, *Physis Secc. C,* 37, 341, 1977.
70. **Vidal, O. R., Giacomozzi, R. O., and Riva, R.,** Los cromosomas de la subfamília Dynastinae (Coleoptera, Scarabaeidae). I. Inversion pericéntrica en *Diloboderus abderus* (Sturm) 1962, *Physis Secc. C,* 37, 303, 1977.
71. **Vidal, O. R. and Giacomozzi, R. O.,** Los cromosomas de la subfamília Dynastinae (Coleoptera, Scarabaeidae). II. Las bandas C en *Enema pan* (Fabr.), *Physis Secc. C,* 38, 113, 1978.
72. **Yadav, J. S. and Pillai, R. K.,** Chromosomes of Rutelinae. I. Genus *Anomala* Sam. (Scarabaeidae: Coleoptera), *Nucleus* 18, 156, 1975.
73. **Yadav, J. S. and Pillai, R. K.,** Karyotypic studies on five species of Melolonthinae (Scarabaeidae: Coleoptera), *Nucleus,* 19, 195, 1976.
74. **Yadav, J. S. and Pillai, R. K.,** Karyotype orthoselection in *Trox* (Troginae: Scarabaeidae), *Chrom. Inf. Serv.,* 20, 8, 1976.
75. **Yadav, J. S. and Pillai, R. K.,** Chromosome studies on four species of Dynastinae and Cetoniinae (Scarabaeidae: Coleoptera), *Zool. Anz.,* 198, 1977.
76. **Yadav, J. S. and Pillai, R. K.,** Karyological investigations on seven species of Coprinae (Scarabaeidae: Coleoptera), *Caryologia,* 30, 255, 1977.
77. **Yadav, J. S. and Pillai, R. K.,** Cytotaxonomy of *Trox* Fabr. (Troginae: Scarabaeidae: Coleoptera), *Curr. Sci.,* 47, 393, 1978.
78. **Yadav, J. S. and Pillai, R. K.,** Evolution of karyotypes and phylogenetic relationships in Scarabaeidae (Coleoptera), *Zool. Anz.,* 202, 105, 1979.
79. **Yadav, J. S., Pillai, R. K., and Karamjeet,** Chromosome numbers of Scarabaeidae (Polyphaga: Coleoptera), *Col. Bull.,* 33, 309, 1979.
80. **Barabás, L. and Bežo, M.,** Chromosome count in some representatives of the family Chrysomelidae (Coleoptera), *Biológia,* 33, 621, 1978.
81. **Barabás, L. and Bežo, M.,** A contribution to the cytotaxonomy of leaf beetles (Coleoptera, Chrysomelidae), *Biológia,* 34, 845, 1979.

82. **Kacker, R. K., Kulkarni, P. P., and Singh, A.,** Chromosome number in *Aulacophora intermedia* Jacoby (Coleoptera: Chrysomelidae) with reference to multiple sex chromosomes, *Occ. Pap. Zool. Surv. India,* 1982.

83. **Lokki, J., Saura, A., Lankinen, P., and Suomalainen, E.,** Genetic polymorphism and evolution in parthenogenetic animals. V. Triploid *Adoxus obscurus* (Coleoptera: Chrysomelidae), *Genet. Res.,* 28, 27, 1976.

84. **Petitpierre, E.,** A chromosome survey of five species of Cassidinae (Coleoptera: Chrysomelidae), *Cytobios,* 18, 135, 1977.

85. **Petitpierre, E.,** Chromosome numbers and sex-determining systems in fourteen species of Chrysomelinae (Coleoptera, Chrysomelidae), *Caryologia,* 31, 219, 1978.

86. **Petitpierre, E.,** Further cytotaxonomical and evolutionary studies on the genus *Timarcha* Latr. (Coleoptera: Chrysomelidae), *Genet. Iber.,* 28, 57, 1976.

87. **Petitpierre, E.,** Chromosome number and sex-determining system in four species of *Galeruca* Geoffr. (Coleoptera, Chrysomelidae), *Chrom. Inf. Serv.,* 25, 4, 1978.

88. **Petitpierre, E.,** Chromosome studies on primitive Chrysomelids. I. A survey of six species of Criocerinae (Coleoptera, Chrysomelidae), *Cytobios,* 28, 179, 1980.

89. **Petitpierre, E.,** New data on the cytology of *Chrysolina* (Mots.) and *Oreina* Mots. (Coleoptera, Chrysomelidae), *Genetica,* 54, 265, 1981.

90. **Petitpierre, E. and Jolivet, P.,** Phylogenetic position of the American *Timarcha* Latr. (Coleoptera, Chrysomelidae) based on chromosomal data, *Experientia,* 32, 157, 1976.

91. **Hsiao, T. H. and Hsiao, C.,** Chromosomal analysis of *Leptinotarsa* and *Labidomera* beetles (Coleoptera: Chrysomelidae), *Genetica,* 60, 139, 1983.

92. **Sharma, G. P. and Sood, V. B.,** Chromosome number and sex determination mechanism in thirty species of Chrysomelidae (Coleoptera), *Nat. Acad. Sci. Lett.,* 1, 351, 1978.

93. **Sharma, G. P. and Sood, V. B.,** Chromosome studies on *Oides* Weber (Galerucinae: Chrysomelidae), *Chrom. Inf. Serv.,* 26, 26, 1979.

94. **Sharma, G. P. and Sood, V. B.,** Chromosome polymorphism in *Cassida syrtica* Boh. (Coleoptera: Chrysomelidae), *Cytobios,* 25, 17, 1979.

95. **Postiglioni, A. and Brum-Zorrilla, N.,** Bandas C en los cromosomas de tres especies de Stolaine Cassidines (Coleoptera, Chrysomelidae), *Reunión Anual de la Sociedad Argentina de Genética,* Castelar, 1975, 26.

96. **Postiglioni, A., Brum-Zorrilla, N., Wettstein, R., and Alemán, E.,** Regiones organizadoras (NORs) y nucléolo en la meiosis de *Chelymorpha variabilis* (Boheman) (Coleoptera-Chrysomelidae), *XI Congr. Annu. Sociedad Argentina de Genética,* Mar del Plata, 1980, 21.

97. **Postiglioni, A. and Brum-Zorrilla, N.,** Localización de regiones organizadoras nucleolares (NORs) en otra especie con sistema sexual Xy_p *Calligraphs polyspila* (Coleoptera, Chrysomelidae, Chrysomelinae), *Res. Com. J. Ci. Natur. (Uruguay),* 1981, 32.

98. **Postiglioni, A. and Brum-Zorrilla, N.,** Sistemas sexuales $Xy_p/X_p neo Xneo Y_p$ en Coleoptera, su estructura y associación durante meiosis, *Arch. Biol. Med. Exp.,* 78, 1981.

99. **Wettstein, R.,** Unusual mechanisms of chromosome pairing in Arthropoda, in *International Cell Biology,* Schweiger, H. G., Ed., Springer-Verlag, Berlin, 1980.

100. **Vidal, O. R., Giacomozzi, R. O., and Riva, R.,** Los cromosomas de *Typophorus nigritus* Fab. (Coleoptera, Chrysomelidae, Eumolpinae). Polimorfismo Xy (Xy_p)/Xyy (Xyy_p), *Physis Secc. C,* 37, 177, 1977.

101. **Virkki, N,** Prophase of spermatocyte I in Oedionychina (Coleoptera), *J. Agric. Univ. P.R.,* 60, 661, 1976.

102. **Virkki, N.,** Brief notes on the cytology of neotropical Coleoptera. 3. "*Luperodes antillarum* Blake" = *Lysathia ludoviciana* (Fall); *J. Agric. Univ. P.R.,* 63, 100, 1979.

103. **Virkki, N.,** Response of an Oedionychina (Coleoptera) karyotype to acute gamma radiation, *J. Agric. Univ. P.R.,* 63, 116, 1979.

104. **Virkki, N.,** Banding of Oedionychina (Coleoptera: Alticinae) chromosomes: C- and Ag- bands, *J. Agric. Univ. P.R.,* 67, 221, 1983.

105. **Virkki, N. and Kimura, M.,** Distribution of nuclear pores and perinuclear dense substances in spermatocytes of some Oedionychina fleabeetles, *BioSystems,* 10, 213, 1978.

106. **Yadav, J. S.,** Post-reductional sex-determining mechanism and chromosomal polymorphism in *Diapromorpha turcica* Fabr. (Chrysomelidae: Coleoptera), *Zool. Anz.,* 197, 125, 1976.

107. **Yadav, J. S. and Pillai, R. K.,** On the cytology of Eumolpinae (Chrysomelidae: Coleoptera), *Zool. Polon.,* 27, 621, 1980.

108. **Yadav, J.S. and Pillai, R. K.,** Karyological notes on four species of Cassidinae Coleoptera: Chrysomelidae), *Genen Phaenen,* 18, 55, 1975.

109. **Garaud, P. and Lecher, P.,** Etude biometrique du caryotype de la bruche du haricot (*Acanthoscelides obtectus,* Coleopteres, Bruchidae), *Can. J. Genet. Cytol.,* (in press).

110. **Takenouchi, Y.,** A hexaploid parthenogenetic weevil, *Blosyrus japonicus* Sharp with 68 chromosomes (Coleoptera: Curcuiionidae), *Chrom. Inf. Serv.,* 19, 24, 1975.

111. **Takenouchi, Y.,** Chromosome survey in twelve species of bisexual weevils (Coleoptera: Curculionidae) in Kyushu, Japan. I., *La Kromosomo,* 2, 1, 1976.
112. **Takenouchi, Y.,** A study on the chromosomal dimorphism in *Catapionus gracilicornis* Roelofs (Curculionidae: Coleoptera), *Jpn. J. Genet.,* 51, 279,1976.
113. **Takenouchi, Y.,** The Xyy$_p$ sex-determining mechanism in an anthribid beetle, *Chrom. Inf. Serv.,* 20, 5, 1976.
114. **Takenouchi, Y.,** Chromosome complement of *Pseudorhycotes insignis* Lewis (Coleoptera: Brentidae), *Jpn. J. Genet.,* 51, 277, 1976.
115. **Takenouchi, Y.,** A study of polyploidy in races of Japanese weevils (Coleoptera: Curculionidae), *Genetica,* 46, 327, 1976.
116. **Takenouchi, Y.,** A tripartite first metaphase plate found in a parthenogenetic weevil, *Callirhopalus minimus* Roelofs (Coleoptera: Curculionidae), *J. Hokkaido Univ. Educ. II B,* 28, 1, 1977.
117. **Takenouchi, Y.,** A further chromosome study on races of two reportedly Japanese polyploid partheniogenetic weevils (Coleoptera: Curculionidae), *J. Hokkaido Univ. Educ. II B,* 29, 1, 1978.
118. **Takenouchi, Y.,** A chromosome study of the parthenogenetic rice weevil, *Lissorhoptus oryzophilus* Kuschel (Coleoptera: Curculionidae) in Japan, *Experientia,* 34, 444, 1978.
119. **Takenouchi, Y.,** A chromosome study on a parthenogenic weevil *Blosyrus japonicus* Sharp in Oshima Peninsula, Hokkaido (Curculionidae: Coleoptera), *J. Hokkaido Univ. Educ. II B,* 29, 1, 1979.
120. **Takenouchi, Y.,** The chromosomes of thirty-seven weevil species from Japan (Coleoptera: Curculionidae), *Entomol. Gener.,* 6, 7, 1980.
121. **Takenouchi, Y.,** Experimental study on the evolution of parthenogenetic weevils (Coleoptera: Curculionidae), *J. Hokkaido Univ. Educ. II B,* 1, 1980.
122. **Takenouchi, Y.,** A diploid parthenogenetic race of the weevil species *Catapionus gracilicornis* from Japan (Coleoptera: Curculionidae), *Entomol. Gener.,* 6, 367, 1980.
123. **Takenouchi Y.,** Chromosome numbers of Japanese weevils of Curculionoidea (Coleoptera). II, *Seibutsu-kyozai,* 16, 155, 1981.
124. **Takenouchi, Y.,** A chromosome study of eggs produced by trial crosses between parthenogenetic *Catapionus gracilicornis* females and bisexual *Catapionus* sp. males, *Zool. Mag.,* 90, 39, 1981.
125. **Takenouchi, Y.,** A chromosome study on the new parthenogenetic weevil, *Myosides pyrus* Sharp (Coleoptera: Curculionidae), *J. Hokkaido Univ. Educ. II B,* 31, 55, 1981.
126. **Takenouchi, Y.,** A chromosome study on two new polyploid parthenogenetic weevils in Kyushu, Japan, *Genetica,* 55, 137, 1981.
127. **Takenouchi, Y.,** A further chromosome survey on a parthenogenetic weevil, *Callirhopalus bifasciatus* Roelofs (Coleoptera: Curculionidae), *Genetica,* 55, 141, 1981.
128. **Takenouchi, Y.,** On the chromosomes of the bisexual weevils of the genus *Catapionus* (Curculionidae: Coleoptera), *Genetica,* 55, 147, 1981.
129. **Takenouchi, Y., Okamoto, H., and Sugawara, H.,** A study on the influence of low temperatures on the eggs of the tetraploid parthenogenetic *Catapionus gracilicornis* Roelofs (Curculionidae: Coleoptera), *J. Hokkaido Univ. Educ. II B,* 32, 1, 1981.
130. **Lokki, J., Saura, A., Lankinen, P., and Suomalainen, E.,** Genetic polymorphism and evolution in parthenogenetic animals. VI. Diploid and triploid *Polydrosus mollis* (Coleoptera: Curculionidae), *Hereditas,* 82, 209, 1976.
131. **Lokki, J., Saura, A., Lankinen, P., and Suomalainen, E.,** Evolution of parthenogenetic animals, *Luonnon Tutkija,* 81, 104, 1977 (Finnish, with English Summary).
132. **Saura, A., Lokki, J., Lankinen, P., and Suomalainen, E.,** Genetic polymorphism and evolution in parthenogenetic animals. III. Tetraploid *Otiorrhynchus scaber* (Coleoptera: Curculionidae), *Hereditas,* 82, 79, 1976.
133. **Saura, A., Lokki, J., Lankinen, P., and Suomalainen, E.,** Genetic polymorphism and evolution in parthenogenetic animals. IV. Triploid *Otiorrhynchus salicis* Ström (Coleoptera: Curculionidae), *Entomol. Scand.,* 7, 1, 1976.
134. **Saura, A., Lokki, J., and Suomalainen, E.,** Selection and genetic differentiation in parthenogenetic populations, *Lecture Notes in Biomathem.,* 19, 381, 1977.
135. **Suomalainen, E.,** Evolution of parthenogenetic insects, *Evol. Biol.,* 9, 209, 1976.
136. **Sharma, G. P. and Gill, T. K.,** Chromosome studies on eight species of *Myllocerus* (Cur. Col.), *4th All India Cell Biol. Conf.,* C 86, 1980.
137. **Sharma, G. P. and Pal, V.,** Chromosomal polymorphism in *Lapidospyris demissus* (Coleoptera: Curculionidae), *Cytobios,* 21, 171, 1978.
138. **Sharma, G. P., Gill, T. K., and Pal, V.,** Chromosomes in the Curculionoid beetles (Coleoptera, Curculionoidea), *Col. Bull.,* 34, 361, 1980.
139. **Sharma, G. P. and Pal, V.,** Chromosomal studies on *Apoderus sissu* (Attelabidae), *Curr. Sci.,* 49, 527, 1980.

140. **Sharma, G. P. and Pal, V.,** Some karyological observations on *Metialma* sp. (Coleoptera: Curculionidae), *Curr. Sci.,* 49, 565, 1980.
141. **Sharma, G. P. and Pal, V.,** Karyological studies on two species of the family Curculionidae (Coleoptera), *Haryana Univ. J. Res.,* 11, 40, 1981.
142. **Sharma, G. P. and Pal, V.,** A curculionid weevil with the lowest chromosome number (Curculionidae: Coleoptera), *Curr. Sci.,* 50, 42, 1981.
143. **Ennis, T. J.,** Low chromosome number and postreductional XO in *Gelus californicus* (LeC.) (Coleoptera: Curculionidae), *Can. J. Genet. Cytol.,* 14, 851, 1972.
144. **Lue, P. S., Watson, J. E., and Gilliland, F. R., Jr.,** Karyology of the boll weevil, *Ann. Entomol. Soc. Am.,* 66, 801, 1973.
145. **Lue, P. S., Watson, J. E., and Gilliland, F. R. Jr.,** Karyotypic determination in the boll weevil, *J. Hered.,* 67, 308, 1976.
146. **Lanier, G. N.,** Cytotaxonomy of *Dendroctonus,* in *Application of Genetics and Cytology in Insect Systematics and Evolution* (Proc. Symp. Natl. Mtg. Entomol. Soc. Am., Atlanta, December 1st and 2nd, 1980), Stock, M.W., Ed., Forest Wildland Range Experiment Station, Moscow, Idaho, 1981. 33.
147. **Virkki, N.,** Further Chrysomelids with an X + Y sex chromosome system: Megalopodinae, *Hereditas,* 98, 209, 1983.
148. **Ferreira, A., Cella, D. M., Mesa, A., and Virkki, N.,** Cytology and systematical position of stylopids (= Strepsiptera), *Hereditas,* in press.
149. **Comings, D. E. and Avelino, E.,** DNA loss during Robertsonian fission in studies of the tobacco mouse, *Nature (London) New Biology,* 237, 199, 1972.
150. **Virkki, N. and Reyes-Castillo, P.,** Cytotaxonomy of Passalidae (Coleoptera), *An. Esc. Nac. Ci. Biol. México,* 19, 49, 1972.
151. **Smith, S. G.,** The chromosomes of some Chrysomelid Coleoptera: Diabroticites, *Chromosomes Today,* 2, 41, 1972.
152. **White, M. J. D.,** *Animal Cytology and Evolution,* Cambridge University Press, London, 1973.
153. **Smith, S. G.,** Adaptive chromosomal polymorphism in *Chilocorus stigma, Proc. Genet. Soc. Canada,* 2, 40, 1967.
154. **Smith, S. G.,** Cytogenetic pathways in beetle speciation, *Can. Entomol.,* 94, 941, 1962.
155. **Smith, S. G.,** Natural hybridisation in the Coccinellid genus *Chilocorus, Chromosoma,* 18, 380, 1966.
156. **Smith, S. G.,** *Chilocorus similis,* Rossi (Coleoptera: Coccinellidae): Disinterment and case history, *Science,* 148, 1614, 1965.
157. **Pathak, S., Hsu, T. C., and Arrighi, F. E.,** Chromosomes of *Peromyscus* (Rodentia, Cricetinae). IV. The role of heterochromatin in karyotype evolution, *Cyto. Cell Genet.,* 12, 315, 1973.
158. **Smith, S. G. and Virkki, N.,** *Animal Cytogenetics: Coleoptera,* John, B., Ed., Borntraeger, Berlin, 1978, 181.
159. **Manna, G. K. and Smith, S. G.,** Chromosomal polymorphism and interrelationships among bark weevils of the genus *Pissodes, Nucleus,* 2, 179, 1959.
160. **Smith, S. G. and Takenouchi, Y.,** Unique incompatibility system in a hybrid species, *Science,* 138, 36, 1962.
161. **Smith, S. G. and Takenouchi, Y.,** Chromosomal polymorphism in *Pissodes* weevils: Further on incompatibility in *P. terminalis, Can. J. Genet. Cytol.,* 11, 761, 1969.
162. **Smith, S. G.,** Chromosomal polymorphism in North American *Pissodes* weevils: structural isomerism, *Can. J. Genet. Cytol.,* 12, 506, 1970.
163. **Piza, S. Toledo de,** Some interesting aspects of the male meiosis in an elaterid beetle, *Caryologia,* 11, 72, 1958.
164. **Smith, S. G.,** Chromosomal polymorphism and interrelationships in *Pissodes* weevils: additional cytogenetic evidence of synonymy, *Can. J. Genet. Cytol.,* 15, 83, 1973.
165. **Lanier, G. N.,** Biosystematics of North American *Ips* (Coleoptera: Scolytidae). Hopping's group IX, *Can. Entomol.,* 102, 1139, 1970.
166. **Weber, F.,** Beitrag zur Karyotypanalyse der Laufkäfergattung *Carabus* L. (Coleoptera), *Chromosoma,* 18, 467, 1966.
167. **Weber, F.,** Zur Polymorphie eines Chromosomes innerhalb von Populationen des Laufkäfers *Carabus auronitens* Fabr., *Experientia,* 34, 257, 1967.
168. **Hopkins, A. D.,** Technical papers on miscellaneous forest insects. I. Contributions toward a monograph on the barkweevils of the genus *Pissodes, USDA Bur. Entom. Tech. Pap.,* Ser. 20, 1-68, 1911.
169. **Smith, S. G. and Virkki, N.,** *Animal Cytogenetics: Coleoptera,* John, B., Ed., Borntraeger, Berlin, 1978, 102.
170. **John, B. and Lewis, K.R.,** Nucleolar controlled segregation of the sex chromosomes in beetles, *Heredity,* 15, 431, 1960.

171. **Wahrman, J., Nezer, R., and Freund, O.,** Multiple sex chromosome mechanisms with "segregation bodies", *Chromosomes Today,* 4, 434, 1973.

172. **Quack, B. and Noel, B,** XY-chromosome pair in mouse and human spermatocytes, visualized by silver staining, *Nature,* 267, 431, 1977.

173. **Noel, B., Quack, B., and Benezech, M.,** Le bivalent sexuel des Mammiferes observé par marquage argentique au stade pachytene, *Ann. Genet.,* 21, 83, 1978.

174. **Smith, S. G.,** Nucleolar organisation and chromocentre formation in *Chilocorus, Chromosomes Today,* 2, 41, 1969.

175. **Monesi, V.,** Chromosome activities during meiosis and spermiogenesis, *J. Reprod. Fertil. Suppl.,* 13, 1, 1971.

176. **Virkki, N.,** Chromosome relationships in some North American Scarabaeoid beetles, with special reference to *Pleocoma* and *Trox. Can. J. Genet. Cytol.,* 9, 107, 1967.

177. **Lanier, G. N. and Raske, A. G.,** Multiple sex chromosomes and configuration polymorphism in the *Monochamus scutellatus oregonensis* complex (Coleoptera: Cerambycidae), *Can.J. Genet. Cytol.,* 12, 947, 1970.

178. **Wagenaar, E.,** End-to-end chromosome attachments in mitotic interphases and their possible significance to meiotic chromosome pairing, *Chromosoma,* 26, 410, 1969.

179. **Shchapova, A.I.,** Karyotype pattern and the chromosome arrangement in the interphase nucleus, *Tsitologiya,* 13, 1157, 1971.

180. **Lewin, R.,** Do chromosomes talk?, *Science,* 214, 1334, 1981.

181. **John, B. and Shaw, D. D.,** Karyotype variation in dermestid beetles, *Chromosoma,* 20, 371, 1967.

182. **Shaw, D. D.,** Selection for supernumerary y-chromosomes in *Dermestes maculatus* (Coleoptera: Dermestidae), *Can. J.Genet. Cytol.,* 10, 54, 1968.

183. **Smith, S. G. and Brower, J. H.,** Chromosome numbers of stored-product Coleoptera, *J. Kansas Entomol. Soc.,* 47, 317, 1974

184. **Lewis, K. R. and John, B.,** The organization and evolution of the sex multiple, in *Blaps mucronata, Chromosoma,* 9, 69, 1957.

185. **Virkki, N.,** Evolution of extremely complicated sex chromosome systems in *Blaps* (Coleoptera), *J. Agric. Univ. P.R.,* 48, 140, 1974.

187. **Vaio, E. S. de and Postiglioni, A.,** Stolaine Cassidines (Coleoptera, Chrysomelidae) with Xy_p sex chromosomes and derivative system $X_pneoXneoY_p$, *Can. J. Genet. Cytol.,* 16, 433, 1974.

188. **Virkki, N.,** Sex chromosomes and karyotypes of the Alticidae (Coleoptera), *Hereditas,* 64, 267, 1970.

189. **Luykx, P.,** *Cellular Mechanisms of Chromosome Distribution,* Academic Press, New York, 1970.

190. **Dietz, R.,** Bau und Funktion des Spindelapparats, *Naturwissenschaften,* 56, 237, 1969.

191. **Virkki, N.,** Orientation and segregation of asynaptic multiple sex chromosomes in the male *Omophoita clerica* Erichson (Coleoptera, Alticidae), *Hereditas,* 57, 275, 1967.

192. **Lorbeer,** Die Zytologie der Lebermoose mit besonderer Berücksichtigung allgemeiner Chromosomenfragen. I. Distanzkonjugation, *J. Wiss. Bot.,* 80, 567, 1934.

193. **Virkki, N., and Purcell, C. M.,** Four pairs of chromosomes: the lowest number in Coleoptera, *J. Hered.,* 56, 71, 1965.

194. **Wahrman, J.,** A carabid beetle with only eight chromosomes, *Heredity,* 21, 154, 1966.

195. **Gotô, E., and Yosida, T. H.,** Sex chromosomes of *Rhaphidopalpa femoralis* Motsch. (Coleoptera, Chrysomelidae), *La Kromosoma,* 17-19, 674-676, 1953.

196. **Jeannel, R.,** *Introduction to Entomology,* Hutchinson, London, 1960.

197. **Blackwelder, R. E.,** Checklist of the Coleopterous insects of Mexico, Central America, the West Indies, and South America, *U.S. Nat. Mus. Bull.,* 185 (Pt. 3), 343, 1945.

198. **Crowson, R. A.,** The phylogeny of Coleoptera, *Ann. Rev. Entomol.,* 5, 111, 1960.

199. **Hughes-Schrader, S.,** Reproduction in *Achroschismus wheeleri* Pierce, *J. Morph.,* 39, 157, 1924.

200. **Scott, A. C.,** Haploidy and aberrant spermatogenesis in a Coleopteran, *Micromalthus debilis* Le Conte, *J. Morph.,* 59, 485, 1936.

201. **Scott, A. C.,** Paedogenesis in the Coleoptera, *Z. Morph. Ökol. Tiere,* 33, 633, 1938.

202. **Scott, A. C.,** Reversal of sex production in *Micromalthus, Biol. Bull.,* 81, 420, 1941.

203. **Pringle, J. A.,** A contribution to the knowledge of *Micromalthus debilis* LeC. (Coleoptera), *Trans. Roy. Entom. Soc. London,* 87, 271, 1938.

204. **Herfs, A.,** Studien an dem Steinnussborkenkäfer, *Coccotrupes tanganus* Eggers, *Höfchen-Briefe,* p. 1, 1950.

205. **Entwistle, P. F.,** Inbreeding and arrhenotoky in the ambrosia beetle *Xyleborus compactus* Eichh. (Coleoptera: Scolytidae), *Proc. Roy. Entomol. Soc. London,* 39, 83, 1964.

206. **Brader, L.,** Étude de la relation entre le Scolyte des rameaux du cafféier, *Xyleborus compactus* Eichh. (*X. morstatti* Hag.) et sa plante hôte, *Meded. Landbowhogeschool Wagen.* 64, 470, 1964.

207. **Kaneko, T. and Takagi, K.,** Biology of some Scolytid ambrosia beetles attacking tea plants. IV, *Jap. J. Appl. Entomol. Zool.,* 9, 303, 1965.

208. **Takagi, K. and Kaneko, T.,** Biology of some Scolytid ambrosia beetles attacking tea plants, *Appl. Entomol. Zool.,* 1, 29, 1966.

209. **Takenouchi, Y. and Takagi, K.,** A chromosome study of two parthenogenetic Scolytid beetles, *Annot.Zool. Jap.,* 40, 105, 1967.

210. **Lawrence, J. F.,** Biology of the parthenogenetic fungus beetle *Cis fuscipes* Mellié (Coleoptera: Cisidae), *Breviora,* 258, 1, 1967.

211. **Suomalainen, E.,** Polyploidy in parthenogenetic Curculionidae, *Hereditas,* 26, 51, 1940.

212. **Suomalainen, E.,** Beiträge zur Zytologie parthenogenetischer Insekten.I.Coleoptera, *Ann. Acad. Sci. Fenn.,* 54, 1, 1940.

213. **Takenouchi, T.,** A chromosome study on bisexual and parthenogenetic races of *Scepticus insularis* Roelofs (Curculionidae, Coleoptera), *Can. J. Genet. Cytol.,* 10, 945, 1968.

214. **Suomalainen, E.,** Evolution in parthenogenetic Curculionidae, *Evol. Biol.,* 3, 261, 1969.

215. **Suomalainen, E.,** Polyploidy in parthenogenetic beetles, *Proc. Xth Intern. Congr. Genet.,* 2, 283, 1958.

216. **Suomalainen, E.,** Die Polypoidie bei dem parthenogenetischen Blattkäfer *Adoxus obscurus* L. (Coleoptera, Chrysomelidae), *Zool. Jb. Syst.,* 92, 183, 1965.

217. **Robertson, J. C.,** The chromosomes of bisexual and parthenogenetic species of *Calligrapha* (Coleoptera: Chrysomelidae) with notes on sex ratio, abundance and egg number, *Can. J. Genet. Cytol.,* 8, 695, 1966.

218. **Suomalainen, E.,** Parthenogenese und Polyploidie bei Rüsselkäfern (Curculionidae), *Hereditas,* 33, 425, 1947.

219. **Lanier, G. N.,** Unpublished results, College of Environmental Science and Forestry, State University of New York, Syracuse, N.Y., 1982.

220. **Sanderson, A.,** The cytology of a diploid bisexual spider beetle, *"Ptinus clavipes"* Panzer and its triploid gynogenetic form *"mobilis"* Moore, *Proc. Roy. Soc. Edinb. Sec. B,* 67, 333, 1960.

221. **Sanderson, A. and Jacob, J.,** Artificial activation of the egg in a gynogenetic spider beetle, *Nature,* 179, 1300, 1957.

222. **Suomalainen, E.,** Significance of parthenogenesis in the evolution of insects, *Ann. Rev. Entomol.,* 7, 349, 1962.

223. **Suomalainen, E., Saura, A, and Lokki, J.,** Evolution of parthenogenetic insects, *Evol. Biol.,* 9, 209, 1976.

224. **Takenouchi, Y.,** A further study on the chromosomes of the parthenogenetic weevil, *Listroderes costirostris,* Schönherr, from Japan, *Cytologia,* 34, 360, 1969.

225. **Suomalainen, E. and Saura, A.,** Genetic polymorphism and evolution in parthenogenetic animals. I : Polyploid Curculionidae, *Genetics,* 74, 489, 1973.

226. **Lokki, J.,** Genetic polymorphism and evolution in parthenogenetic animals. VII. The amount of heterozygosity in diploid populations, *Hereditas,* 83, 57, 1976.

227. **Lokki, J.,** Genetic polymorphism and evolution in parthenogenetic animals. VIII. Heterozygosity in relation to polyploidy, *Hereditas,* 83, 65, 1976.

228. **Suomalainen, E.,** On morphological differences and evolution of different polyploid parthenogenetic weevil populations, *Hereditas,* 47, 309, 1961.

229. **Smith, S. G.,** Parthenogenesis and polyploidy in beetles, *Am. Zool.,* 11, 341, 1971.

230. **Serrano, J.,** New chromosome numbers of Spanish Caraboidea (Coleoptera, Adephaga), *Genét. Ibér.,* in press.

231. **Kowalczyk, M.,** Chromosomes of *Pterostichus cupreus* and *Pterostichus vulgaris* (= *Pterostichus melanarius* Ill.*sensu* Lindroth, 1957) (Coleoptera, Carabidae), *Fol. Biol.,* 24, 231, 1976.

232. **Rozek, M.,** Karyological studies on Bembidiinae (Coleoptera, Carabidae). I. Chromosome number and sex-determining mechanism system in six species of the genus *Bembidion* Latr., *Fol. Biol.,* 29, 119, 1981.

233. **Rozek, M., and Warchalowska-Śliva, E.,** Karyological studies on Trechinae (Coleoptera, Carabidae). I. Chromosomes of *Trechus pilisensis* Csiki and *Trechus pulchellus* Putz., *Fol. Biol.,* 28, 313, 1980.

234. **Gambardella, L. A. de, and Vaio, E. S. de,** Estudio morfológico, biológico y cariológico de *Chauliognathus scriptus* (Germ.) (Coleoptera, Cantharidae), *Rev. Biol. Urug.,* 6, 69, 1978.

235. **Srivastava, M. D. L.,** Unpublished data, Department of Zoology, Allahabad University, Allahabad, India, 1982.

236. **Vidal, O. R.,** Cariotipos infrequentes en Coleoptera, *Actas IV Congr. Latinoam. Genét.,* 1, 15, 1980.

237. **Petitpierre, E.,** Karyometric differences among nine species of the genus *Chrysolina* Mots. (Coleoptera, Chrysomelidae), *Can. J. Genet. Cytol.,* 25, 33, 1983.

238. **Petitpierre, E.,** Chromosomal findings on twenty-two species of Chrysomelinae (Chrysomelidae: Coleoptera), *Chrom. Inf. Serv.,* 32, 22, 1982.

239. **Alegre, C. and Petitpierre, E.,** Chromosomal findings on eight species of European *Cryptocephalus,* *Experientia,* 38, 774, 1982.

240. **Segarra, C. and Petitpierre, E.,** Preliminary data on the chromosomes of European Alticinae, *Spixiana,* Suppl. 7, 29, 1982.

241. **Jolivet, P. and Petitpierre, E.,** Sélection trophique et evolution chromosomique chez les Chrysomelinae (Col. Chrysomelidae), *Acta Zool. Pathol. Antverp.,* 66, 59, 1976.

242. **Jolivet, P. and Petitpierre, E.,** Les plantes-hotes connues des *Chrysolina* (Col. Chrysomelidae). Essai sur les types de sélection trophique, *Ann. Soc. Entomol. France,* 12, 123, 1976.

243. **Panzera, F.,** Estudio cromosómico de *Botanochara* sp., Manuscript, Universidad Mayor de la República, Montevideo, Uruguay, 1982.

244. **Vaio, E. S. de, Dergam, J. A., and González, L. E.,** Tres especies de género *Botanochara* (Coleoptera, Chrysomelidae, Cassidinae), *V., J. Argent. Zool.,* Córdoba, 1978.

245. **Panzera, F.,** Estudio citogenetico de *Botanochara bonariensis* (Coleoptera - Chrysomelidae), Manuscript, Universidad Mayor de la República, Montevideo, Uruguay, 1982.

246. **Mazzella, M. C.,** Estudio cromosómico de *Stolas lacordairei* (Coleoptera, Chrysomelidae), Manuscript, Universidad Mayor de la República, Montevideo, Uruguay, 1982.

247. **Panzera, F., Mazzella, M., and Vaio, E. S. de,** Estudio de dos especies de Cassidinos (Coleoptera, Chrysomelidae) pertenecientes a la tribu Stolaini con cromosomas sexuales muy distintos, V. *Congr. Latinoam. Genét.,* Viña del Mar, p. 75, 1981.

248. **Vaio, E. S. de,** Unpublished results, Universidad Mayor de la República, Montevideo, Uruguay, 1982.

249. **Wise, D., Wright, J. E. and McCoy, J. R.,** Meiotic chromosomes of the boll weevil, *J. Hered.,* 75, 234, 1982.

250. **Ferreira, A., Cella, D.M., Ramos Tardivo, J., and Virkki, N.,** Two pairs of chromosomes: A new record for Coleoptera, *Heredity* (in press).

251. **Mesa, A. and Fontanetti, S.,** Personal communication. Instituto de Biociências, Universidade Estadual Paulista, Rio Claro, São Paulo, Brazil, 1983.

252. **Blackwelder, R. E.,** Checklist of the coleopterous insects of Mexico, Central America, The West Indies, and South America, *U.S. Nat. Mus. Bull.,* 185 (Pt. 2), 189, 1944.

253. **Takenouchi, Y., Suomalainen, E., Saura, A., and Lokki, J.,** Genetic polymorphism and evolution in parthenogenetic animals. XII. Observations on Japanese polyploid Curculionidae (Coleoptera), *Jpn. J. Genet.,* 58, (in press).

254. **Angus, R. B.,** Separation of two species standing as *Helophorus aquaticus* (L.) (Coleoptera, Hydrophilidae) by banded chromosome analysis, *System. Entomol.,* 7, 265, 1982.

255. **Kacker, R. K.,** Studies on the chromosomes of the Indian Coleoptera. VI. Chromosome number and sex determining mechanism in 15 species of Coleoptera, *Newsl. Zool. Surv.India,* 2, 48, 1976.

Chapter 4

CHROMOSOMES IN EVOLUTION OF NEMATODES

A. C. Triantaphyllou

TABLE OF CONTENTS

I. INTRODUCTION

Nematodes are probably the most numerous multicellular animals on earth. They constitute a highly diversified group comprising approximately 25 thousand described species. They are found in a wide variety of habitats, and on this basis they are characterized as free-living soil, fresh-water, and marine forms — saprophagous or predaceous, as well as animal-parasitic and plant-parasitic forms (Table 1). Such ecological diversification is, in many cases, instructive of the direction and the pathway of evolution of nematodes. Thus, free-living nematodes feeding on bacteria are considered to be primitive and closely related to ancestral forms from which algal and fungal feeders, plant-parasitic, and animal-parasitic forms have evolved. Systematists have attempted to make nematode classification as natural as possible, in order to reflect evolutionary relationships. They have considered primarily evolution of the excretory system and secondarily modifications of the digestive system and other anatomical features. Biological and ecological criteria have not been considered at all, because of the scarcity of information in this field.

The study of the chromosomes of nematodes was initiated late in the nineteenth century. Early investigations elucidated some basic cytological and biological phenomena that were incomprehensible at that time. Working with the horse roundworm, *Parascaris equorum*, van Beneden[1] in 1883 demonstrated the process of "meiosis", involving the reduction of the chromosomes from the diploid ($2n$) to the haploid (n) state and the re-establishment of the diploid number during "fertilization" by the fusion of sperm/egg pronuclei. The processes of "chromosome fragmentation" and "chromatin diminution" occurring in somatic cells during embryogenesis of ascarid nematodes were discovered later by Boveri.[2] Nematodes were also the second animal group, after the insects to be studied with regard to chromosomal mechanisms of sex determination. Such studies led to the discovery of the multiple X mechanism in ascarids.

Most of the early cytological and cytogenetic work with nematodes involved animal-parasitic and some free-living soil forms.[3-5] Cytological work with plant-parasitic forms was initiated in the late 1950s and proceeded in slow pace up to now.[6] The total number of nematode species for which some cytological information is available represents less than 1% of the total described species. An attempt is made in this article to give a comprehensive list of the nematodes that have been studied cytologically and to evaluate the significance of this work in relation to nematode evolution.

II.THE CHROMOSOMAL COMPLEMENT

A. General Features

Practically all information about the chromosomal complement of nematodes refers to meiotic chromosomes during gametogenesis, especially during the first maturation division of gametocytes. Occasionally, mitotic chromosomes have been studied in oogonial and spermatogonial cells in the germinal zone of male and female gonads, and in a few cases in early blastomeres during embryogenesis. In general, mitotic chromosomes have not helped much in the study of nematode karyotypes.Variation in the degree of contraction of such chromosomes during the mitotic cycle often influences their size and morphology. Because of cell constancy in nematodes, very few somatic cell divisions occur during postembryogenesis and they are always unfavorable for karyotype analysis.

The use of colchicine or colcemid to stabilize the mitotic chromosomes with regard to chromatin condensation at a prometaphase stage could help in karyotype characterization, but these methods have rarely been used.[7] Somatic cell culturing could facilitate the study of nematode chromosomes, but establishing such cultures has been unsuccessful so far. Finally, such modern differential staining methods as Giemsa banding techniques, quinacrine

Table 1
LIFE HABITS AND PREDOMINANT
HABITATS OF MAJOR TAXONOMIC
CATEGORIES OF NEMATODA

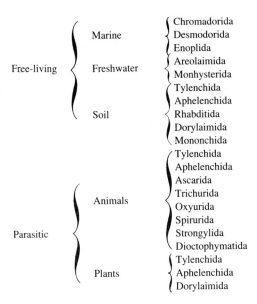

Free-living	Marine	Chromadorida
		Desmodorida
		Enoplida
	Freshwater	Areolaimida
		Monhysterida
	Soil	Tylenchida
		Aphelenchida
		Rhabditida
		Dorylaimida
		Mononchida
Parasitic	Animals	Tylenchida
		Aphelenchida
		Ascarida
		Trichurida
		Oxyurida
		Spirurida
		Strongylida
		Dioctophymatida
	Plants	Tylenchida
		Aphelenchida
		Dorylaimida

fluorescent banding, and others, so successful in karyotype analysis of many organisms, have not been employed in the study of nematode chromosomes.

Mitotic chromosomes, as observed during gonial and cleavage divisions, are small, 1 to 5 μm long, with no striking morphological features (Figures 1 and 2). At prometaphase and metaphase, mitotic chromosomes appear as single rods with little physical evidence of the fact that they consist of two chromatids (Figure 3). This appearance is in contrast to the meiotic chromosomes, in which the chromatids of each chromosome are distinct. Only in some nematodes with large chromosomes has the karyotype been described precisely.

Meiotic chromosomes have been studied during maturation of oocytes and spermatocytes. Prometaphase I chromosomes are the most favorable for determining the chromosome number in most species, but telophase I and metaphase II figures are also favorable in some species. At prometaphase I, the bivalent chromosomes appear as tetrads, i.e., consist of two single chromosomes which, in most cases, are in an end-to-end orientation, and each one consists of two distinct chromatids oriented parallel to each other (Figures 4 and 5).[8] In some nematodes with long chromosomes, cross configurations and loops indicating the presence of more than one chiasma are occasionally observed. In some nematodes with compound chromosomes, meiotic configurations have been difficult to interpret and have caused quite a controversy.[9] Similarly, the meiotic configurations of other nematodes have been interpreted in an unconventional manner, because of their peculiar appearance and behavior during prophase I.[10] Part of the difficulty in interpretation may be due to the diffuse nature of the centromere. The lack of a localized centromere probably causes precocious separation of sister chromatids and their parallel association during prometaphase. It seems also that centromeric activity at metaphase and anaphase I is often limited to one end of the chromosomes, thus resulting in an end-to-end orientation of the chromosomes forming the bivalents. At anaphase II, however, centromeric activity extends along the entire side of each chromosome and thus the chromosomes (chromatids) migrate with their broad side toward the poles.[8,11]

A definite centromere with localized activity has been reported in some trichurids.[3,12]

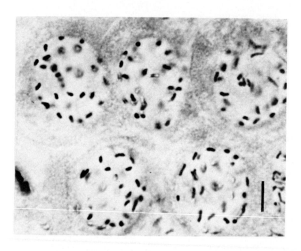

FIGURE 1.   Mitotic chromosomes of *Anguina tritici* in sperma-togonia, at late prophase. Bar = 5 μm. (From Triantaphyllou, A. C. and Hirschmann, H., *Nematologica,* 12, 437, 1966. With permission.)

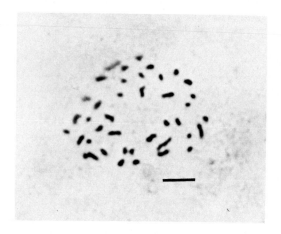

FIGURE 2.   Mitotic chromosomes of *Meloidogyne incognita* in an oogonium at metaphase. Bar = 3 μm.(From Triantaphyllou, A. C., *J. Nematol.,* 13, 95, 1981. With permission.)

Similarly, multicentric, compound chromosomes occur in some ascarids and possibly in some spirurids.[11] Physical evidence of the diffuse nature of centromeres in nematodes has been provided by electron microscope studies of mitotic chromosomes in *Meloidogyne*[13] and in *Caenorhabditis elegans.*[133]

B. Chromosome Numbers and Amount of DNA

The chromosome number has been determined in approximately 200 species of nematodes. For many of these, the chromosome number is the only cytological information available and has been used to interpret relationships among members of certain taxonomic categories, or to characterize races within species.[14] Most nematode groups include members with a haploid complement of six chromosomes. This may be the basic chromosome number that conceivably was common in ancestral forms of nematodes. Deviations occur within many

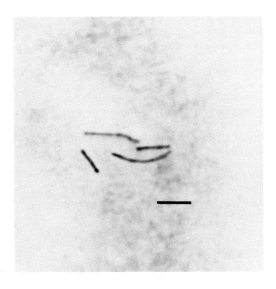

FIGURE 3. Mitotic chromosomes of *Strogyloides papillosus* at metaphase during the second cleavage division. Bar = 5 μm. (From Triantaphyllou, A. C. and Moncol, D. J., *J. Parasitol.*, 63, 961, 1977. With permission.

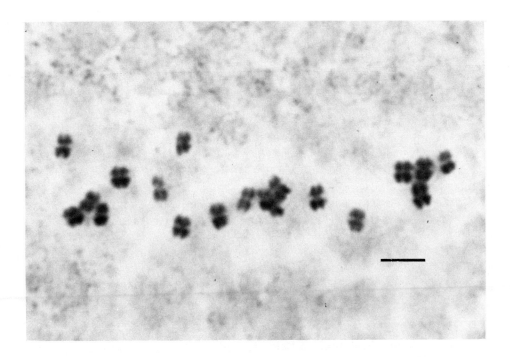

FIGURE 4. Bivalent chromosomes of *Anguina tritici* in a primary oocyte at metaphase I. Bar = 3 μm. (From Triantaphyllou, A. C. and Hirschmann, H., *Nematologica*, 12, 437, 1966. With permisson.)

groups and include increases or decreases in the haploid number through chromosomal rearrangements, and the establishment of polyploidy and possibly aneuploidy. The haploid chromosome number varies from $n = 1$, as in the free-living nematode *Diploscapter coronata*,[15] to $n = 19$ in the plant-parasitic nematodes *Anguina tritici*[16] and *Meloidogyne*

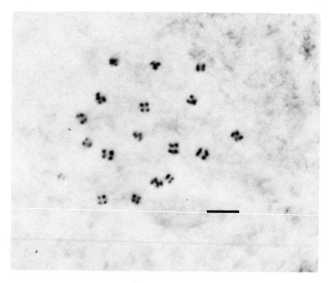

FIGURE 5.   Bivalent chromosomes (tetrads) of *Meloidogyne graminis* in a primary oocyte at metaphase I. Bar = 3 μm. (From Triantaphyllou, A. C., *J. Nematol.*, 5, 87, 1973. With permission.)

microtyla.[14] Higher chromosome numbers are encountered in nematodes considered to be polyploid in nature, e.g., $n = 24$ in *Ditylenchus destructor*,[17] and $3 n = 54$ in *M. arenaria*.[18] Most pronounced deviations from the basic number of $n = 6$ are observed in highly evolved groups, such as the Ascarida, Trichurida, and Strongyloididae among animal-parasitic nematodes, and the Tylenchida and some Dorylaimida which comprise plant-parasitic nematodes. Thus, there apparently is a correlation between the general cytological evolution in these groups and specialization toward parasitism.

The relative amount of DNA per nucleus has been determined cytophotometrically in very few nematode species. Such studies have demonstrated the polyploid nature of some *Meloidogyne* and *Heterodera* species.[19] Similarly, they demonstrated that cells of some tissues of the nematode body are indeed polyploid.[20]

Chemical analysis and renaturation kinetic studies of DNA have given an estimate of 0.8×10^8 base pairs or 0.088 pg of the DNA content of the haploid genome of *Caenorhabditis elegans*.[21] According to Sulston and Brenner,[21] this estimate is only 20 times as much as that of *E. coli*, and apparently the smallest amount of DNA observed in any animal. However, in the nematode *Panagrellus silusiae*, taxonomically related to *Caenorhabditis*, the haploid DNA content, as determined cytophotometrically, was found to be more than four times that of *C. elegans*.[20,21] An even larger amount of DNA appears to be present in nuclei of *Ascaris lumbricoides*.[21] Comparative DNA studies may be useful in elucidating relationships between major taxonomic groups of nematodes, as they may reveal that certain steps in DNA amplification have occurred during nematode evolution.

C. Sex Chromosomes and Mode of Reproduction

Understanding evolution of the chromosomal mechanism of sex determination and of the methods of reproduction is instructive in interpreting evolution in several nematode groups. Sex chromosomes were recognized in nematodes as early as 1910 and the study of reproduction had been initiated even earlier.[6]

Most nematodes are bisexual, amphimictic (cross-fertilizing). The female is the homogametic sex and the male the heterogametic sex. Thus, the chromosomal mechanism of sex determination is XX ♀ — XO ♂. An XX ♀ — XY ♂ mechanism is common in some groups and is believed to have evolved secondarily from the previous one. Examples of this

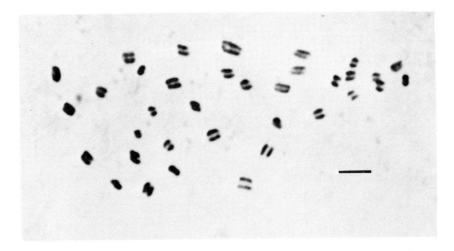

FIGURE 6.   Univalent chromosomes (dyads) of *Heterodera trifolii* in an oocyte at metaphase I. Bar = 3 μm.

type of evolution involve species of *Strongyloides*,[22] *Trichuris*,[23,24] and others. In many plant-parasitic nematodes, males and females have indistinguishable chromosomal complements. Often, these nematodes are assumed to be of the XX ♀ — XY ♂ type, with the Y chromosome being indistinguishable from the X chromosomes or the autosomes. However, since chromosomal differences cannot be demonstrated, sex determination in these nematodes may involve genic differences only. Indeed, simple genic systems that influence sex determination have been discovered in *Caenorhabditis elegans*.[25] They involve one or a few genes located in one or several autosomes.

Hermaphroditism has been demonstrated only in some rhabditid and a few aphelenchid nematodes. Nematode hermaphrodites are bisexual individuals which produce in the same gonad (ovotestis) sperm initially, and eggs later on (protandric hermaphrodites). The sperm usually fertilizes the eggs of the same individual (automixis). There are no recognizable sex chromosomes in hermaphrodites. However, males, often present in hermaphroditic species, have one chromosome less than hermaphrodites or females of the same species. The missing chromosome is an X chromosome and, therefore, the mechanism of sex determination is XX ♀ or ⚥ — XO ♂. Whenever males are present, amphimictic reproduction may occur in addition to automixis.[26]

Some ascarid nematodes exhibit more complex mechanism of sex determination, having not only one but a set of two to eight sex chromosomes which during division migrate as a unit toward the same pole.[3] Thus, a multiple X — O and a multiple X — Y mechanism have evolved, suggesting a distinct line of evolution of ascarids.

Parthenogenetic and gynogenetic species are found in many groups of nematodes and are believed to have evolved from bisexual — amphimictic and hermaphroditic forms. The cytological details of the various methods of reproduction have been discussed.[6] However, brief definitions of gynogenesis and parthenogenesis are presented here.

''Gynogenesis'' is a type of pseudofertilization. A sperm enters an oocyte and activates it for further development. However, the sperm nucleus degenerates later in the cytoplasm without fusing with the egg pronucleus.

Parthenogenesis occurs as meiotic or mitotic, depending on the type of maturation of the gametocytes. Thus, in ''meiotic parthenogenesis'' maturation of oocytes follows the conventional meiotic cycle which involves pairing of homologous chromosomes and formation of bivalents (Figures 4 and 5). Two maturation divisions that follow lead to the formation of two polar nuclei and the egg pronucleus with the haploid chromosomal complement. The

Table 2
CHROMOSOMES AND MODE OF REPRODUCTION IN NEMATODES

Taxon	Haploid chromosome number[a]	Sex chromosomes ♀ — ♂	Mode of reproduction	Ref.
ADENOPHOREA				
Dorylaimida				
Longidoridae				
Xiphinema mediterraneum	5		Parthenogenesis (meiotic)	27, 28, 63
Xiphinema diversicaudatum	(10)		Amphimixis	27
Xiphinema index	(10)		Parthenogenesis (meiotic)	27, 29
Longidorus elongatus	7		Parthenogenesis (meiotic)	27, 63
Longidorus macrosoma	7		Parthenogenesis (meiotic)	27, 63
Longidorus vineacola			Parthenogenesis (meiotic)	27, 63
Mermithidae				
Paramermis contorta	6		Amphimixis	30
Hexamermis albicans	8			31
SECERNENTEA				
Rhabditida				
Rhabditidae				
Rhabditis aberrans	9(♂) 18(2n)♀		Gynogenesis? Apomixis	64
Rhabitis aberrans	9 9;8	XX—XO		26
Rhabditis anomala	20 (2-n)♀		Gynogenesis Apomixis	32
Rhabditis espera	7;6	XX—XO		65
Rhabditis guerneyi	5			66
Rhabditis monohystera	10		Gynogenesis (meiotic)	67
Rhabditis maupasi	7;6		Amphimixis	32
Rhabditis pellio	7;6	XX—XO	Amphimixis	32, 65, 68
Rhabditis pellio (mutant)	7		Gynogenesis	68
Rhabditis terricola	7;6	XX—XO	Amphimixis	65
Rhabditis ''XX''	12(2n)♀ variable		Gynogenesis? Apomixis	66
Rhabditis ''XIX''	5		Parthenogenesis (pseudomeiotic)	66
Pelodera strongyloides	11;10	XX—XO	Amphimixis	5
Caenorhabditis briggsae	6;5	XX—XO	Automixis Amphimixis (rarely)	69
Caenorhabditis dolichura	6♀ 6;5♂	XX—XO	Automixis Amphimixis	5, 70
Caenorhabditis elegans	6♀ 6;5♂	XX—XO	Automixis	5, 70
Caenorhabditis elegans (polyploid)	12♂ 12;11	XX—XO		71
Mesorhabditis belari	10		Gynogenesis Amphimixis (rarely)	5, 66

Table 2
CHROMOSOMES AND MODE OF REPRODUCTION IN NEMATODES

Taxon	Haploid chromosome number[a]	Sex chromosomes ♀ — ♂	Mode of reproduction	Ref.
Diploscapter coronata	1		Apomixis	15
Panagrolaimidae				
Panagrellus redivivus	5;4	XX—XO	Amphimixis	72
Panagrellus redivivoides	5;4	XX—XO	Amphimixis	73
Panagrolaimus rigidus	4		Amphimixis	5
Paroigolaimella bernesis	6		Amphimixis	74
Diplogasteridae				
Diplogaster sp.	6			75
Fictor anchiocoprophaga	6		Amphimixis	74
Mononchoides changi	7;6	XX—XO	Amphimixis	76
Neodiplogaster pinicola	7;6	XX—XO	Amphimixis	73
Rhabdiasidae				
Rhabdias bufonis	6 ♀		Automixis	4, 77
	6;5	XX—XO	Amphimixis	
Rhabdias fülleborni	6;5	XX—XO	Amphimixis	78
Rhabdias ranae	6 ♀		Automixis	79
	6;5	XX—XO	Amphimixis	
Strongyloididae				
Strongyloides ratti	3;2	XX—XO	Amphimixis	34, 35
			Gynogenesis + apomixis	
Strongyloides papillosus	2	XX—XY	Gynogenesis + apomixis	22,36,80
Strongyloides ransomi	2	XX—XY	Gynogenesis + apomixis	22
Strongylida				
Metastrongylina				
Chabertia ovina	6		Amphimixis	38
Cyclostomum (Strongylus) tetracanthus	6			3, 4
Dictyocaulus filaria	6;5	XX—XO	Amphimixis	4, 40, 82
Dictyocaulus viviparus	6;5	XX—XO	Amphimixis	4, 40, 65
Metastrongylus elongatus	6;5	XX—XO	Amphimixis	40
Oesophagostomum columbianum	6;5	XX—XO	Amphimixis	11, 38
Oesophagostomum venulosum	6			4, 134
Ophiostomum mucronatum	6;5	XX—XO	Amphimixis	82
Stephanurus dentatus	6;5	XX—XO	Amphimixis	83
Strongylus equinus	6;5	XX—XO	Amphimixis	4
Stronglylus vulgaris	6;5	XX—XO	Amphimixis	4
Strongylina				
Ancylostoma caninum	6;5	XX—XO	Amphimixis	11
Trichostrongylina				
Bunostomum phlebotomum	6;5	XX—XO	Amphimixis	82
Filaroides mustelarum	8		Amphimixis	37
Haemonchus contortus	6;5	XX—XO	Amphimixis	11, 39
Marshallagia marshalli	6;5	XX—XO	Amphimixis	84
Nematodirus spathiger	6;5	XX—XO	Amphimixis	82
Nematospira turgida	6;5	XX—XO	Amphimixis	3
Ostertagia circumcincta	6;5	XX—XO	Amphimixis	82
Oswaldocruzia filiformis	6;5	XX—XO	Amphimixis	85
Oswaldocruzia goezei	6;5	XX—XO	Amphimixis	86

Table 2 (continued)
CHROMOSOMES AND MODE OF REPRODUCTION IN NEMATODES

Taxon	Haploid chromosome number[a]	Sex chromosomes ♀ — ♂	Mode of reproduction	Ref.
Spirurida				
Spirurina				
Acuaria spiralis	6;5	XX—XO	Amphimixis	3
Aspiculuris kazakstanica	6;5	XX—XO	Amphimixis	11
Cystidicola farionis	6;5	XX—XO	Amphimixis	87
Cystidicola stigmatura	6;5	XX—XO	Amphimixis	88
Cystodicola cristivomeri	6;5	XX—XO	Amphimixis	88
Mastophorus muris	5;4	XX—XO	Amphimixis	3
Physaloptera indiana	5;4	XX—XO	Amphimixis	11
Physaloptera turgida	5;4	XX—XO	Amphimixis	3
Filarina				
Dipetalonema witei	6;5	XX—XO		132
Dirofilaria immitis	5;4	XX—XO	Amphimixis	89
Dirofilaria immitis	5	XX—XY	Amphimixis	12
Litomosoides carinii	5;4	XX—XO	Amphimixis	89, 90
Onchocerca volvulus	5;4(?)			90
Setaria digitata	6;5	XX—XO	Amphimixis	12
Setaria equina	6;5	XX—XO	Amphimixis	91
Setaria labiatopapillosa	3;2	XX—XO	Amphimixis	41
Setaria servi	6;5	XX—XO	Amphimixis	91
Wuchereria bancrofti	5(?)			90
Camallanina				
Camallanus baylisi	5;4	XX—XO	Amphimixis	92
Camallanus kachugae	6;5	XX—XO	Amphimixis	11
Oxyurida				
Heterakina				
Ascaridia dissimilis	5;4	XX—XO	Amphimixis	93
Ascaridia galli	5;4	XX—XO	Amphimixis	93
Cosmocerca kashmirensis	8		Amphimixis	42
Cruzia tentaculata	6;5	XX—XO	Amphimixis	3
Heterakis dispar	5;4	XX—XO	Amphimixis	94
Heterakis gallinarum	5;4	XX—XO	Amphimixis	94, 11
Heterakis papillosa	5;4	XX—XO	Amphimixis	3
Heterakis spumosa	6;4	4X—(2X)O	Amphimixis	3
Subulura distans	5;4	XX—XO	Amphimixis	11
Subulura minetti	5;4	XX—XO	Amphimixis	11
Spironoura onama	6;5	XX—XO	Amphimixis	11
Oxyurina				
Gyrinicola batrachiensis	4		Haplodiploidy	136
	6,7,8(2n)		Parthenogenesis (meiotic)	136
Heth mauriesi	8;7	XX-XO		137
Ichthyocephalus sp.	5;4	XX-XO		137
Mehdiella uncinata	5		Haplodiploidy	137
Mehdiella microstoma	5		Haplodiploidy	137
Passalurus ambiguus	5;6	XX-XO	Amphimixis	4, 97
Syphacia (Oxyuris) equi	5			95
Syphacia obvelata	8;7	XX-XO	Amphimixis	3
Syphacia obvelata	4			96
Syphacia (Oxyuris) equi	5			95
Syphacia obvelata	8;7	XX—XO	Amphimixis	3
Syphacia obvelata	4			96
Tachygonetria conica	5		Haplodiploidy	137
Tachygonetria dentata	5		Haplodiploidy	137

Table 2 (continued)
CHROMOSOMES AND MODE OF REPRODUCTION IN NEMATODES

Taxon	Haploid chromosome number[a]	Sex chromosomes ♀ — ♂	Mode of reproduction	Ref.
Tachygonetria macrolaimus	5		Haplodiploidy	137
Tachygonetria numidica	5		Haplodiploidy	137
Tachygonetria longicollis	5		Haplodiploidy	137
Trichurida				
Trichinella domestica	3;2	XX—XO	Amphimixis	98
Trichinella nativa	3;2	XX—XO	Amphimixis	98
Trichinella nelsoni	3;2	XX—XO	Amphimixis	43, 98
Trichinella pseudospiralis	3;2	XX—XO	Amphimixis	98, 99
Trichinella spiralis	3;2	XX—XO	Amphimixis	98-101
Trichosomoides crassicauda	4;3	XX—XO	Amphimixis	3
Trichuris globulosa	3	XX—XY	Amphimixis	11, 23
Trichuris muris	3	XX—XY	Amphimixis	24
Trichuris myocastoris	3	XX—XY	Amphimixis	24, 102
Trichuris ovis	3	XX—XY	Amphimixis	11, 23
Trichuris trichiura	4;3	XX—XO	Amphimixis	11, 23
Ascarida				
Ascaris lumbricoides	12♀		Amphimixis	45
Ascaris lumbricoides	24;19	10X—(5X)O	Amphimixis	3, 104
Ascaris suum	8♀		Amphimixis	45
Ascaris suum (ovis)	24;19	10X—(5X)O	Amphimixis	44, 105
Contracaecum clavatum	12		Amphimixis	4
Contracaecum incurvum	21	16X—(8X)Y	Amphimixis	4, 103
Contracaecum osculatum			Amphimixis	106
Contracaecum spiculigerum	5;4	XX—XO	Amphimixis	4
Parascaris equorum	1—3	XX—XO	Amphimixis	3
Toxocara canis	18;12	12X—(6X)O	Amphimixis	3
Toxocara canis	10	XX—XY	Amphimixis	107
Toxocara leonina		XX—XO	Amphimixis	107
Toxocara mystax	10	XX—XY	Amphimixis	3
Toxocara mystax (cati)	10;9	4X—(2X)Y	Amphimixis	107
Toxocara triquetra or *vulpis*	12;10	4X—(2X)O	Amphimixis	3
Aphelenchida				
Aphelenchidae				
Aphelenchus avenae	8		Amphimixis and parthenogenesis (meiotic)	47
Aphelenchoididae				
Aphelenchoides besseyi	3		Amphimixis	54, 55
Aphelenchoides composticola	3;2 8(2n)	XX—XO	Amphimixis	55, 108
Aphelenchoides fragariae	2		Amphimixis	54
Aphelenchoides ritzemabosi	4		Amphimixis	54, 55
Aphelenchoides tuzeli	3(2n?)		Parthenogenesis	55
Seinura celeris	3♀		Amphimixis	46
Seinura oliveirae	3♀		Amphimixis	46
Seinura oxura	6♂		Automixis	46
Seinura steineri	6♂		Automixis	46
Seinura tenuicaudata	6♀		Amphimixis	109
Tylenchida (Criconematina)				
Hemicriconemoides sp.	5		Amphimixis	6
Cacopaurus pestis	5(6?)		Amphimixis	110
Tylenchulus semipenetrans	5		Amphimixis and parthenogenesis (meiotic)	110

Table 2 (continued)
CHROMOSOMES AND MODE OF REPRODUCTION IN NEMATODES

Taxon	Haploid chromosome number[a]	Sex chromosomes ♀ — ♂	Mode of reproduction	Ref.
Tylenchida (Tylenchina)				
Pratylenchidae				
Pratylenchus penetrans	5♀		Amphimixis	48, 115
Pratylenchus penetrans	6♀		Amphimixis	116
Pratylenchus vulnus	6(♀,♂)		Amphimixis	48
Pratylenchus coffeae	7(♀,♂)		Amphimixis	48
Pratylenchus scribneri	6♀		Parthenogenesis (meiotic)	48
Pratylenchus scribneri	25—26(2n)		Apomixis	48
Pratylenchus zeae	26(2n)		Parthenogenesis	115
Pratylenchus zeae	21—26(2n)		Apomixis	48
Pratylenchus neglectus	20(2n)		Apomixis	48
Pratylenchus brachyurus	30—32(2n)		Apomixis	48
Radopholus similis (banana race) (citrus race)	4		Amphimixis and parthenogenesis (meiotic)	49
Hoplolaimidae	5			
Helicotylenchus dihystera	30, 34, 38(2n)		Apomixis	50
Helicotylenchus erythrinae	5♀		Amphimixis	50
Rotylenchus buxophilus	8♀			6
Rotylenchus robustus	8		Amphimixis	6, 113
Hoplolaimus columbus	50(2n)		Apomixis	Triantaphyllou (unpublished)
Hoplolaimus galeatus	10(♀,♂)		Amphimixis	6
Scutellonema cavenessi	8		Amphimixis	117
Rotylenchulus reniformis	9♀		Amphimixis	6, 118
Dolichodoridae				
Tylenchorhynchus claytoni	8♀		Amphimixis	6
Tylenchorhynchus lamelliferus	8		Amphimixis	113
Tylenchorhynchus husingi	8		Amphimixis	113
Tylenchorhynchus icarus	8		Amphimixis	113
Tylenchorhynchus brevidens	8		Amphimixis and parthenogenesis (meiotic)	113
Trophurus imperialis	8		Amphimixis	113
Macrotrophurus arbusticola	8—9		Amphimixis	113
Belonolaimidae				
Belonolaimus longicaudatus	8♂		Amphimixis	114
Belonolaimus maritimus	8♂		Amphimixis	114
Telotylenchus ventralis	16(2n?)		Amphimixis	113
Anguinidae				
Anguina agrostis	18			52
Anguina agrostis	6			51
Anguina graminis	9			51
Anguina graminophila	9			52
Anguina graminophila	9			51
Anguina millefolii	9—10			51
Anguina tritici	19		Amphimixis	16
Anguina plantaginis	27		Amphimixis	Triantaphyllou (unpublished)
Anguina sp. from *Carex* spp.	18—19			52
Paranguina agropyri	9			51

Table 2 (continued)
CHROMOSOMES AND MODE OF REPRODUCTION IN NEMATODES

Taxon	Haploid chromosome number[a]	Sex chromosomes ♀ — ♂	Mode of reproduction	Ref.
Subanguina radicicola	9—12			51
Ditylenchus destructor	44—48(2n)			17
Ditylenchus dipsaci	12(♀,♂)		Amphimixis	6, 111
Ditylenchus dipsaci (from *Picris* sp., *Taraxacum, Hieracium, Plantago,* and *Vicia*) (from red clover and narcissus)	19—28 12,13,14			53, 56 112
Ditylenchus sonchophila	26—28			53
Ditylenchus triformis	30—40(2n)			6
Heteroderidae				
Cactodera betulae	12,13		Parthenogenesis (meiotic)	58
Globodera mexicana	9♀		Amphimixis	57
Globodera pallida	9♀		Amphimixis	Triantaphyllou (unpublished)
Globodera rostochiensis	9♀			119, 120
Globodera tabacum	9♀		Amphimixis	57
Globodera virginiae	9♀		Amphimixis	57
Globodera solanacearum	9♀		Amphimixis	57
Heterodera avenae	9♀		Amphimixis	121, 122
Heterodera carotae	9♀		Amphimixis	57
Heterodera cruciferae	9♀		Amphimixis	57, 121, 122
Heterodera fici	9		Amphimixis	57, 113
Heterodera galeopsidis	32(2n)		Apomixis	60
Heterodera gambiensis	9		Amphimixis	123
Heterodera glycines	9,(18)		Amphimixis	59, 124
Heterodera goettingiana	9♀			121, 122
Heterodera graminophila	9		Amphimixis	57
Heterodera lespedezae	27(2n)		Apomixis	60
Heterodera leuceilyma	9		Amphimixis	57
Heterodera longicolla	9		Amphimixis	57
Heterodera oryzae	9		Amphimixis	61
Heterodera sacchari	27(2n)		Apomixis	61
Heterodera schachtii	9♀		Amphimixis	57, 121, 122, 125
Heterodera mothi	9		Amphimixis	126
Heterodera trifolii	24, 26, 27, 28, 33, 34, 35(2n)		Apomixis	60, 113, 121
Heterodera trifolii	9		Amphimixis	113
Heterodera weissi	9♀		Amphimixis	57
Heterodera sp. (from *Rumex crispus*)	24(2n)		Apomixis	60
Sarisodera africana	9		Amphimixis(?)	127
Meloidodera floridensis	26,27(2n)			62
Meloidogynidae				
Meloidogyne arenaria (diploid form) (triploid form)	34—37(2n) 51—54(2n)		Apomixis	18
Meloidogyne carolinensis	18		Amphimixis	14
Meloidogyne exigua	18		Amphimixis and parthenogenesis (meiotic)	6

Table 2 (continued)
CHROMOSOMES AND MODE OF REPRODUCTION IN NEMATODES

Taxon	Haploid chromosome number[a]	Sex chromosomes ♀ — ♂	Mode of reproduction	Ref.
Meloidogyne graminicola	18(♀,♂)		Amphimixis and parthenogenesis (meiotic)	128
Meloidogyne graminis	18(♀,♂)		Amphimixis and parthenegenesis (meiotic)	129
Meloidogyne hapla (race A)	14, 15, 16, 17 (♀,♂)		Amphimixis and parthenogenesis (meiotic)	8, 130
Meloidogyne hapla	(34)		Amphimixis	14
Meloidogyne hapla (race B)	30—32 43,45,48(2n)		Apomixis	8, 130
Meloidogyne incognita				6, 7
(race A)	40—46(2n)		Apomixis	
(race B)	32—36(2n)		Apomixis	
Meloidogyne javanica	42—48(2n)		Apomixis	14, 81
Meloidogyne megatyla	18		Amphimixis	14
Meloidogyne microtyla	19		Amphimixis	14
Meloidogyne naasi	18(♀,♂)		Amphimixis and parthenogenesis (meiotic)	128
Meloidogyne ottersoni	18♀		Amphimixis and parthenogenesis (meiotic)	129
Meloidogyne (Hypsoperine) spartinae	7		Amphimixis	6, 131

Numbers separated by a semicolon indicate a heterogametic condition: e.g., 6;5 indicates that the male produces sperm with 6 and 5 chromosomes — the female produces eggs with 6 chromosomes. Numbers in parentheses indicate polyploid forms.

diploid chromosome number is re-established usually by fusion of the egg pronucleus with the second polar nucleus. In "mitotic parthenogenesis"(apomixis) maturation of oocytes consists of a single mitotic division resulting in the formation of a single polar nucleus and the egg pronucleus. No pairing of homologous chromosomes takes place (Figure 6); therefore, metaphase I figures have univalent chromosomes (dyads) in the diploid number.

A single case of haplodiploidy has been reported in the oxyuroid nematode *Gyrinicola batrachiensis,* parasitic on the tadpole stage of various anurans.[136] Some strains of this nematode have no males and the females reproduce by mitonic parthenogenesis. Other strains, however, have variable numbers of males which are haploid and appear to develop from unfertilized eggs produced by amphimictic diploid females. More recently, haplodiploidy was confirmed in several species of the genera *Tachygonetria* and *Mehdiella* of oxyuroid nematodes.[137]

III. CHROMOSOMES AND EVOLUTION OF CERTAIN NEMATODE GROUPS

The available cytological information about nematodes is very limited and often refers to chromosome numbers only. Evaluation of such information is very difficult. Furthermore, a number of references occur in journals of various languages or bulletins of limited circulation, and could be searched only from abstracts. Therefore, the primary effort in this section has been to list the available information (Table 2). Conclusions about the evolutionary significance of cytogenetic data are drawn whenever possible.

A. Adenophorea

Information about the chromosomes of adenophorean nematodes is available only for four genera of the order Dorylaimida. No studies have been conducted with members of any of the other orders, namely Dioctophymatida, Enoplida, Chromadorida, Mononchida, and Monhysterida.

The genera *Xiphinema* and *Longidorus* of the Dorylaimida include exclusively plant-parasitic members. *Xiphinema mediterraneum* is diploid with $n = 5$ chromosomes, whereas *X. diversicaudatum* and *X. index* are considered to be tetraploid with $n = 10$. All have very small, rod shaped, almost spherical chromosomes, less than 0.5 μm in length. During gametogenesis they undergo regular meiosis, but only *X. diversicaudatum* reproduces by amphimixis. The other two species reproduce by meiotic parthenogenesis.[27-29]

All three *Longidorus* species have $n = 7$ chromosomes of small size, similar to those of *Xiphinema*. Meiosis takes place during maturation of gametocytes and reproduction appears to be by facultative meiotic parthenogenesis,[27,29] i.e., by parthenogenesis when males are absent in a population and by amphimixis when males are present.

Paramermis contorta of the family Mermithidae that comprises many insect-parasitic forms, has $n = 6$ chromosomes, four telocentric and two metacentric.[30] One of them is exceptionally long (5 μm), another is very small (1 μm), and the rest are intermediate. Although males in this species constitute only 15% of a population, reproduction is obligatorily amphimictic. *Hexamermis albicans* has $n = 8$ chromosomes but all are of intermediate length (1.5 to 2.5 μm) and without a differentiated centromeric region.[31]

Because of limited information, it is not possible to make any comments about the role of chromosomes in the evolution of Adenophorea. A haploid number of 6, the most common chromosome number in Secernentea, occurs in the amphimictic *Paramermis contorta* of the Adenophorea. The karyotype of this species, however, is not typical of the karyotypes of other nematodes. The extreme variation in size of the chromosomes of *Paramermis contorta* indicates that certain chromosomal rearrangements have occurred in the evolution of its karyotype.

The Adenophorea are assumed to be less advanced evolutionarily compared to Secernentea. However, the members of the Dorylaimida studied cytologically represent the most advanced forms of this order. They are specialized plant parasites, and as such, they have undergone extensive anatomical and biological adaptations. Karyotypic modifications and the establishment of polyploidy, especially in association with parthenogenetic reproduction, as observed in *Xiphinema* and *Longidorus*, are cytogenetic phenomena expected to occur in conjunction with evolution toward plant parasitism. The same type of evolution is seen in many groups of secernentean plant parasites.

Only meiotic parthenogenesis has been detected in the genera *Xiphinema* and *Longidorus* indicating that their evolution toward parthenogenesis is rather recent. Similarly, adaptation of these dorylaimids toward plant parasitism is recent as revealed by the type of parasitism which represents the least advanced form (ectoparasitism), and by the scarcity of plant-parasitic forms among the Adenophorea.

B. Secernentea

The Secernentea have been studied cytogenetically much more extensively than the Adenophorea, apparently because they comprise practically all the animal- and plant-parasitic nematodes in addition to the most common soil-inhabiting forms.

1. Rhabditida

Cytological studies include primarily free-living soil nematodes of the families Rhabditidae, Panagrolaimidae, and Diplogasteridae. The most common chromosomal complement in these families has $n = 6 \pm 1$, small (1 to 3 μm) chromosomes, with two X chromosomes in the females and hermaphrodites and one X chromosome in the males. Larger chromosome numbers of $2n = 18$ to 21 may represent polyploid forms derived through hybridization. Thus, *Rhabditis anomala* (♀ $2n = 20$) is assumed to be a triploid evolved from the closely related *R. pellio* or *R. maupasi* ($n = 7;6$) following fertilization of an unreduced egg ($2n = 14$) with a reduced sperm of $n = 6$.[32] *Pelodera strongyloides* has approximately twice as many chromosomes ($n = 11; 10$) as other related species, but, as in the case of other nematodes assumed to be polyploid, it shows only bivalent chromosomes at metaphase I and reproduces primarily by amphimixis.[5] *Diploscapter coronata* with only one pair of chromosomes has the smallest chromosome number in Nematoda,[15] and its relationship to ancestral forms from which it may have evolved is not known.

All members of the families Panagrolaimidae and Diplogasteridae studied are amphimictic and show only slight diversification in chromosomal complement. The family Rhabditidae, on the other hand, shows extensive cytogenetic diversification with the establishment of polyploidy in association with parthenogenetic or gynogenetic reproduction (meiotic or mitotic). The most striking aspect of cytogenetic evolution in the family Rhabditidae, however, is the establishment of hermaphroditism — possibly secondarily from amphimixis.[33]

Cytological studies in the Rhabditida include also animal-parasitic forms of the families Rhabdiasidae and Strongyloididae in which parasitic generations alternate with free-living generations (heterogony). Three species of Rhabdiasidae are cytogenetically similar to rhabditids in that their haploid karyotype consists of six small-size chromosomes, and they have an XX ♀ or ♂ — XO ♂ sex mechanism. Hermaphrodites are also present in this family, as in Rhabditidae.

The family Strongyloididae appears to represent a more advanced state of karyotypic evolution in the order Rhabditida. *Strongyloides ratti* has one half the chromosomes of other rhabditids.[34,35] Reduction of the haploid number may be the result of chromosomal fusions, as indicated from the larger size of the chromosomes. *S. ratti* maintains the same XX ♀ — XO ♂ sex mechanism characteristic of all rhabditids. The free-living generations have males and females and reproduce by amphimixis or gynogenesis, whereas the parasitic generations are apomictic. The modified karyotype of *S. ratti* has been further evolved in *S. papillosus* and *S. ransomi* where the X chromosome probably has fused with an autosome to give rise to a neo-X chromosome and an XX ♀ — XY ♂ mechanism of sex determination.[22,36]

2. Strongylida

Cytogenetically, the Strongylida is the most stable order of nematodes. All strongylids studied thus far cytologically are bisexual, amphimictic with a haploid chromosome number of 6, and a chromosomal mechanism of sex determination of XX ♀ — XO ♂. Only *Filaroides mustelarum*,[37] has $n = 8$, but confirmation of this early report may be needed. It is indeed difficult to interpret the lack of cytogenetic diversification in such a large group of nematodes. The chromosomes within the karyotypes of most species are of uniform size (2 to 4 μm) and morphology (rod shaped). Even species of different genera may have indistinguishable karyotypes; e.g., *Chaberia ovina* and *Oesophagostomum columbianum*.[38] On the other hand, the bovine strain of *Haemonchus contortus* can be differentiated by its larger X chromosomes (8 μm) from the ovine strain of the same species which has normal size (3 μm) X chromosomes.[39] A multiple or diffuse type of centromere has been reported for at least some species[11,40] and may be a general feature of strongylid chromosomes.

3. Spirurida

There is only limited cytogenetic and chromosomal diversification in this order. Approximately one half of the species studied have a haploid chromosome complement of 6 and the rest have $n = 5$. *Setaria labiatopapillosa* has a haploid chromosome number of only 3, but two of its chromosomes are twice the size of those of other species of the genus. Podgornova et al.[41] suggested that $n = 3$ may be the primitive karyotype from which the higher chromosome numbers of the other members of genus have been derived. However, it is more likely that this exceptional case has derived from a basic karyotype of $n = 6$ through chromosomal fusions resulting in a reduced number.

The chromosomal mechanism of sex determination is XX ♀ — XO ♂ in all Spirurida, except in *Dirofilaria immitis*, in which some forms appear to have acquired a mechanism of XX ♀ — XY ♂.[12]

Chromosome fragmentations may occur in some spirurids during early cleavage divisions,[3] so that somatic cells may have approximately twice as many chromosomes as the zygote nucleus. This behavior may be facilitated by the multicentric nature of the chromosomes as reported[11] for *Physaloptera*, *Spironoura* and *Camallanus* spp. Furthermore, chromatin diminution has been observed during the early cleavage divisions in *Physaloptera indiana*, and chromosomal polymorphism involving one chromosome pair has been reported in *Camallanus baylisi*.[11]

4. Oxyurida

Cytogenetically, the order Oxiurida is quite uniform. Most of the species studied have a haploid chromosome number of 5 (some 6) and a chromosomal mechanism of sex determination of XX ♀ — XO ♂. Karyotypes are similar between species, and the chromosomes within each karyotype are uniform and of small size. *Cosmocerca kashmirensis* with $n = 8$ and two pairs of long chromosomes is an exception.[42] *Syphacia ovelata* may also have $n = 8$, but further documentation is needed. *Gyrinicola batrachiensis* and several species of the genera *Tachygonetria* and *Mehdiella* are the only nematodes known to exhibit haplodiploidy.[136,137]

5. Trichurida

The taxonomic position of the Trichurida and their phyletic relationships to other nematodes are not clear as yet. Their karyotype, consisting of three or four small chromosomes, deviates considerably from the karyotype of more primitive nematodes, suggesting an extensive karyotypic evolution, probably associated with specialization toward animal parasitism.

Trichuris trichiura and *Trichosomoides crassicauda*, with $n = 4$ small chromosomes and an XX ♀ — XO ♂ mechanism of sex determination, may represent the most ancestral forms within this order.[23] Other *Trichuris* species with $n = 3$ and an XX ♀ — XY ♂ sex mechanism probably have evolved from the previous karyotype following chromosomal rearrangements that resulted in the transfer of a large part of the X chromosome to an autosome. Evidence supporting this interpretation is the persistence in the latter species of supernumerary microchromosomes.[23]

In the genus *Trichinella*, which has $n = 3$ and XX ♀ — XO ♂ sex mechanism, reduction in chromosome number may have resulted from fusion of two autosomes. Indeed, one of the autosomes of the *Trichinella* karyotype is significantly longer than the others.[43] The fact that the various *Trichinella* strains, recognized recently as distinct species, have very similar or identical karyotypes supports their common, but unique cytological derivation.

Trichurids have chromosomes with a localized centromere,[3,23,24] resembling the Adenophorea, under which they were classified until recently, and are different from most Secernentea which have chromosomes with a diffuse type of centromere. Still, the nature of the centromere (localized or diffuse) needs to be investigated in many more nematodes, before its evolutionary significance can be evaluated.

6. Ascarida

Among all animal-parasitic forms, the Ascarida has undergone the most extensive cyto-genetic diversification. Cytological features that make the ascarids unique among other nematodes are the presence of compound, multicentric chromosomes, the evolution of a multiple X mechanism of sex determination, and the occurrence of chromosome fragmentation and chromatin diminution in early cleavages.

The large, compound chromosomes are present only in germ-line cells. They undergo fragmentation and give rise to a number of smaller chromosomes in somatic cells during cleavage.[3] It is believed that the compound chromosomes have evolved by the end-to-end association of smaller chromosomes which may have their own localized centromeres, or possibly a diffuse type of centromere. Consequently, ascarids with compound chromosomes have evolved from ancestors with simple chromosomes. This type of evolution must have occurred several times, in different phyletic lines, as evidenced by the presence of members with compound chromosomes in each one of the four genera of ascarids studied thus far.

The simple chromosomes of ascarids are small (1 to 3 μm), like those of other orders of animal-parasitic nematodes. A localized centromere has been observed in the set of five sex chromosomes of *Ascaris suum*,[44] but a diffuse type of centromere is suspected in other cases.

The most primitive cytological situation has been encountered in *Contracaecum spiculigerum* which has $n = 5$ and possibly $n = 8$ chromosomes and an XX — XO mechanism of sex determination.[3,4] Eight simple chromosomes have been reported also in a strain of *Ascaris suum*.[45] However, most species of ascarids have larger haploid chromosome numbers ($n = 10, 12, 18, 21, 24$) and show an increase in total chromosome volume. The total amount of DNA probably is larger in such species, but much of the extra DNA may represent heterochromatin. It is possible that certain species of ascarids are polyploid. However, the increased number of X chromosomes present in some species cannot correspond to the degree of ploidy. In some species, e.g., *Toxocara triquetra*, the 4 X — (2 X)O sex mechanism may indicate tetraploidy, but in other species, e.g., *T. canis*, the 12 X — (6 X)O situation may indicate chromosome fragmentation, or other chromosomal mutations leading to chromosome number increases.

7. Aphelenchida

This order has been treated by many taxonomists as a suborder of the Tylenchida. It comprises some specialized plant-parasitic forms, but the majority are predaceous soil inhabitants or insect parasites and show distinct evolution from that of tylenchids. Cytogenetically also they exhibit a distinct line of evolution. The genus *Seinura* includes species with haploid numbers of 6 and 3 chromosomes.[46] *S. tenuicaudata*, a bisexual amphimictic species with $n = 6$, may be regarded cytogenetically as the most primitive.[6] Other bisexual amphimictic species, e.g., *S. celeris* and *S. oliveira*, with $n = 3$ may have evolved through chromosome fusions resulting in reduction of the haploid number. Still other species *(S. oxura, S. steineri)* have maintained the ancestral ($n = 6$) chromosomal complement but have evolved toward hermaphroditism. This is the second case of demonstrated hermaphroditism in nematodes and, as in the Rhabditida, it is suspected to have evolved secondarily from bisexuality.[14] The only other nematode group suspected to exhibit hermaphroditism is the Criconematina of the Tylenchida for which, however, no cytological data are available.

Reductions in the haploid chromosome number are observed also in members of the genus *Aphelenchoides* which includes some specialized plant-parasitic forms.

In contrast to this pattern of evolution, an increased haploid complement with 8 or 9 chromosomes is observed in the least specialized, soil-inhabiting, fungus feeder, *Aphelenchus avenae* which, however, has evolved toward meiotic parthenogenesis.[47]

8. Tylenchida

The recognition, during the last 25 years, that members of the order Tylenchida are causing considerable damage to agricultural crops around the world stimulated an intensive study in various areas, including cytogenetics. As a result, a considerable amount of cytogenetic information has become available, especially about the most important plant-parasitic groups.[6,14] These nematodes have undergone extensive cytogenetic diversification involving karyotypic changes, establishment of polyploidy, and evolution toward parthenogenetic reproduction. The topic has been reviewed recently,[14] and only a brief account will be presented here.

Within the Tylenchida, three species of Criconematina have a haploid complement of 5 chromosomes and reproduce by amphimixis or parthenogenesis. Nothing can be said about cytological evolution.

In the Tylenchina a larger number of species from many families provide a much clearer cytogenetic picture. Unfortunately, no cytological information is available about the least specialized, alga- and fungus-feeding members of the family Tylenchidae which, possibly, have chromosomal complements more representative of the ancestral forms of tylenchid nematodes. The available information shows that the most specialized forms such as members of the families Anguinidae, Heteroderidae and Meloidogynidae, tend to have more chromosomes and possibly more DNA per haploid nucleus than other, less specialized tylenchids.

Small chromosome numbers of $n = 6 \pm (1$ or 2$)$ are found in some species of the families Pratylenchidae, Hoplolaimidae, Dolichodoridae, and Belonolaimidae which, primarily, comprise migratory plant-parasitic forms. Thus, *Pratylenchus penetrans, Radopholus similis,* and *Helicotylenchus erythrinae* are amphimictic and have a haploid number of five chromosomes of small size.[48-50] Other members of the same families have $n = 6$, 7, or 8 chromosomes and also reproduce by amphimixis. *Hoplolaimus galeatus* with $n = 10$, i.e., twice the haploid number of *Helicotylenchus erythrinae,* may represent a polyploid state of the same karyotype.[6] Members with large, haploid numbers have chromosomes of equal size, or larger than those of species with smaller haploid numbers. This fact may indicate that aneuploidy and polyploidy, rather than chromosomal fissions, are responsible for haploid number increases. Definitely, polyploidy has been established in parthenogenetic species of *Helicotylenchus dihystera, Pratylenchus scribneri, P. zeae, Hoplolaimus columbus,* and others.[48,50]

With only few exceptions, members of the families Dolichodoridae and Belonolaimidae have $n = 8$ chromosomes and are amphimictic. This cytogenetic stability may indicate their recent evolution and the close phyletic relationship of the species within each family.

A more extensive cytogenetic diversification apparently has occurred in the family Anguinidae, where the most frequent haploid chromosome number is nine in *Anguina* and twelve in *Ditylenchus.*[51-53] The *Ditylenchus* chromosomes are slightly smaller than those of *Anguina;* therefore, increase in number through fragmentation is possible. Larger haploid chromosome numbers observed in some amphimictic species, e.g., *A. tritici* with $n = 19$, *A. plantaginis* with $n = 27$, *Ditylenchus dipsaci* (some races) with $n = 19$ to 27, probably represent polyploids.[14,53,59] However, such polyploid forms cytologically behave like diploids, i.e., they form only bivalents during the first maturation division.

The family Heteroderidae is relatively stable cytogenetically. Most of its members are bisexual and amphimictic with a haploid chromosomal complement of nine.[57] There are only slight differences in chromosome size between some genera. An increased haploid number ($n = 10$ to 13) has been observed in *Cactodera betulae* which, however, reproduces primarily by meiotic parthenogenesis.[58] Polyploidy combined with amphimixis has been found only in one population of *Heterodera glycines.*[59] Several other cases of polyploidy and aneuploidy, e.g., in *H. trifolii, H. sacchari, Meloidodera floridensis,* etc., are associated with apomictic reproduction.[60-62] They have not played a significant role in the evolution of the family.

Only the *H. trifolii* parthenogenetic species complex, which exhibits extensive chromosomal variation, has been successful biologically as judged by its wide distribution and successful parasitism.

Finally, the Meloidogynidae cytogenetically represents the most evolved family of plant-parasitic nematodes. With a haploid complement of 18 in amphimictic species of *Meloidogyne*, it has the largest haploid chromosome number among all nematodes. Some decreases in the haploid number appear to have occurred in *M. hapla* ($n = 17$ to 14) which can also reproduce facultatively by meiotic parthenogenesis.[8] Increases in chromosome number, including the establishment of polyploidy, have occurred in forms that have acquired the capacity to reproduce by mitotic parthenogenesis, e.g., *M. incognita, M. arenaria,* and *M. javanica.* These polyploid parthenogenetic forms are considered to be the most important nematode parasites of plants. Evolution toward polyploidy in association with parthenogenesis evidently has contributed to successful parasitism and enabled these nematodes to extend their ecological niche and become worldwide in distribution.[135]

IV. CONCLUSIONS

Nematodes constitute a highly diversified group of organisms. They are adapted to a wide variety of life styles and exist in every conceivable habitat on earth. Nematodes feeding on bacteria and other low forms of life (algae, fungi) are considered to be the least evolved biologically, whereas most evolved are those adapted to parasitic life on animals and plants.

Cytogenetically most nematodes are diploid, amphimictic, with an XX—XO mechanism of sex determination and a haploid complement of approximately six chromosomes of small (1 to 5 μm) size. This cytogenetic situation is common among primitive nematodes, but it is observed also in many specialized animal-parasitic and some plant-parasitic forms. Deviations, indicating cytogenetic evolution, are encountered in some groups, as follows:

Hermaphroditism, demonstrated cytologically in some rhabditids and aphelenchoidids, appears to have evolved secondarily from bisexuality.

Decreases in chromosome numbers ($n = 2$ to 3), probably resulting from chromosome fusions, are observed in the family Strongyloididae and the order Trichurida of animal-parasitic forms. Chromosome fusions involving sex chromosomes probably have led to the establishment of an XX—XY sex mechanism in both these groups. Decreases in chromosome number also have occurred in some specialized plant-parasitic forms of the Aphelenchida.

Increases in the haploid chromosomal complement, possibly involving some cases of polyploidy, are common in two orders of nematodes. In the Ascarida, which includes highly specialized animal-parasitic forms, cytological evolution involves increases of the haploid chromosome number from six to twelve and up to 24; establishment of a multiple X—O and multiple X—Y sex mechanisms; and evolution of compound, multicentric chromosomes with subsequent chromosome fragmentation and chromatin diminution during early cleavage. In the Tylenchida, which includes most plant-parasitic nematodes, cytogenetic evolution involves increases in the haploid chromosomal complement from six to nine, twelve, and eighteen, and the establishment of polyploidy, usually in association with apomictic reproduction.

Less than 1% of the described nematode species have been studied cytologically; therefore, these conclusions are tentative. Many nematode groups have not been studied at all and others are represented only by a small number of species. More extensive and in-depth cytological studies are needed. Still, it can be stated that nematodes, in general, have been conservative in chromosomal evolution. Striking modifications in chromosomal complement have occurred primarily in specialized groups among animal and plant-parasitic forms. In many cases, such evolution has been accompanied by the establishment of parthenogenetic reproduction.

REFERENCES

1. **Beneden, E. van,** Rechereches sur la maturation de l'oeuf et la fecondation, *Arch. Biol.,* 4, 265, 1883.
2. **Boveri, T.,** Über Differenzierung der Zellkerne während der Furchung des Eies von *Ascaris megalocephala, Anat. Anz.,* 2, 688, 1887.
3. **Walton, A. C.,** Studies on nematode gametogenesis, *Z. Zellen Gewebelehre,* 1,167, 1924.
4. **Walton, A. C.,** Gametogenesis, in *An Introduction to Nematology,* Sect. II, Part 1, Chitwood, B. G, Ed., M. B. Chitwood, Babylon, New York, 1940, 205.
5. **Nigon, V.,** Modalités de la reproduction et déterminisme du sexe chez quelques nématodes libres, *Ann. Scien. Naturelles, Zool.,* 11, 1, 1949.
6. **Triantaphyllou, A. C.,** Genetics and cytology, in *Plant Parasitic Nematodes,* Vol. 2, Zuckermann, B. M., Mai, W. F., and Rohde, R. A., Eds., Academic Press, New York, 1971, 51.
7. **Triantaphyllou, A. C.,** Oogenesis and the chromosomes of the parthenogenetic root-knot nematode *Meloidogyne incognita, J. Nematol.,* 13, 95, 1981.
8. **Triantaphyllou, A. C.,** Polyploidy and reproductive patterns in the rootknot nematode *Meloidogyne hapla, J. Morphol.,* 118, 403, 1966.
9. **Lin, T. P.,** The chromosomal cycle in *Parascaris equorum (Ascaris megalocephala):* oogenesis and diminution, *Chromosoma,* 6, 175, 1954.
10. **Nigon, V. and Brun, J.,** L'évolution des structures nucléaires dans l'ovogenése de *Caenorhabditis elegans* Maupas, 1900, *Chromosoma,* 7, 129, 1955.
11. **Goswami, U.,** Karyological studies on fifteen species of parasitic nematodes, *Res. Bull. (Sci.), Panjab Univ.,* 28, 111, 1979.
12. **Sakaguchi, Y., Kihara, S., and Tada, I.,** The chromosomes and gametogenesis of *Dirofilaria immitis, Jpn. J. Parasitol.,* 29, 377, 1980.
13. **Goldstein, P. and Triantaphyllou, A. C.,** The ultrastructure of sperm development in the plant-parasitic nematode *Meloidogyne hapla, J. Ultrastr. Res.,* 71, 143, 1980.
14. **Triantaphyllou, A. C. and Hirschmann, H.,** Cytogenetics and morphology in relation to evolution and speciation of plant-parasitic nematodes, *Annu. Rev. Phytopathol.,* 18, 333, 1980.
15. **Hechler, H. C.,** Postembryonic development and reproduction in *Diploscapter coronata* (Nematoda: Rhabditidae), *Proc. Helminthol. Soc. Wash.,* 35, 24, 1968.
16. **Triantaphyllou, A. C. and Hirschmann, H.,** Gametogenesis and reproduction in the wheat nematode, *Anguina tritici, Nematologica,* 12, 437, 1966.
17. **Ladygina, N. M. and Barabashova, V. N.,** The genetic and physiological compatibility and the karyotypes of stem nematodes, *Parazitologiya,* 10, 449, 1976 (in Russian).
18. **Triantaphyllou, A. C.,** Polyploidy and parthenogenesis in the root-knot nematode *Meloidogyne arenaria, J. Morphol.,* 113, 489, 1963.
19. **Lapp, N. A. and Triantaphyllou, A. C.,** Relative DNA content and chromosomal relationships of some *Meloidogyne, Heterodera,* and *Meloidodera* spp. (Nematoda: Heteroderidae), *J. Nematol.,* 4, 287, 1972.
20. **Sin, W. C. and Pasternak, J.,** Number and DNA content of nuclei in the free-living nematode *Panagrellus silusiae* at each stage during postembryonic development, *Chromosoma,* 32, 191, 1970.
21. **Sulston, J. E. and Brenner, S.,** The DNA of *Caenorhabditis elegans, Genetics,* 77, 95, 1974.
22. **Triantaphyllou, A. C. and Moncol, D. J.,** Cytology, reproduction, and sex determination of *Strongyloides ransomi* and *S. papillosus, J. Parasitol.,* 63, 961, 1977.
23. **Goswami, U.,** Some cytogenetical aspects of genus *Trichuris* (Nematoda), *Curr. Sci.,* 47, 368, 1978.
24. **Dmitrieva, T. I., Podgornova, G. P. and Shlikas, A. V.,** Karyotypes of *Trichocephalus muris* (Schrank, 1788) and *T. myocastoris* (Enigk, 1933), *Acta Parasitol. Lituanica,* 17, 47, 1979 (in Russian).
25. **Nelson, G. A., Lew, K. K. and Ward, S.,** Intersex, a temperature-sensitive mutant of the nematode *Caenorhabditis elegans, Develop. Biol.,* 66, 386, 1978.
26. **Nigon, V.,** Développement et reproduction des nématodes, in *Traité de Zoologie,* Vol. 4, Grassé, P., Ed., Masson & Cie., Paris, 1965, 218.
27. **Dalmasso, A.,** Cytogenetics and reproduction in *Xiphinema* and *Longidorus,* in *Nematode Vectors of Plant Viruses,* Lamberti, F., Taylor, C. E. and Seinhorst, J. W., Eds., Plenum Publ. Co., New York, 1975, 139.
28. **Dalmasso, A. and Younés, T.,** Étude de la gamétogenèse chez *Xiphinema mediterraneum, Nematologica,* 16, 51, 1970.
29. **Dalmasso, A. and Younès, T.,** Ovogenèse et embryogenèse chez *Xiphinema index* (Nematoda: Dorylaimida), *Ann. Zool. Écol. Anim.,* 1, 265, 1969.
30. **Parenti, U.,** Fecondazione e corredo cromosomico di *Paramermis contorta* parassita di *Chironomous tentans, Atti. Accad. Naz. Lincei,* 32, 699, 1962.
31. **Goncharova, S. N. and Romanenko, L. N.,** Preliminary studies on the chromosomes of *Hexamermis albicans* (Mermithidae), *Byull. Vses, Inst. Gel'mintol.,* 17, 30, 1976 (in Russian).

32. **Goodchild, C. G. and Irwin, G. H.,** Occurrence of nematodes *Rhabditis anomala* and *R. pellio* in oligochaetes *Lumbricus rubellus* and *L. terrestris, Trans. Am. Microsc. Soc.,* 90, 231, 1971.
33. **Triantaphyllou, A. C. and Hirschmann, H.,** Reproduction in plant and soil nematodes, *Annu. Rev. Phytopathol.,* 2, 57, 1964.
34. **Nigon, V. and Roman, E.,** Le déterminisme du sexe et le développement cyclique de *Strongyloides ratti, Bull. Biol. France Belg.,* 86, 404, 1952.
35. **Abe, Y. and Tanaka, H.,** Studies on the chromosomes of *Strongyloides ratti* Sandground, 1925, *Jpn. J. Parasitol,* 14, 520, 1965.
36. **Albertson, D. G., Nwaorgu, O. C, and Sulston, J. E.,** Chromatin diminution and a chromosomal mechanism of sexual differentiation in *Strongyloides papillosus, Chromosoma,* 75, 75, 1979.
37. **Carnoy, J. B.,** La cytodiéfèśe de l'oeuf chez quelques nématodes, *La Cellule,* 3, 1, 1887.
38. **LeJambre, L. F.,** The chromosome numbers of *Oesophagostomum columbianum* Curtice and *Chabertia ovina* Fabricius (Nematoda: Strongylata), *Trans. Am. Microsc. Soc.,* 87, 105, 1968.
39. **Bremner, K. C.,** Cytological studies on the specific distinctness of the ovine and bovine 'strains' of the nematode *Haemonchus contortus* (Rudolphi) Cobb (Nematoda: Trichostrongylidae), *Aust. J. Zool.,* 3, 312, 1955.
40. **Dmitrieva, T. I.,** The chromosomes in mitosis and gametogenesis of 2 nematodes from the genus *Dictyocaulus,* in *Fauna, sistem, biol ekologiya gel'mintov i ikh promezhutochnykh khozyaev (Respublikanskii Sbornik), Gor'kii USSR,* 1977, 33 (in Russian).
41. **Podgornova, G. P., Shol', V. A, and Gubaidulin, N. A.,** Karyological study of *Setaria* of ungulates, in *Zhiznennye tsikly, ekologiya i morfologiya gel'mintov zhivotnykh Kazakhstana,* Alma-Ata, USSR, "Nauka", 1978, 157 (in Russian).
42. **Fotedar, D. N., Duda, P. L., and Raina, M. K.,** The chromosomes of *Cosmocerca kashmirensis* Fotedar 1959 (Oxyuroidea: Cosmocercidae), *Chromos. Inform. Serv.,* 14, 16, 1973.
43. **Mutafova, T. and Komandarev, S.,** On the karyotype of a laboratory *Trichinella* strain from Bulgaria, *Z. Parasitenkd.,* 48, 247, 1976.
44. **Mutafova, T.,** Morphology and behavior of sex chromosomes during meiosis in *Ascaris suum, Z. Parasitenkd.,* 46, 291, 1975.
45. **Kurashvili, B. E., Pkhakadze, G. M., and Shengelia, F. V.,** Cytological study of *Ascaris lumbricoides* and *A. Suum, Genetika,* 5, 170, 1965 (in Russian).
46. **Hechler, H. C. and Taylor, D. P.,** The life histories of *Seinura celeris, S. oliveirae, S. oxura,* and *S. steineri* (Nematoda: Aphelenchoididae), *Proc. Helminthol. Soc. Wash.,* 33, 71, 1966.
47. **Triantaphyllou, A. C. and Fisher, J.,** Gametogenesis in amphimictic and parthenogenetic populations of *Aphelenchus avenae, J. Nematol.,* 8, 168, 1976.
48. **Roman, J. and Triantaphyllou, A. C.,** Gametogenesis and reproduction of seven species of *Pratylenchus, J. Nematol.,* 1, 357, 1969.
49. **Huettel, R. N. and Dickson, D. W.,** Karyology and oogenesis of *Radopholus similis* (Cobb) Thorne, *J. Nematol.,* 13, 16, 1981.
50. **Triantaphyllou, A. C. and Hirschmann, H.,** Cytology and reproduction of *Helicotylenchus dihystera* and *H. erythrinae, Nematologica 13, 575, 1967.*
51. **Krall, E. L. and Aomets, E. K.,** Cytological and genetical aspects of the evolution of *Anguina,* causing the formation of galls on grasses, in *Materialy. VI. Pribaltiiskoi Nauchno-Koordinatsionnoi Konferentsii po Voprosam Parazitologii,* Akademiya Nauk Litovskoi SSR, Institut Zoologii i Parazitologii, 1973, 82 (in Russian).
52. **Solov'eva, G. I. and Gruzdeva, L. I,** Karyotypes of plant-parasitic nematodes of the genus *Anguina* Scopoli, 1799, *Parazitologiya,* 11, 366, 1977 (in Russian).
53. **Barabashova, B. N.,** Karyotypes of stem eelworms of wild plants, *Parazitologiya,* 13, 257, 1979 (in Russian).
54. **Cayrol, J. C. and Dalmasso, A.,** Affinités interspécifiques entre trois nématodes des feuilles *(A. fragariae, A. ritzemabosi* et *A. besseyi), Cah ORSTOM Sér Biol.,* 10, 215, 1975.
55. **B'Chir, M. M. and Dalmasso, A.,** Meiosis and mitotic chromosome numbers in certain species of the genus *Aphelenchoides, Rev. Nematol.,* 2, 249, 1979.
56. **Sturhan, D.,** *Ditylenchus dipsaci* — doch ein Artenkomplex?, *Nematologica,* 16, 327, 1970.
57. **Triantaphyllou, A. C.,** Oogenesis and the chromosomes of twelve bisexual species of *Heterodera* (Nematoda: Heteroderidae), *J. Nematol.,* 7, 34, 1975.
58. **Triantaphyllou, A. C.,** Oogenesis and reproduction of the birch cyst nematode, *Heterodera betulae, J. Nematol,* 2, 399, 1970.
59. **Triantaphyllou, A. C. and Riggs, R. D.,** Polyploidy in an amphimictic population of *Heterodera glycines, J. Nematol.,* 11, 371, 1979.
60. **Triantaphyllou, A. C. and Hirschmann, H.,** Cytology of the *Heterodera trifolii* parthenogenetic species complex, *Nematologica,* 24, 418, 1978.

61. **Netscher, C.,** L'ovogénèse et la reproduction chez *Heterodera oryzae* et *H. sacchari* (Nematoda: Heteroderidae), *Nematologica,* 15, 10, 1969.

62. **Triantaphyllou, A. C.,** Oogenesis and the chromosomes of the cystoid nematode, *Meloidodera floridensis, J. Nematol.,* 3, 183, 1971.

63. **Dalmasso, A.,** La gamétogenèse des genres *Xiphinema* et *Longidorus* (Nematoda:Dorylaimida), *C. R. Acad. Sc. Paris,* 270, 824, 1970.

64. **Krüger, E.,** Fortpflanzung und Keimzellenbildung von *Rhabditis aberrans* n. sp., *Z. Wiss. Zool.,* 105, 87, 1913.

65. **Kröning, F.,** Studien zur Chromatinreifung der Keimzellen, Die Tetradenbildung und die Reifetelungen bei einigen Nematoden, *Arch. Zellforsch,* 17, 63, 1923.

66. **Belar, K.,** Über den Chromosomenzyklus von parthenogenetischen Erdnematodes, *Biol. Zentrbl,* 43, 513, 1923.

67. **Nigon, V.,** Le déterminisme du sexe et la pseudogamie chez un nématode parthénogénétique, *Rhabditis monohystera* Bütschli, *Bull. Biol. France Belg.,* 81, 33, 1947.

68. **Hertwig, P.,** Beobachtungen über die Fortpflanzungsweise und die systematische Einteilung der Regenwurmnematoden *Z. Wiss. Zool.,* 119, 539, 1922.

69. **Nigon, V. and Dougherty, E. C.,** Reproductive patterns and attempts at reciprocal crossing of *Rhabditis elegans* Maupas, 1900, and *Rhabditis briggsae* Dougherty and Nigon, 1949, (Nematoda: Rhabditidae), *J. Exp. Zool.,* 112, 485, 1949.

70. **Honda, H.,** Experimental and cytological studies on bisexual and hermaphrodite free-living nematodes, with special reference to problems of sex, *J. Morphol. Physiol.,* 40, 191, 1925.

71. **Nigon, V.,** Polyploidie experimentale chez un nematode libre, *Rhabditis elegans Maupas, Bull. Biol. France Belg.,* 85, 187, 1951.

72. **Hechler, H. C.,** Reproduction, chromosome number, and postembryonic development of *Panagrellus redivivus* (Nematoda: Cephalobidae), *J. Nematol.,* 2, 355, 1970.

73. **Hechler, H. C.,** A brief report on the chromosome number of *Neodiplogaster pinicola* and *Panagrellus redivivoides, J. Nematol.,* 4, 243, 1972.

74. **Pillai, J. K. and Taylor, D. P.,** Biology of *Paroigolaimella bernensis* and *Fictor anchiocoprophaga* (Diplogasterinae) in laboratory culture *Nematologica,* 14, 159, 1968.

75. **Mulvey, R. H.,** Oogenesis in several free-living and plant-parasitic nematodes, *Can. J. Zool.,* 33, 295, 1955.

76. **Hechler, H. C.,** Chromosome number and reproduction in *Mononchoides changi* (Nematode: Diplogasterinae), *J. Nematol.,* 12, 125, 1970.

77. **Schleip, W.,** Das Verhalten des Chromatins bei *Angiostomum (Rhabdonema) nigrovenosum.* Ein Beitrag zur Kenntnis der Beziehungen zwischen Chromatin und Geschlechtsbestimmung, *Arch. Zellforsch.,* 7, 87, 1911.

78. **Dreyfus, A.,** Contribucao para o estudo do cyclo chromosomico e da determinacao do sexo de *Rhabdias fülleborni, Biol. Geral.,* 1, 1, 1937.

79. **Runey, W. M., Runey, G. L. and Lauter, F. H.,** Gametogenesis and fertilization in *Rhabdias ranae* Walton 1929. I. The parasitic hermaphrodite, *J. Parasitol.,* 64, 1008, 1978.

80. **Zaffagnini, F.,** Parthenogenesis in the parasitic and free-living forms of *Strongyloides papillosus* (Nematoda, Rhabdiasoidea), *Chromosoma,* 40, 443, 1973.

81. **Triantaphyllou, A. C.,** Oogenesis in the root-knot nematode *Meloidogyne javanica, Nematologica,* 7, 105, 1962.

82. **Gonzales, J. C. and Malmann, M. C.,** Cromosomas de algumas espécies de nematódeos parasitos de ovinos e bovinos no Rio Grande do Sul, Brasil, *Revista de Medicina Veterinaria, São Paulo,* 6, 132, 1970.

83. **Tromba, F. G. and Steele, A. E.,** The chromosomes of *Stephanurus dentatus* (Nematoda: Strongyloidea), *J. Parasitol.,* 43, 590, 1957.

84. **Valero, A. and Pretel, A.,** Karyological studies on *Marshallagia marshalli,* Nematoda Trichostrongylidae, *Rev. Iber. Parasitol.,* 39, 119, 1979.

85. **John, B.,** The chromosomes of zooparasites. II. *Oswaldocruzia filiformis* (Nematoda: Trichostrongylidae), *Chromosoma,* 9, 61, 1957.

86. **Podgornova, G. P,** Chromosome complexes during the maturation of the gametes of *Oswaldocruzia goezei* Skryabin & Shults, 1952, *Ekologicheskaya i Eksperimental'naya Parazitologiya,* 1, 140, 1975 (in Russian).

87. **Mulsow, K.,** Chromosomenverhältnisse bei *Ancyracanthus cystidicola, Zool. Anz.,* 38, 484, 1911.

88. **Boyes, J. W. and Anderson, R. C,** Meiotic chromosomes of *Cystidicola stigmatura* and *C. cristivomeri* (Nematoda: Spiruroidea), *Can. J. Genet. Cytol.,* 3, 231, 1961.

89. **Taylor, A. E. R.,** The spermatogenesis and embryology of *Litomosoides carinii* and *Dirofilaria immitis, J. Helminthol.,* 34, 3, 1960.

90. **Miller, M. J.,** Observations on spermatogenesis on *Onchocerca volvulus* and *Wuchereria bancrofti, Can. J. Zool.,* 44, 1003, 1966.

91. **Podgornova, G. P., Trofimenko, V. Ya., Dmitrieva, T. I., Lomakin, V. V., Shlikas, A. V. and Shol',
V. A,** The karyology of some Adenophorea and Secernentea nematodes, *Trudy Gel'mintologicheskoi La-
boratorii (Gel'minty zhivotnykh i rastenii)*, 29, 112, 1979 (in Russian).

92. **Goswami, U.,** An unequal pair of chromosomes in *Camallanus baylisi* (Nematoda), *Cytologia*, 39, 321,
1974.

93. **Mutafova, T.,** Comparative cytological studies of mitotic and male meiotic karyotype of *Ascaridia dissimilis*
(Vigueras, 1931) and *Ascaridia galli* (Schrank, 1788), *Z. Parasitenkd.*, 48, 239, 1976.

94. **Gulick, A.,** Über die Geschlechtschromosomen bei einigen Nematoden nebst Bemerkungen über die Be-
deutung dieser Chromosomen, *Arch. Zellforsch.*, 6, 339, 1911.

95. **Sultanov, M. A., Tashkhodzhaev, P. I., Zimin, Yu. M., Chinenkov, V. A. and Ismailov, T.,** The
biology of *Oxyuris equi* (Schrank, 1788), *Doklady Akademii Uzbekskoi SSR (Uzbekistan SSR Fanlar Aka-
demiyasining, Dokladlari)*, 12, 48, 1977 (in Russian).

96. **Vogel, R.,** Zur Kenntnis der Fortpflanzung, Eireifung, Befruchtung und Furchung von *Oxyuris obvelata*
Bremser, *Zool. Jahrb. (Zool.)*, 42, 243, 1926.

97. **Meves, F.,** Über Samenbildung und Befruchtung bei *Oxyuris ambigua*, *Arch. Mikr. Anat.*, 94, 135, 1920.

98. **Pen'kova, R. A. and Romanenko, L. N.,** A study of *Trichinella* chromosomes, *Trudy Vsesoyuznogo
Instituta Gel'mintologii im. K. I. Skryabina*, 20, 133, 1973 (in Russian).

99. **Penkova, R. A. and Romanenko, L. N.,** Studies of karyotypes of *Trichinella spiralis* and *T. pseudospiralis*,
in *4th Int. Congr. Parasitol.*, August 19-26, 1978, Warsaw, Poland, Short commun., Section C., 1978,
144 (in Russian).

100. **Geller, E. R. and Gridasova, L. F.** The karyotype and ovogenesis of *Trichinella spiralis*, *Med. Parazitol.
Parazit. Bolezni.*, 43, 324, 1974 (in Russian).

101. **Thomas, H.,** Beiträge zur Biologie und mikorskopischen Anatomie von *Trichinella spiralis* Owen 1835,
Z.Tropenmed. Parasitol., 16, 148, 1965.

102. **Jenkins, T., Larkman, A., and Funnell, M.,** Spermatogenesis in a trichuroid nematode, *Trichuris muris*.
I.Fine structure of spermatogonia, *Int. J. Invertebr. Reprod.*, 1, 371, 1979.

103. **Goodrich, H. B.,** The germ cells in *Ascaris incurva*, *J. Exp. Zool.*, 21, 61, 1916.

104. **Edwards, C. L.,** The idiochromosomes in *Ascaris megalocephala* and *Ascaris lumbricoides*, *Arch. Zellforsch.*, 5, 422, 1910.

105. **Vassilev, I. and Mutafova, T.,** Comparative studies on the karyotype of *Ascaris suum* and *Ascaris ovis*,
Z. Parasitenkd., 43, 115, 1974.

106. **Monné, L.,** On cyclical alterations in the Feulgen staining of nuclei during the development of nematodes,
Parasitology, 53, 273, 1963.

107. **Chinenkov, V. A.,** The structure of the chromosomal apparatus of *Toxocara mystax* (*T. cati*) (Seder, 1800)
Stiles, 1907 and *T. canis* (Werner, 1782) Stiles, 1905, *Tr. (Pervogo) Mosk. Med. Inst.*, 84, 118, 1975 (in
Russian).

108. **Brun, J. L. and Younès, T.,** Gamétogenèse et déterminisme du sexe chez le nématode mycophage
Aphelenchoides composticola, *Nematologica*, 15, 591, 1969.

109. **Hechler, H. C.,** Description, developmental biology, and feeding habits of *Seinura tenuicaudata* (de Man)
J. B. Goodey, 1960 (Nematoda: Aphelenchoididae), a nematode predator, *Proc. Helminthol. Soc. Wash.*,
30, 182, 1963.

110. **Dalmasso, A., Macaron, J., and Bergé, J. B.,** Modalités de la reproduction chez *Tylenchulus semipentrans*
et chez *Cacopaurus pestis* (Nematoda—Criconematoidea), *Nematologica*, 18, 423, 1972.

111. **Barabashova, V. N.,** Karyotypical pecularities of some forms of stem nematodes of the collective species
Ditylenchus dipsaci, *Parazitologiya*, 8, 408, 1974 (in Russian).

112. **Barabashova, V. N,** The karyotype of the stem nematode of red clover and narcissi, *Byulleten' Vsesoyuznogo Instituta Gel'm intologii im. K. I. Skryabina*, 15, 24, 1975 (in Russian).

113. **Bergé, J., Dalmasso, A., and Ritter, M.,** Nouvelles données phylogéniques sur les Tylenchoidea (Nematoda), *C. R. Acad. Sc. Paris*, 276, 3307, 1973.

114. **Robbins, R. T. and Hirschmann, H.,** Variation among populations of *Belonolaimus longicaudatus*, *J.
Nematol.*, 6, 87, 1974.

115. **Hung, Chia-Ling and Jenkins, W. R.,** Oogenesis and embryology of two plant-parasitic nematodes,
Pratylenchus penetrans and *P. zeae*, *J. Nematol.*, 1, 352, 1969.

116. **Thistlethwayte, B.,** Reproduction of *Pratylenchus penetrans* (Nematoda: Tylenchida), *J. Nematol.*, 2, 101,
1970.

117. **Demeure, Y., Netscher, C. and Quénéhervé, P.,** Biology of the plant-parasitic nematode *Scutellonema
cavenessi* Sher, 1964: Reproduction, development and life cycle, *Rev. Nématol.*, 3, 213, 1980.

118. **Nakasono, K.,** Role of males in reproduction of the reniform nematode, *Rotylenchulus* spp. (Tylenchida:
Hoplolaimidae), *Appl. Ent. Zool.*, 1, 203, 1966.

119. **Riley, R. and Chapman, V.,** Chromosomes of the potato root eelworm, *Nature*, 180, 662, 1957.

120. **Cotten, J.,** Observations on the cytology of the potato-root eelworm, *Heterodera rostochiensis* Wollen-
weber, *Nematologica*, (Suppl. II), 123, 1960.

121. **Mulvey, R. H.,** Oogenesis in some species of *Heterodera* and *Meloidogyne* (Nematoda: Heteroderidae), in *Nematology — Fundamentals and Recent Advances,* Sasser, J. N. and Jenkins, W. R., Eds., University of North Carolina Press, Chapel Hill, 1960, 323.
122. **Cotten, J.,** Cytological investigations in the genus *Heterodera, Nematologica,* 11, 337, 1965.
123. **Merny, G. and Netscher, C,** *Heterodera gambiensis* n. sp. (Nematoda: Tylenchida) parasite du mil et du sorgho en Gambie, *Cah. ORSTOM, Sér. Biol,* 11, 209, 1976.
124. **Triantaphyllou, A. C. and Hirschmann, H.,** Oogenesis and mode of reproduction in the soybean cyst nematode, *Heterodera glycines, Nematologica,* 7, 235, 1962.
125. **Mulvey, R. H.,** Chromosome number in the sugar beet nematode *Heterodera schachtii* Schmidt, *Nature,* 180, 1212, 1957.
126. **Masood, A. and Husain, S. I.,** Oogenesis and mode of reproduction in cyperus cyst nematode, *Heterodera mothi, Acta Bot. Indica,* 6, 141, 1978.
127. **Luc, M., Germani, G., and Netscher, C.,** Description de *Sarisodera africana* n. sp. et considerations sur les relations entre les genres *Sarisodera* Wouts & Sher, 1971 et *Heterodera* A. Schmidt, 1871 (Nematoda: Tylenchida), *Cah. ORSTOM, Sér. Biol.,* 21, 35, 1973.
128. **Triantaphyllou, A. C.,** Gametogenesis and the chromosomes of two root-knot nematodes, *Meloidogyne graminicola* and *M. naasi, J. Nematol.,* 1, 62, 1969.
129. **Triantaphyllou, A. C.,** Gametogenesis and reproduction of *Meloidogyne graminis* and *M. ottersoni* (Nematoda: Heteroderidae), *J. Nematol.,* 5, 84, 1973.
130. **Triantaphyllou, A. C.,** Cytogenetics of root-knot nematodes, in *Root-knot Nematodes (Meloidogyne species), Systematics, Biology and Control,* Lamberti, E. and Taylor, C. E., Eds., Academic Press, New York, 1979, 477
131. **Fassuliotis, G. and Rau, G. J.,** Observations on the embryogeny and histopathology of *Hypsoperine spartinae* on smooth cordgrass roots, *Spartina alterniflora, Nematologica,* 12, 90, 1966.
132. **Terry, A., Terry, R. J., and Worms, M. J.,** *Dipetalonema witei,* filarial parasite of the jird, *Meriones libycus.* II. The reproductive system, gametogenesis and development of microfilaria, J. Parasitol., *47, rom.* 703, 1961.
133. **Albertson, D. G. and Thomson, J. N.,** The kinetochores of *Caenorhabditis elegans, Chromosoma (Berlin)* 86, 409, 1982.
134. **Carnoy, J. B.,** La cytodiérèse de l'oeuf chez quelques nématodes, *La Cellule,* 3, 5, 1887.
135. **Triantaphyllou, A. C.,** Cytogenetic aspects of nematode evolution, in *Concepts in Nematode Systematics,* Stone, A. R., Platt, H. M. and Khalil, L. F., Eds., Systematics Association Special Volume 22, Academic Press, London, 1983, 55.
136. **Adamson, M. L.,** Studies on gametogenesis in *Gyrinicola batrachiensis* (Walton, 1929) (Oxyuroidea: Nematoda), *Can. J. Zool.,* 59, 1368, 1981.
137. **Adamson, M. L.,** personal communication.

Chapter 5

CHROMOSOMES AND EVOLUTION IN PTERIDOPHYTES

Trevor G. Walker

TABLE OF CONTENTS

I. INTRODUCTION

The history of pteridophyte cytology is a relatively short one due to difficulties inherent in the material and the need to develop suitable techniques to deal with them. Although a few accurate chromosome counts had been made in the past,[1] together with descriptions and interpretations of cytological processes,[2,3] the subject dates in effect from 1950. This year saw the publication of Manton's classic *Problems of Cytology and Evolution in the Pteridophyta* in which the technical problems of examining meiosis had been successfully overcome by a modified squash technique involving quite considerable manual pressure.[4] This enabled the high numbers of chromosomes customarily met with in this group to be counted with complete accuracy instead merely of the close estimates which had been the general rule with paraffin wax sections.

Root tips for mitotic counts initially formed fairly intractable material because of the exceptionally tough nature of the intercellular bonding. Good hydrolysis was, however, obtained by Meyer by boiling in acetocarmine.[5] Although a very effective method, it was exceedingly unpleasant to use on any scale, and the process of hydrolysis was immeasurably improved by the introduction of snail gastric juice as a hydrolyzing agent at normal temperatures.[6] Purified enzymes for this purpose can now be obtained commercially.

While pteridophyte cytology has paralleled in many respects that of the considerably older angiosperm cytology, the order of progress and the relative emphasis given have been very different in these two groups of plants. Thus, the majority of chromosome counts have been established on meiotic rather than mitotic material. One of the most compelling reasons for this is the very high chromosome number characteristic of the pteridophytes, where it is considerably easier to count the gametic rather than the doubled sporophytic number. A further advantage is that the pairing behavior of the chromosomes can be observed as well, and many instances of hybrids and apomicts have been noted which would otherwise have gone undetected in somatic nuclei. The restricted seasonal reproductive period in most flowering plants is normally less of a problem in the pteridophytes as here the fixation period available tends to last for a very much longer time. Almost contemporaneous with the determination of chromosome numbers, biosystematic studies involving the synthesis and analysis of hybrids went apace, although much of this is outside the scope of this chapter except when the results have a direct bearing on chromosome behavior.

Very shortly after the appearance of Manton's book, research on pteridophyte cytology developed in many countries of the world in both tropical and subtropical zones in addition to the temperate ones. Pteridophyte floras have been extensively studied in Britain, parts of continental Europe, Canada, the U.S., Mexico, Jamaica, Trinidad, Madeira, the Canary Islands, West Africa, Sri Lanka, S. India, the Himalayan region, Malaya, Taiwan, Japan, New Zealand, and very recently in Australia (see Walker[8] and in Lovis[9]). In addition, small but important samples from Costa Rica and Brazil have been published.[10,11] Work in progress involves the pteridophyte floras of Indonesia, Papua New Guinea, and Sarawak.[12] The only large areas for which very few data are available are mainland China and South America as a whole, although Trinidad may be regarded as a small part of the latter in this context. As a consequence the pteridophytes in general and the ferns in particular have provided a basis whereby whole floras from many different parts of the world can be compared with one another. A further and most important feature is that virtually all the results are based upon material of known wild origin, the very few exceptions being clearly indicated as such, thereby avoiding many of the pitfalls which beset some of the earlier investigations in the angiosperms.

II. BASIC CHROMOSOME NUMBERS

Estimates of the total numbers of taxa present in the pteridophytes vary from some 10,000 to 13,000, of which all but about 1,000 at a generous estimate are ferns. Approximately 200 non-fern taxa have been cytologically examined and some 2,500 ferns (including much personal unpublished data). In addition, several hundred different hybrid combinations have been studied. Thus some 20% of taxa are cytologically known to a greater or lesser degree belonging to approximately three quarters of the total number of genera. Of the unexamined genera, the vast majority are either monotypic or only comprise two or three species and are almost invariably tropical, very rare, or in very isolated regions. Lists of chromosome numbers may be found in Chiarugi,[13] Fabbri,[14,15] and Löve et al.[16] In the last mentioned work all numbers are converted to the presumed $2n$ values regardless of the form of the original citation and also excluding sterile hybrids. A list of basic chromosome numbers in the pteridophytes is presented in Table 1.

Probably the most striking feature of the pteridophytes is the high level of the chromosome numbers almost universally encountered, with only a very few exceptions. These exceptions for the most part occur in plants having a heterosporous life cycle, namely *Selaginella*, *Isoetes* and the aquatic ferns. The low basic numbers of 7, 8, 9, and 10 in *Selaginella*, 11 (10) in *Isoetes*, 9 in *Salvinia*, 10, 13 in *Pilularia*, and the higher but clearly derived numbers of 19 in *Regnellidium* and 20 in *Marsilea* contrast strongly with those in the great majority of homosporous members. Thus, in the non-fern members the lycopods have base numbers of 23 and 34 or 39 established (although unpublished work shows several additional numbers[12]), in *Psilotum* and *Tmesipteris* $x = 52$, while in *Equisetum* $x = 108$. Similarly, in the ferns the vast majority of genera or part-genera have basic chromosome numbers that range from $x = 22$ to 60 with three very pronounced peaks at $x = 29/30$, $x = 36$, and $x = 40/41$. The striking contrast between the heterosporous and the homosporous ferns leads to the question as to the origin of the high numbers. High numbers are considered to have been derived from low ones by a process of very ancient multiplication — palaeopolyploidy — followed by diploidization. Its antiquity is indicated by the prevalence of a particular basic number in all species of a genus throughout the world and frequently with no trace of what the original number may have been. Thus, *Equisetum* has universally been found to have $x = 108$. Various attempts have been made to postulate an ancestral number for this genus, e.g., the $x = 9$ favored by Löve et al.[16] Such attempts are futile at present, particularly as 108 has a very large number of factors and can be divided in a multiplicity of ways, even assuming it represents a straightforward multiplication in the first place.

Evidence for original low basic numbers rests on data derived from several sources. Mention has already been made of the low values found at the present day in the heterosporous groups. It would seem to be very strange if these were fundamentally different from all other pteridophytes. In the filmy ferns belonging to the family Hymenophyllaceae, numbers such as $n = 32, 33, 34, 36$, or their multiples abound and at first sight this looks like a straightforward aneuploid series. However, closer inspection reveals that some are polyploids on different low base numbers. Thus $n = 11$ has been reported in *Hymenophyllum peltatum* and $n = 13$ in *H. tunbrigense* and it would be surprising if some at least of the many lines with $n = 36, 72, 108$, and 144 did not ultimately refer back to $x = 12$.[17,4] Similarly in *Mecodium* in the same family numbers such as $2n = 42, 56$, and 84 suggest the existence of a series based on 7.

Two more lines of a more indirect nature have been proposed, one based on genetic data and the other on chromosomal morphology. In the former, Klekowski presents figures suggesting the presence of duplicated loci in *Osmunda*,[18] indicating that the universal $n = 22$ in this genus has been derived from an ancestral base number ($= b$) of $b = 11$. Several Japanese workers have analyzed excellent photographs of somatic cells of a number of species. They claim that the chromosomes can be divided into a number of distinct groups which give a clue as to the ancestral number. Thus, since the chromosomes of *Asplenium* and *Camptosorus* with $x = 36$ are of six types, each of which can be further split into two

Table 1

BASIC CHROMOSOME NUMBERS OF PTERIDOPHYTES

Order; family; subfamily	Base numbers	Genus
Psilotales		
Psilotaceae	52	*Psilotum*
	52	*Tmesipteris*
Lycopodiales		
Lycopodiaceae	34	*Lycopodium*
	35,39	*Lycopodiella*
	?17,23	*Diphasium*
		n = c. 34-36
	?	*Huperzia*
		n = c. 128-132,136,165-178
	?	*Phylloglossum*
		2*n* = c. 502
Selaginellales		
Selaginellaceae	7,8,9,10	*Selaginella*
Isoetales		
Isoetaceae	[10],11	*Isoetes*
	?	*Stylites*
		2*n* = c. 50
Equisetales		
Equisetaceae	108	*Equisetum*
Ophioglossales		
Ophioglossaceae	45 [46]	*Botrychium*
	94	*Helminthostachys*
	120	*Ophioglossum*
Marattiales		
Marattiaceae	40	*Angiopteris*
	—	*Archangiopteris*
	—	*Protomarattia*
	[39],40	*Marattia*
	40	*Macroglossum*

Order; family; subfamily	Base numbers	Genus
Adiantaceae (continued)	29,30	*Pellaea*
	29,30	*Dryopteris*
	—	*Ormopteris*
	29 or 30	*Saffordia*
	30	*Trachypteris*
	[26],[27],29	*Anogramma*
	29,30	*Pityrogramma*
	30	*Trismeria*
	30	*Hemionitis*
	30	*Gymnopteris*
	30	*Bommeria*
	30	*Paraceterach*
	29*	*Pterozonium*
	—	*Syngrammatopteris*
	—	*Jamesonia*
	29	*Eriosorus*
	29	*Nephopteris*
	—	*Syngramma*
	29	*Craspedodictyum*
	29	*Toxopteris*
	—	*Austrogramme*
	22	*Taenitis*
	—	*Platytaenia*
	—	*Holttumiella*
	30	*Coniogramme*
	29	*Asplenopsis*
	—	*Rheopteris*
	—	*Cerosora*
Vittarioideae	29,30	*Adiantum*
	30	*Antrophyum*

Family / Subfamily	Genus	*n*
Marattiales (continued)	*Danaea*	40
	Christensenia	40
Filicales		
Osmundaceae	*Osmunda*	22
	Todea	22
	Leptopteris	22
Plagiogyriaceae	*Plagiogyria*	25?,33?
	n = 66,75, c.100,125	
Schizaeaceae	*Schizaea*	?
	n = 77,94,96,103,134,c. 154,c. 270	
	Actinostachys	?
	n = 140,325 ± 30,350-370	
	Lygodium	28,29,30
	Anemia	38
	Mohria	38
Parkeriaceae	*Ceratopteris*	39,40
Platyzomataceae	*Platyzoma*	38
Adiantaceae		
Adiantoideae	*Actiniopteris*	29
	Afropteris	—
	Ochropteris	—
	Anopteris	29
	Onychium	29
	Cryptogramma	30
	Llavea	29
	Neurosoria	?
	n = c.53-54*	
	Cheilanthes	26*,27*, 28,29,30
	Adiantopsis	30
	Aleuritopteris	29,30
	Aspidotis	30
	Cheiloplecton	—
	Mildella	29
	Notholaena	29,30
	Sinopteris	—
	Negripteris	—
Adiantoideae	*Polytaenium*	30
	Scoliosorus	—
	Anetium	30
	Hecistopteris	—
	Ananthacorus	30
	Vittaria	30
	Monogramma	—
	Vaginularia	30
Pteriodoideae	*Pteris*	29
	Idiopteris	27
	Copelandiopteris	—
	Neurocallis	29
	Acrostichum	30
Loxsomaceae	*Loxsoma*	50
	Loxsomopsis	46
Hymenophyllaceae	*Hymenoglossum*	—
	Serpyllopsis	—
	Rosenstockia	—
	Hymenophyllum	11,13,17*,18,21, 22
	Buesia	—
	Meringium	21,22,26,28
	Eupectinum	—
	Myriodon	21
	Sphaerocionium	36
	Apteropteris	36
	Craspedophyllum	12*
	Hemicyatheon	—
	Mecodium	21,[27],28,36
	Cardiomanes	36
	Trichomanes	32
	Lacosteopsis (pro sectione)	36
	Crepidomanes	36
	Polyphlebium	36
	Reediella	36
	Abrodictyum	36

Table 1 (continued)
BASIC CHROMOSOME NUMBERS OF PTERIDOPHYTES

Order; family; subfamily	Base numbers	Genus
Hymenophyllaceae (continued)	36	*Pleuromanes*
	36	*Gonocormus*
	28?,33,36	*Selenodesmium*
	32	*Davalliopsis*
	32	*Cephalomanes*
	36	*Callistopteris*
	36	*Nesopteris*
	34	*Didymoglossum*
	34	*Microgonium*
	32	*Lecanolepis*
	32,36	*Achomanes* (pro sectione)
	?	*Neuromanes*
		n = 26,32,c. 36
	—	*Odontomanes*
	32,36	*Lacostea*
	32	*Trigonophyllum* (pro sectione)
		Homoeotes
	32	*Feea*
	32	*Ragatelus*
	32	*Acarpacrium* (pro sectione)
Hymenophyllopsidaceae	—	*Hymenophyllopsis*
Stromatopteridaceae	—	*Stromatopteris*
Gleicheniaceae	20,22	*Gleichenia*
	28?	*Diplopterygium*
	n = 56	
	34	*Sticherus*
	39	*Dicranopteris*

Order; family; subfamily	Base numbers	Genus
Microsorioideae (continued)	36,37	*Arthromeris*
	—	*Polypodiopteris*
Pleopeltoideae	[22,23,25,26]34, 35,36,37,39,[47]	*Pleopeltis*
		Microgramma
	37	*Anapeltis*
	36,37	*Craspedaria*
	37	*Solanopteris*
	35	*Marginariopsis*
	37	*Neurodium*
	36	*Neolepisorus*
	36	*Lemmaphyllum*
	36	*Drymotaenium*
	36	*Paragramma*
	33,35	*Belvisia*
	36	*Dicranoglossum*
		Neocheiropteris
	—	*Niphidium*
	37	*Pessopteris*
	37	*Campyloneurum*
	35	*Dictymia*
	37	*Phlebodium*
		Synammia
	35,36,37	*Polypodium*
	37	*Goniophlebium*
	—	*Thylacopteris*
Grammitidaceae	35,37	*Marginaria*
	36,37	*Grammitis*
	32,33,36,37	*Xiphopteris*
	37	*Ctenopteris*

Family / Subfamily	Genus	
Gleicheniaceae (continued)	*Acropterygium*	43
Matoniaceae	*Matonia*	26
	Phanerosorus	25
Cheriopleuriaceae	*Cheriopleuria*	57?
	n = c. 57	
Dipteridaceae	*Dipteris*	33
Polypodiaceae		
Drynarioideae	*Drynaria*	36,37
	Photinopteris	36
	Merinthosorus	36
	Aglaomorpha	36
	Dryostachyum	—
	Holostachyum	—
	Drynariopsis	36,37
	Pseudodrynaria	37
	Thayeria	36,37
Platycerioideae	*Platycerium*	37
	Pyrrosia	36,37
	Saxiglossum	—
	Drymoglossum	—
Microsorioideae	*Microsorum*	36,37
	Dendroconche	36,37
	Phymatosorus	37
	Lecanopteris	36
	Colysis	—
	Podosorus	—
	Diblemma	—
	Leptochilus	36
	Paraleptochilus	—
	Dendroglossa	—
	Christiopteris	—
	Pycnoloma	—
	Grammatopteridium	—
	Oleandropsis	—
	Holcosorus	—
	Crypsinus	33,35
	Selliguea	37

Family / Subfamily	Genus	
Grammitidaceae (continued)	*Calymmodon*	—
	Acrosorus	—
	Amphoradenium	37
	Prosaptia	37
	Glyphotaenium	37
	Oreogrammitis	—
	Nematopteris	—
	Scleroglossum	33
	Cochlidium	37
	Hyalotricha	35,36
	Loxogramme	37
	Anarthopteris	
Metaxyaceae	*Metaxya*	95 or 96
Lophosoriaceae	*Lophosoria*	65
Cyatheaceae	*Cyathea* (incl. *Trichopteris, Nephelea, Alsophila, Sphaeropteris*)	69,[70]
	Dicksonia	65
	Cystodium	56
Thyrsopteriadaceae	*Thyrsopteris*	—
	Culcita subg. *Culcita*	66
	Culcita subg. *Calochlaena*	56*
	Cibotium	68
Dennstaedtiaceae		
Dennstaedtioideae	*Dennstaedtia*	30,31,32,33,34, c. 44,46,47
	Microlepia	40,42,43,44
	Oenotrichia	41*
	Leptolepia	c.47
	Hypolepis	?26,29,39,45*, 50,52
	Paesia	26
	Pteridium	26

Table 1 (continued)
BASIC CHROMOSOME NUMBERS OF PTERIDOPHYTES

Order; family; subfamily	Base numbers	Genus
Denstaedioideae (continued)	?24 or 48	*Histiopteris*
	—	*Lepidocaulon*
	50	*Lonchitis*
	38	*Blotiella*
	?	*Saccoloma*
	$n = 63 \pm 2$	
	47	*Orthiopteris*
	$n = 188$	
Monachosoroideae	56	*Monachosorum*
Lindsaeoideae	34,c.	*Lindsaea*
	40,41,42,44,47, c. 50	
	42	*Ormoloma*
	38,39,41,47,50	*Sphenomeris*
	c. 48	*Odontosoria*
	?	*Tapeinidium*
	$n = c. 120$	
Thelypteridaceae	—	*Xyropteris*
	35	*Thelypteris*
	29	*Amauropelta*
	30	*Phegopteris*
	31	*Pseudophegopteris*
	36	*Cyclogramma*
	27,31(c. 36)	*Parathelypteris*
	32	*Coryphopteris*
	31	*Macrothelypteris*
	34	*Oreopteris*
	(31),35(36)	*Metathelypteris*
	36	*Cyclosorus*
	36	*Trigonospora*

Order; family; subfamily	Base numbers	Genus
Athyrioideae	40	*Athyrium*
	40,41	*Cornopteris*
	40	*Rhachidosorus*
	41	*Diplazium* (incl. *Callipteris*, *Diplaziopsis*)
	40	*Anisocampium*
	31	*Hemidictyum*
	40	*Gymnocarpium*
	42	*Cystopteris*
	[33,38,39],41	*Woodsia*
	40	*Lunathyrium*
	—	*Adenoderris*
	—	*Cheilanthopsis*
	40,41	*Hypodematium*
	40	*Kuniwatsukia*
	—	*Trichoneuron*
Tectarioideae	41	*Crenitis*
	40	*Ctenitopsis*
	41	*Lastreopsis*
	41	*Psomiocarpa*
	—	*Atalopteris*
	41	*Pleocnemia*
	41	*Arcypteris*
	41	*Pteridrys*
	—	*Dryopolystichum*
	40	*Tectaria*
	—	*Pseudotectaria*
	40	*Hemigramma*
	40	*Quercifilix*

Thelypteridaceae (continued)	*Pronephrium*	36
	Mesophlebion	36
	Plesioneuron	36
	Glaphyropteridopsis	36
	Chingia	36
	Haplodictyum	—
	Nannothelypteris	—
	Stegnogramma	36
	Steiropteris	36
	Glaphyropteris	36
	Sphaerostephanos	36
	Ampelopteris	36
	Goniopteris	c. 36
	Meniscium	36
	Pneumatopteris	—
	Christella	—
	Amphineuron	36
Aspleniaceae Asplenioideae	*Asplenium*	36,[40]
	Camptosorus	36
	Phyllitis	36
	Ceterach	36
	Ceterachopsis	36
	Pleurosorus	36
	Pleurosoriopsis	36
	Loxoscaphe	c. 36
	Diellia	36
	Holodictyum	—
	Antigramma	—
	Diplora	36
	Schaffneria	?38
	Boniniella	*n* = 76
Dryopteridaceae Onocleoideae	*Matteuccia*	[39],40
	Onoclea	37
	Onocleopsis	40

Tectarioideae (continued)	*Cionidium*	40
	Tectaridium	—
	Fadyenia	40
	Pleuroderris	40
	Hypoderris	40
	Amphiblestra	—
	Dictyoxiphium	40
	Camptodium	40
	Stenosemia	40
	Heterogonium	40
	Cyclopeltis	41
	Didymochlaena	41
Dryopteridoideae	*Peranema*	41
	Diacalpe	41
	Polystichum	41
	Papuapteris	—
	Plecosorus	—
	Cyclodium	(c. 42)
	Cyrtomium	41
	Cyrtomidictyum	41
	Cyrtogonellum	—
	Phanerophlebia	41
	Arachniodes	41
	Polystichopsis	41
	Lithostegia	—
	Polybotrya	41
	Maxonia	41
	Dryopteris	41
	Nothoperanema	41
	Stigmatopteris	41
	Acrophorus	41
	Stenolepia	—
	Bolbitis	—
Lomariopsidaceae	*Egenolfia*	41
	Thysanosoria	41
	Arthrobotrya	(c. 40)
	Teratophyllum	41

Table 1 (continued)
BASIC CHROMOSOME NUMBERS OF PTERIDOPHYTES

Family	Genus	Number	Family	Genus	Number
Lomariopsidaceae (continued)	Lomagramma	41	Blechnaceae	Nephrolepis	41
	Lomariopsis	[16,31,39],41		Blechnum	28,29,30,31,32, 33,34,36
	Elaphoglossum	41		Salpichlaena	40
	Peltapteris	40		Doodia	32
Davalliaceae				Brainea	33
Davallioideae	Humata	40		Sadleria	33
	Trogostolon	—		Woodwardia	34
	Scyphularia	—		Anchistea	35
	Parasorus	—		Chieniopteris	—
	Davallia	40		Lorinseria	35
	Davallodes	40		Pteridoblechnum	27*
	Paradavallodes	40		Stenochlaena	37
	Araiostegia	40	Marsileaceae	Marsilea	20
	Leucostegia	41		Regnellidium	19
	Gymnogrammitis	—		Pilularia	10,13
Oleandroideae	Rumohra	41			
	Oleandra	40, 41	Salviniaceae	Salvinia	9
	Arthropteris	41	Azollaceae	Azolla	22
	Psammiosorus	—			

Note: (1) The list of pteridophyte genera is based on that of Crabbe, Jermy and Michel,[33] with some modifications. (2) Orders have been kept to a minimum. Thus all ferns with the exception of Ophioglossales and Marattiales are subsumed as Filicales. (3) Believed aneuploid derivatives indicated by []; cases requiring confirmation, etc., indicated by (). (4) Entries marked with an asterisk are new base numbers (personal communication) to be published in a cytotaxonomic survey of the pteridophytes of Australia by Mary D. Tindale and S. K. Roy.

groups, the suggestion made was that $b = 12$.[19-21] The same answer was arrived at for *Lemmaphyllum* and *Pyrrosia*, also with $x = 36$.[22] Osmundaceae, with $x = 22$, was considered to be derived from $b = 11$.[23,24] More involved explanations had to be devised to account for the situation in such diverse genera as *Diplazium* with $2n = 82$ and *Matteucia* with $2n = 80$,[25] *Pteris* with $2n = 58$ and 116,[26] *Hymenophyllum* with $2n = 42$ and *Mecodium* with $2n = 42$, 56, and 84.[27] All these invoked multiplication of ancestral genomes with $b = 11$, followed by the loss of a number of chromosomes. While the quality of the preparations is excellent, nevertheless karyotyping ferns presents great technical difficulties, and the evidence for the conclusions of these workers is by no means convincing as discussed by Duncan and Smith.[28]

Wagner and Wagner suggest that most pteridophyte basic numbers are derived from b = 9, 10, or 11 which have reached various levels of ploidy,[29] sometimes combined with simple aneuploidy. For example, *Osmunda* with $x = 22$ is 2×11, *Pteris* with $x = 29$ is $3 \times 10 (-1)$, *Asplenium* with $x = 36$ is 4×9, and *Botrychium* with $x = 45$ is 5×9. While their general tenor is that palaeopolyploidy has resulted in present day base numbers which represent different ancient ploidy levels, the whole scheme is too simplistic and ignores the part that dibasic numbers may have played in aneuploid changes which have arrived at a number which is a multiple of one of the so-called basic ones by coincidence.

However, despite any controversy surrounding derivation, the establishment of accurate present day basic chromosome numbers has been of considerable importance in studies on pteridophytes. In no other large group of organisms have cytological data played such an important part in the formulation or revision of various systems of classification and in influencing ideas concerning interrelationships. The various systematic revisions which have appeared since 1950 such as those of Pichi Sermolli,[30,31] Nayar,[32] Crabbe et al.,[33] and Lovis[9] have been considerably influenced by such data. Similar attention has also been paid in monographic works such as those of Holttum on Thelypteridaceae.[34] This is clearly brought out in Table 1 of basic chromosome numbers arranged with modifications according to the scheme of Crabbe et al.[33] The few discrepancies usually point to groups of species in need of further taxonomic study or less frequently to species where drastic alterations in the chromosome complement have occurred such as in *Lomariopsis* and *Hemidictyum*.

Prior to these later systems, probably the work which had the greatest influence this century was Copeland's *Genera Filicum*.[34] His family Pteridaceae was a very large one, encompassing 61 genera. Manton showed that while the latter half of the family (hinging on *Adiantum* and *Pteris*) were very uniform cytologically with base numbers of $x = 29$ or 30, the other half was extremely diverse.[36] This has been formally recognized by the splitting off of Adiantaceae as a separate family in most modern schemes and in treating the remainder as being composed of a number of families, often without close relationships. Similarly, Holttum suggested that Thelypteridaceae bore no relationship to Dryopteridaceae, despite being placed by Copeland in one family Aspidiaceae. Cytology has lent support to Holttum's view in highlighting the contrast between the aneuploid series ranging from $x = 27$ to 36 in Thelypteridaceae to an almost universal $x = 40/41$ in Dryopteridaceae. In his revision of Thelypteridaceae, Holttum recognized 23 genera. Each genus proved to be cytologically homogeneous, having a single typical base number.[34] Since the range of x is from 27 to 36, obviously some genera have the same number, but in the very rare instances where species had a number deviating from that typical of the genus, these proved almost without exception to be those which were doubtfully placed in the scheme in the first instance.[8]

One of the most unsatisfactory groups in present classifications is Dennstaedtiaceae, centred on the many species presently placed in *Dennstaedtia*, *Microlepia*, *Oenotrichia* and *Hypolepis*. A very varied range of chromosome numbers exists here. *Dennstaedtia* has $x = 30$, 31, 32, 33, 34, and c. 44, 46, 47. Lovis suggests that as *Microlepia* partly fills the gap between these two series of numbers with $x = ?40$, 42, 43, and c. 44; this apparent gap may not be real.[9] However, few taxonomists would dispute the fact that this group is in need of monographic study on a world-wide basis. A further extended discussion on the impact of cytology upon the taxonomy and systematics of ferns and of the problems involved may be found in Walker[8] and in Lovis.[9]

While a study of basic chromosome numbers has led to taxonomic revisions, often on a handsome scale, due attention must be paid to morphological features. Although cytology has been extraordinarily successful in solving some problems (or indeed in pointing to the existence of certain unidentified problems) and in highlighting relationships within families, it has so far failed to give much information regarding the possible interrelationships of the families themselves. It can only point to the rather isolated position of some families such as Loxsomaceae which had been previously closely associated with Hymenophyllaceae and Davalliaceae, but in which its two component genera are in different hemispheres, the monotypic *Loxsoma* in New Zealand with $n = 50$ and *Loxsomopsis,* consisting of about three species in tropical South America and with $n = 46$.[37,29] This gap in our knowledge may eventually be filled as methods of chromosomal analysis improve and comparative biochemical studies become more common.

III. ANEUPLOIDY

Aneuploidy occurs at two levels in the pteridophytes, affecting either the basic chromosome number or members of a polyploid series. For the most part at the lower level it involves the loss or gain either of a single chromosome or of a very small number from the basic complement such as in *Lygodium* with $x = 28, 29, 30$, or the more extended series in the genus *Blechnum* with $x = 28, 29, 30, 31, 32, 33, 34, 36$. On the higher taxonomic level of a family, Thelypteridaceae shows an unbroken series from $x = 27$ to 36. In the ancient family Gleicheniaceae the base numbers are $x = 20, 22, 34, 39, 43$, and 56. How they arose is open to conjecture. Two hypotheses have been advanced by Mehra and by Sorsa.[38,39] The former postulated that these numbers derived from original base numbers of 7, 10, 13, and 17 by straight polyploidy, while the latter author derives the present numbers from only two original ones, viz., 11 and 17, by a more complex route which assumes the formation of two dibasic polyploids, followed by loss of chromosomes. Lovis has pointed out various objections.[9] A third possibility follows the simplest (although not necessarily the correct) route, and that is to consider that the $n = 56$, which so far is established on counts for only two species of *Diplopterygium,* may represent the tetraploid condition based on $x = 28$. Thus the sequence may run $x = 20, 22, 28, 34, 39, 43$ — a more obviously aneuploid series. The family is composed of five genera or subgenera, each of which is characterized by a single basic number except for *Gleichenia* s.s. for which both $x = 20$ and 22 have been reported. Because of the very characteristic and unique assemblage of characters, Gleicheniaceae represents a very natural family. Holttum notes that some of the Cretaceous fossils from Greenland show more-or-less transitional stages between the features of *Gleichenia* and *Diplopterygium,* not shown by living members of the family.[40] It may be that certain lines evolved having their own characteristic basic chromosome numbers with the gradual extinction of types which were intermediate in both morphology and chromosome numbers, giving rise to the present situation.

Occasionally a departure from the normal is seen in families with a very stable cytology. Thus, Dryopteridaceae, with the exception of the small subfamily Onocleoideae, almost universally has a base number of $x = 40$ or 41. In *Woodsia,* however, which belongs here, the lower numbers of $x = 33, 38$, and 39 appear in addition to the $x = 41$ typical of the family as a whole. A similar case of loss of chromosomes has occurred in the monotypic Central and South American fern genus *Hemidictyum*. At various times it has been associated with *Diplazium* but has a number of peculiarities, including a chromosome number of $n =$

31 which is very different from the $x = 41$ characteristic of *Diplazium*.[8] Lovis considered this to be a somewhat peculiar thelypterid fern but accepts the view that it has its nearest affinities in *Diplazium*.[9] This is confirmed by various lines of morphological evidence derived from both sporophyte and gametophyte.[12]

In Aspleniaceae the very stable genus *Asplenium* based on $x = 36$ has two species which deviate from this number in an upward direction, representing a gain of chromosomes. *A. unilaterale* has two levels of ploidy based on $n = 40$ and with a tetraploid having $n = 80$.[41-44] *A. repandulum* from Mexico has been recorded by Smith and Mickel as having a gametic number of either 39 or 40.[45] Similarly, in Blechnaceae, basic chromosome numbers range from 28 to 36 (with the exception of 35) but a single species shows $n = 40$.[8] This is the very isolated plant in terms of habit, etc. which is usually placed in the monotypic genus *Salpichlaena*. Unlike all other members of the family, this plant has very long climbing fronds built on a very similar pattern to that seen in *Lygodium* in the Schizaeaceae.

It is usually not possible to be able to say whether specific aneuploid changes represent loss or gain of chromosomes but in *Woodsia*, loss seems the only logical answer, and gain in *Salpichlaena*. Similarly in the very ancient eusporangiate fern family Marattiaceae, with only two exceptions, all the members investigated belonging to five genera throughout its entire range in the eastern and western hemisphere show $x = 40$. The exceptions are two species of *Marattia* based on $x = 39$, while *M. alata* from Jamaica and *M. weinmannifolia* from Mexico show the $x = 40$ typical of the family.[46,45] Here again, loss of a chromosome at the basic level is indicated. It is doubtful if such a relatively small change could be detected by karyotype analysis at the present time due to the small size and large numbers of chromosomes involved.

The most spectacular case on record of aneuploidy in pteridophytes is that occurring in some members of the fern genus *Lomariopsis*.[47] In the family Lomariopsidaceae, the genera *Bolbitis*, *Egenolfia*, *Arthrobotrya*, *Teratophyllum*, *Lomagramma*, *Elaphoglossum*, and *Peltapteris* are all based on $x = 41$ or 40. The central genus *Lomariopsis* also has a number of species which have been cytologically examined from Asia,[47,48] Mauritius,[49] and Trinidad,[12] all based on $x = 41$. In Africa, however, two species showed $2n = 78$ while two others showed very spectacular reductions to $2n = 62$ in *L. rossii* and $2n = 32$ in *L. hederacea*, representing base numbers of 39, 31, and 16, respectively. The last two taxa had extremely uneven-sized chromosomes ranging from the smallest at c. 3μm to the largest at c. 12 μm.[47] These compare with the very much smaller range of from c. 3 μm to c. 6 μm in the much higher-numbered *L. palustris*. The extremely different chromosome sizes in the low-numbered species strongly suggest a repatterning of the chromosomes by very unequal translocations and loss of centromeres.

Adiantum apparently is unique in pteridophytes in the extent to which aneuploidy abounds and is frequently undetectable morphologically. While $x = 29$ and 30 are the accepted base numbers in the genus, frequently in an individual species deviations from the normal number may occur from plant to plant. In Trinidad the diploid sexual species *A. lucidum* yielded seven plants with $2n = 60$ in root tip cells, while a further specimen had an extra chromosome present, giving $2n = 61$.[12,50] Also in Trinidad, *A. tetraphyllum* has both diploid and tetraploid cytotypes. Bidin examined two diploid plants which had the expected $2n = 58$.[50] A further specimen showed $2n = 62$.[12] Similarly, in the tetraploid cytotype of this species, Bidin recorded one plant with $2n = 114$, one with $2n = 117$, and two with $2n = 118$. Indeed, none were found with the "expected" number of $2n = 116$. An extremely large jump in chromosome numbers was again recorded from Trinidad in *A. pulverulentum*. Three plants were straightforward diploids with $n = 30$, $2n = 60$ while a specimen from another locality in the island showed $n = 122$ very clearly,[12] thus representing a jump to the hyper-octoploid level ($8x + 2$). Both the hyper-octoploid and the diploids were virtually indistinguishable on morphological grounds. The above instances appear to represent chance fluctuations in chromosome numbers against which the genus appears to be well buffered. However, in some apomictic forms an aneuploid number may become stabilized and be found in plants covering a large area, suggesting that it may be at some selective advantage under certain conditions. One of the most striking cases in this respect is met with in *Adiantum hispidulum*. Here one of two cytotypes represents an exact pentaploid level of $2n = 150$, while a variant showed $2n = 157$ in Australia.[50] The other cytotype can best be regarded as a hypo-hexaploid with "n" = 171 (instead of 180) and has a very wide distribution, from Southern India,[51] Australia, and Mauritius.[50] A further plant with $2n = $ c. 175 was recorded from New Caledonia.[52] A somewhat similar case is that of *Asplenium flabellifolium* of Australia in which Lovis demonstrated c. 212 univalents at meiosis and that it was an apomict of the Braithwaite type.[53] This was corroborated by Brownsey who recorded either 210 or 211 univalents in New Zealand material.[54]

Ophioglossum is a genus in which only very high chromosome numbers occur and where polyploids exist based on $x = 120$. Manton recorded $n = $ c. 116 for *O. costatum* from Sierra Leone.[55] I examined a number of individual plants from this population and they had several different numbers, namely $n = 116, 117, 118$, and 120. Here the very high chromosome numbers probably involve much multiplication of individual genes so that the loss of a few chromosomes represents very little loss of genetic material. The same explanation can no doubt be applied to the high aneuploid polyploids of *Adiantum* and *Asplenium*.

In a very few cases hybrids have been recorded between species having different basic chromosome numbers, e.g., *Adiantum* × *tracyi* with 59 chromosomes[56,57] originated by *A. pedatum, n = 29*, crossing with *A. jordani*, with $n = 30$.[57,58] In some cases aneuploidy has played a role in the evolution at the high polyploid level. In *Taenitis blechnoides* two cytotypes with $n = 44$ and 110 indicate the possession of a basic number of $x = 22$. Other members of the genus have $n = 108$ and 114, suggesting a possible aneuploid shift at what is essentially the decaploid level.

IV. POLYPLOIDY

A. Occurrence

Palaeopolyploidy has undoubtedly led to the situation met with in the vast majority of living pteridophytes with very high basic chromosome numbers, as has been discussed earlier. Despite such high basic levels, nevertheless they have provided the platform for further increases, often of considerable magnitude, during more modern times. This neopolyploidy is still proceeding at the present time, several examples having appeared in experimental cultures. The combination of palaeo– and neo-polyploidy has resulted in the pteridophytes as a whole having the highest general level of chromosome numbers known in plants. Individually it may approximate to, or surpass, that recorded for any other organism, such as occurs in *Ophioglossum* with $x = 120$ and in which octoploids and decaploids have been recorded.

Neopolyploidy occurs in all the major groups of homosporous pteridophytes with the sole exception of *Equisetum*, in which all members have a gametic number of $n = 108$. In the ferns in particular, only some 40 to 45% of cytotypes are diploid. However, here polyploidy is very uneven both in its occurrence and in the levels attained and does not appear to conform to any set pattern. Thus Osmundaceae and Cyatheaceae with one of the lowest and highest base numbers of $x = 22$ and $x = 69$, respectively, have no recorded natural polyploidy. Equally, members of great antiquity such as Marattiaceae and of relatively modern aspect such as some species-groups of Aspleniaceae show it in abundance. Table 2 gives a cross section of genera having a range in percentage of polyploidy from 0 to 100. In addition, the levels of polyploidy reached vary from genus to genus. Thus, in *Polystichum, Dryopteris,* and *Pteris* only a very small percentage are at a level higher than tetraploid,

Table 2
GRADES OF PLOIDY IN SOME FERN GENERA

Genus	2x	3x	4x	5x	6x	8x	10x	12x	16x	Total no. cytotypes	Polyploidy (%)
Cyathea	54	—	—	—	—	—	—	—	—	54	0
Polystichum	47	3	34	—	—	1	—	—	—	85	45
Dryopteris	64	29	26	1	3	—	—	—	—	123	48
Pteris	61	22	38	—	1	2	—	—	—	124	51
Adiantum	34	4	30	2	4	4	1	2	—	81	58
Asplenium	44	6	110	4	6	22	1	3	1	197	78
Jamesonia-Eriosorus	—	—	—	—	7	—	—	2	—	9	100

while in *Asplenium* approximately a quarter of the polyploids are greater than tetraploids and extend up to the 16-ploid level. The extreme case is in the small South American duo of very closely related genera, *Jamesonia* and *Eriosorus*, in which nothing lower than a hexaploid level has been recorded. Some yet unsolved problems are posed by these facts, such as: what are the constraints on polyploidy; why should some members reach high levels (implying several successive acts of multiplication); and why have most pteridophytes been capable of undergoing palaeopolyploidy while some of them are unable to achieve neopolyploidy?

An outstanding exception to the general rule of high frequency of polyploidy and of high basic chromosome numbers is found in the heterosporous pteridophytes. Klekowski and Baker demonstrated this contrast by plotting gametic chromosome numbers against number of members,[59] the low chromosome numbers belonging predominantly to the heterosporous types. The mean gametic number (13.62) furthermore corresponded very closely to that of seed plants (15.99). Klekowski, in respect of sexually reproducing members, showed that the comparable figure for homosporous pteridophytes was four times as great at 54.[60] However, the mean basic chromosome numbers for the heterosporous and homosporous members were 12.7 and 37.5, respectively, i.e., an increase in basic numbers of homosporous types of only three times. The much larger difference between the mean base and the mean gametic numbers is accounted for by the much greater incidence of polyploidy in homosporous groups, namely, 40% as opposed to 12% in heterosporous members. Thus, on all counts of basic chromosome number, gametic chromosome number, and percentage of polyploidy, heterosporous pteridophytes have dramatically lower values than those for homosporous ones.

Neither palaeopolyploidy nor neopolyploidy has played anything like the predominant role in heterosporous pteridophytes that it has in the homosporous ones, and the question arises as to why this should be so. In sexual members of heterosporous groups, megaspores and microspores are produced which involve separate acts of meiosis, closely comparable with the situation found in sexually reproducing flowering plants. DeWet has pointed out that probably the commonest method of creation of a polyploid is at least a two-stage event,[61] namely a female diploid gamete being produced by meiotic failure and this being fertilized by a normal haploid male gamete, giving rise to a triploid. An unreduced egg of this triploid on fusing with a normal haploid male gamete would then produce a tetraploid. This sequence of events requires two relatively rare events to occur in the nonreduction of the egg cell at the diploid and triploid levels, followed by successful fertilization and the survival of the progeny. Of the approximately 123 heterosporous pteridophytes which have had their chromosome number determined, some 33 are polyploid, i.e., 27%. If the few cytotypes are omitted which are known to be apomictic in *Isoetes* and *Salvinia,* plus those which are almost certainly so, based on such evidence as unbalanced levels of ploidy such as 3x, 5x, etc., then the incidence of polyploidy is reduced to 15%, which is very close to Klekowski's original figure.

In a multicellular organism the occurrence of an abnormal nuclear division leading ultimately to a change in the level of ploidy as a whole is most likely to succeed if the division occurs at the single cell stage of the organism's life. The two stages in a pteridophyte life cycle where the generations start from single cells are in the spores and the zygote. DeWet has indicated that in flowering plants the known cases of a polyploid originating from a zygote are very few indeed.[61] There is no evidence to suggest that this method operates in the pteridophytes. On the other hand, all the indications are that an effective doubling of the chromosome number in spores is responsible. Well-documented cases in *Asplenium* concern diploid hybrids showing failure of pairing of their chromosomes which have given rise to fertile tetraploid plants showing 72 bivalents at meiosis instead of the 72 univalents of their forebears. Such instances occur in *Asplenium platyphyllum* × *rhizophyllum* = *A.* × *ebenoides*,[62] *A. fissum* × *A. viride*,[63] *A. viride* × *A. trichomanes* subsp. *inexpectans* = *A.* × *adulteriniforme*,[64,65] *A. viride* × *A. trichomanes* subsp. *trichomanes* = *A.* × *protoadulterine*,[64,65] *A. viride* × *A. fontanum* = *A.* × *gastonii-gautieri*.[66] It appears likely that *Asplenium trichomanes* × *Camptosorus rhizophyllus* = × *Asplenosorus shawneensis* behaves in a similar way, as Moran noted that the contents of the sporangia varied between 64 aborted spores and 16 unreduced spores.[67] In all these instances doubling of the chromosome numbers of the spores was indicated and Lovis found abundant dividing cells up to and including metaphase I of meiosis but little sign of later stages,[66] suggesting that meiosis was aborting after metaphase I and that restitution nuclei were being formed. When sown, spores from these plants give rise to prothalli producing both male and female gametes which are diploid instead of haploid.

Self fertilization is very common in homosporous ferns, resulting in the formation of a tetraploid within weeks after the production of unreduced spores. In the case of the European hybrids listed above, the period from the moment of hybridization to the production of allotetraploids occupied at most a very few years. In addition to having the overwhelming advantage of homospory most pteridophytes are more-or-less herbaceous and long-lived and have a massive spore output, conditions closely paralleling those defined by Stebbins as being most conducive to the formation of polyploids in angiosperms.[71]

Klekowski has advanced the hypothesis that because homosporous ferns are capable of self-fertilization there is a tendency to homozygosity and consequent loss of genetic variability and that this effect may be offset by multiplication of chromosome sets and by homoeologous pairing between them.[60,68,69] Lovis has criticized the evidence advanced for homeologous pairing.[9] However, regardless of whether or not polyploidy may increase genetic variability, it may be argued that polyploidy in homosporous ferns is a direct consequence of the unique system whereby a spore may double its chromosomal complement in one act and be able to produce a sporophyte by self-fertilization.

Wagner and Wagner have pointed to the existence of many hybrids of what they term the AB type with nonhomologous chromosomes which form univalents at meiosis but to there being many fewer polyploids of the AABB type resulting from a doubling of the chromosomes of such hybrids.[29] Of the European *Asplenium* hybrids, both the sterile diploids *A.* × *adulteriniforme* and *A.* × *protoadulterinum* are well known in nature. Both spontaneously produce unreduced spores from which fully fertile allotetraploids are produced. However, although the tetraploid form of *A.* × *protoadulterinum* is well known in nature under the name of *A. adulterinum*, that of *A.* × *adulteriniforme* has failed to establish itself. Lovis and Reichstein ascribe this failure to a lack of sufficient vigor to compete successfully in the limestone habitats of its parent hybrid.[65] Undoubtedly there are several examples of diploid hybrids of this nature which may be thought to be ideal material for producing alloploids but which have failed to do so in nature. Other hybrids may be dead ends in that the genus seems to be incapable of establishing polyploids in nature despite the hybrid having existed for long periods of time. Wagner et al.[70] record a single clone of the royal fern hybrid *Osmunda* × *rugii* which they estimate from the size of the clone and the known growth-rate to be over eleven centuries old. Similarly, the large vigorous colonies of the three horsetail hybrids in Ireland, namely, *Equisetum* × *trachyodon, E.* × *littorale* and *E.* × *moorei* are probably thousands of years old. In general, experience in working with tropical floras indicates that the presence of diploid hybrids of the AB type is not as universal as Wagner and Wagner suggest, most of the hybrids being triploids, etc.[29] representing backcrosses to one of the parents, while in others the numbers of bivalents or univalents fall far short of the full basic numbers. The cytology of temperate ferns on the whole gives the appearance of being more clear cut in this respect than tropical ones.

Stebbins classified polyploids as consisting of four main types, viz., autopolyploidy, genomic allopolyploidy, segmental allopolyploidy, and autoallopolyploidy,[71] with continuous gradations from one type to another. All these types are present in the pteridophytes but in many cases their cytological behavior requires the synthesizing of hybrids and analysis of meiosis to demonstrate their true nature. For example, most natural fern polyploids show regular pairing of the chromosomes regardless of their origin. The classic example of autopolyploidy in ferns is the synthesized series in *Osmunda regalis,* consisting of autotriploids and autotetraploids. These have been described in detail by Manton and showed the presence of numerous multivalents at meiosis.[4] However, all the European tetraploids of the genus *Asplenium* consistently show 72 bivalents at meiosis and uniformly set good spores. Of the 13 tetraploids analyzed experimentally, no less than 5 are autopolyploid in origin, namely, *A. petrarchae* subsp. *petrarchae, A. billotii, A. septentrionale, A. trichomanes* subsp. *quadrivalens* and *A. ruta-muraria* subsp. *ruta-muraria.* In crosses with unrelated species, bivalents are formed in the hybrids by autosyndesis of the two sets of chromosomes contributed by the tetraploid parent. The fact that no multivalents are formed in the wild tetraploid indicates the presence of a genetic mechanism for the suppression of multivalent formation, possibly similar to that known in wheat. Hence meiosis is regular and indistinguishable from that of a typical genomic allopolyploid. Occasionally traces of the true nature of the species may be found such as in members of nine out of 80 populations of *Asplenium ruta-muraria* subsp. *ruta-muraria* from Central and Eastern Europe.[72] Unlike the majority of these populations with the normal meiotic pattern of 72 bivalents, plants from the nine had both multivalents and univalents present, in addition to some bivalents. In this subspecies there has evidently been selection to restrict synapsis to the formation of bivalents only over most of its range.

The origin of genomic allopolyploids has been considered earlier and experimental analyses strongly suggest that the vast majority of pteridophyte polyploids are of this nature, being in essence hybrids between species showing nonhomology of their genomes, leading initially to the formation of a sterile asynaptic plant to which full chromosome pairing and fertility are restored following chromosome doubling. Assuming that only a relatively small proportion of initial hybrids are suitable potential polyploid material, the existence today of an estimated several thousand neopolyploids presupposes hybridization on a grand scale, and in fact almost every sizeable cytological survey has revealed the presence of hybrids in each pteridophyte flora investigated.

The normal sequence of events in the production of an allopolyploid, of hybridization followed by chromosome doubling, may occur in reverse order in certain circumstances. It was described by Lovis as delayed allopolyploidy and occurs in *Asplenium* × *murbeckii.*[9] This plant is known in the wild from several localities in Europe and has resulted from the hybridization of the two autotetraploid species *A. septentrionale* and *A. ruta-muraria* ssp. *ruta-muraria.* Lovis suggested that the latter two species had genomic formulae of SSSS and RRRR, respectively, and that hence *A.* × *murbeckii* would be SSRR. Pairing of the chromosomes in some populations of *A.* × *murbeckii* was complete due to synapsis of the R and R and the S and S genomes. A similar situation also occurs in *Asplenium* × *clermontiae*, derived from *A. ruta-muraria* subsp. *ruta-muraria* × *A. trichomanes* ssp. *quadrivalens*.[9]

The third type of polyploid, the segmental allopolyploid, in which the constituent genomes are incompletely differentiated, possibly forms the hardest type to detect. This requires the setting up of a fairly elaborate hybridization program such as that reported for the *Adiantum caudatum* complex by Manton et al.[73-75] and summarized in Figure 5 of Walker.[76] A remarkable feature of this small group of 4 tetraploids was that although each showed 60 bivalents at meiosis, one was an autopolyploid, two were allopolyploids, and the remaining one was a segmental allopolyploid.

Autoallopolyploids occur in nature to an unknown extent since few experimental studies have involved species at the hexaploid or higher level of ploidy, but some members of the fern complex based on *Cystopteris fragilis* would appear to provide examples in nature. Hybridization between diploid *Polystichum lonchitis* with the genomic formula LL and allotetraploid *P. aculeatum* (LLSS) has produced the well-known triploid *P.* × *illyricum* (LLS). Vida and Reichstein report the spontaneous production of hexaploid *P.* × *illyricum* which has the genomic formula of LLLLSS and is hence an autoallopolyploid.[77]

A noteworthy feature of pteridophyte cytology is that with few exceptions such as in *Ceratopteris* the chromosomes appear to be very conservative and retain their homologies for considerable lengths of time. This has made the analysis of species complexes possible and it has allowed cytologists to trace relationships on an intercontinental basis in certain species-complexes, e.g., *Polypodium vulgare*,[78-80] *Dryopteris spinulosa*,[81,82] and *Dryopteris dilatata-campyloptera*,[83,84] in addition to the *Adiantum caudatum* complex discussed above which embraced plants from Africa, India, Sri Lanka, Malaya and China.[73-75]

One important way in which polyploids differ from diploid species in their origin is that a polyploid may be created and recreated in different places and at different times. Brownsey has shown that this has happened in *Asplenium lepidum*.[85] This allotetraploid consists of two subspecies, *lepidum* and *haussknechtii*, together with a distinctive variety of *Samarkandense*, of the last-named subspecies. All these have arisen from the same diploid parental taxa, *A. ruta-muraria* subsp. *dolomiticum* and *A. aegeum*, and show a great deal of variation between populations and also of overlapping in the range of variation within the populations themselves. Brownsey considers that different morphological variants of the same two diploid species have given rise to the three main tetraploid types in different geographical locations. The additional interpopulation variation arises because populations on the limestone mountains which form their habitat tend to be broken up into small isolated groups, diverging from one another in morphology.

Extensive use has been made in flowering plants of morphological features which could be shown to vary with ploidy. In the pteridophytes the relative lengths of both stomata and spores have been widely used for this purpose and have enabled extrapolations from herbarium material to be made of features such as the geographical range and relative frequencies of polyploids and diploids, e.g., *Anemia adiantifolia*,[86] *Pellaea*,[87] and *Woodwardia*.[88] However, care must be taken that any supposed correlations between cell size and ploidy are valid and that the comparisons are made between the members of a species-complex and not applied indiscriminately. In a survey of spore size in relation to ploidy levels in Himalayan species of *Asplenium, Athyrium, Cystopteris*, and *Diplazium*,[89] Bir found spore size to be an effective indicator within various species-complexes. But it was impossible to predict the level of ploidy from the mean spore size of a species taken at random. Even within a species stomatal measurements may not be valid, as in diploid *Adiantum pulverulentum* from Trinidad and its hyper-octoploid cytotype, in which the mean measurements varied so widely from one plant to another as to be valueless as a predictor of ploidy level.[12] In brief, each case needs testing before applying the method which may then yield very valuable information.

B. Geographical Distribution

Many attempts have been made to correlate differences in the levels of polyploidy in flowering plants with differences in a wide variety of factors such as geographical latitude, vegetational type, life forms, taxonomic position, etc.[90] A regular increase was noted from North Africa and the Mediterranean area to the extreme north of Europe, the percentage of polyploidy being very approximately equal to the degree of latitude. Various suggestions were made to account for these facts, one being that in general polyploids were hardier than diploids and therefore better adapted to the cold and another that high percentages of polyploidy were more or less a direct outcome of the Pleistocene Ice Age. Such figures need careful scrutiny on a number of counts. Stebbins pointed out that different life-forms differ in their susceptibility to polyploidy and that the relative proportions of life-forms alter with increasing latitude.[71] Furthermore, many statistics were amassed regarding polyploid percentages for different countries, which were often not entirely based on the wild floras of the countries concerned but were extrapolated from chromosome numbers determined on supposedly the same species from another country. Yet another source of confusion lies in the differing opinions regarding the basic chromosome numbers in flowering plants: 7, 8, and 9 are considered to be clearly truly basic numbers, but while some workers would regard 13, 14, 15, 16, etc. as basic, others would regard them as being polyploid in origin.

The ferns have several great advantages over flowering plants in that virtually all authors base their statistics for polyploidy on neopolyploid levels and ignore the palaeopolyploid derivation of the high numbers. Ferns also tend to be composed of perennials of a similar nature and the genera with and without polyploidy tend to have a randomized distribution. Finally, all the chromosome numbers are from wild, clearly localized plants so that the figures are authentic. These factors allow direct comparison of fern floras with one another.

The first fern floras to be treated in this way were those of Britain and of Madeira. Manton found 53% polyploidy in the former and 42% in the latter.[4] This apparently gave support to the theory of higher levels of ploidy in more temperate climates. However, the Madeiran flora is not a simple one as it is composed of two separate floristic elements, one of Oceanic European affinities and the other of tropical African ones. Lovis presents a partial breakdown of the figures showing that the European element contains 35% polyploids while the African has 50%.[9] In the latter, three species reach high levels of ploidy at $8x$, $10x$ and $12x$ levels, respectively, each representing several individual acts of multiplication.

As more tropical floras were investigated (see Table 3), it soon became evident from the figures of c. 60% polyploidy in Sri Lanka,[44] Jamaica,[46,91] and West Tropical Africa[92] that the idea of cold being the principal direct stimulus to polyploidy must be abandoned. Equally, tropical and subtropical conditions per se are not in themselves guarantees of high ploidy levels as seen in Malaya with only 39%. The factors affecting ploidy levels are obviously complex and may differ from region to region. In connection with the European and Madeiran flora the finding of Page of only 25% polyploidy in the Canary Islands is highly significant.[93,94] There is a large body of evidence supporting the view that the flora and fauna of the Canaries are ancient and relict, living in what is essentially a fossil climate. Both organisms and climate appear to be very similar to those existing in North Africa and Southern Europe during Tertiary times. Larsen also found a comparable low percentage (23%) of polyploidy in the native flowering plants of the island.[95,96]

The very close similarity between the figures for Europe and North America is very striking. It is tempting to conclude that to some extent this may be a reflection of their having been part of a single land mass in Eocene times, and detailed biosystematic work indicates common genomes present in the two continents at the present day in *Asplenium*, *Dryopteris*, *Polypodium*, and *Polystichum*.

An apparent anomaly in Table 3 is highlighted by the high polyploidy figured for Sri Lanka and Southern India (60% and 53%, respectively) as compared with those for the

Table 3
PERCENTAGE POLYPLOIDY IN FERN FLORAS

Locality	No. spp. counted	Highest ploidy	Percent polyploidy	Authority
Canaries	28	$6x$	25	93,94
Malaya	101	$8x$	39	97
Madeira	38	$12x$	42	4
Trinidad	117	$8x$	43	76
Japan	392	$8x$	45	In Ref. 140
North America (U.S. and Canada)	192	$10x$	52	Numerous authors
New Zealand	92	$8x$	52	17,37,141, 142,143
Britain	47	$6x$	53	4
Hungary	37	$6x$	54	92
Southern India	108	$8x$	53	51
Jamaica	229	$16x$	60	44,91
West Tropical Africa	94	$12x$	60	92
Ceylon	132	$12x$	60	44
Tristan de Cunha	20	$12x$	65	92
Eastern Himalayas	251		36	38
Western Himalayas	61		23	38
Central Himalayas	40		35	144

Note: Figures have only been quoted where (with the possible exception of West Africa) between 50 and 100% of the fern floras has been examined. The above list excludes Hymenophyllaceae because of the uncertainty of the levels of ploidy.

With permission from Walker, T. G., in *The Experimental Biology of Ferns*, Dyer, A. F., Ed., Academic Press, London, 1979, 87. Copyright: Academic Press Inc. (London) Ltd.

western (23%), central (35%) and eastern (36%) Himalayas. The history and floristics of the subcontinent and the Himalayas are vastly different, South India and Sri Lanka having drifted into their present position while the Himalayas, thrown up by the impact, belong to the Asiatic mainland. The flora of the Himalayas has its strongest affinities with Europe in the west and China in the east.

Jamaica with its high level of polyploidy (60%) is an island with a great wealth of diverse habitats and climatic conditions together with moderately high mountains. It also has had a somewhat disturbed geological history, all features thought to be conducive to polyploidy when taken in conjunction with one another. Although the Canary Islands have very high mountains, there have not been large climatic changes and polyploid percentages are low.

By contrast, tropical Malaya shows only 39% polyploidy, and preliminary figures for Sarawak suggest a similar or even lower figure.[12] Manton suggested that the land mass of which Malaya is a part may support a Tertiary flora, less disturbed by the climatic changes which have stimulated polyploidy elsewhere.[97] In contrast to most other land masses a large portion of what is now Southeast Asia has remained in the low latitudes for most of geological history.[98] The relationship of Japan, also with a relatively low figure, to the mainland mass is less clear but it may also have a somewhat similar pattern.

The temperate oceanic isolated island of Tristan de Cunha with the exceptionally high figure of 65% polyploidy up to the $12x$ level clearly represents a different case.[92] There is a sparcity of species (which are all immigrant) suggesting that autoploidy is at work here to a great extent.

The relatively high percentages in Britain and Hungary may be ascribed to the indirect effects of moving ice sheets of the Ice Age in mixing floras and in altering the ecological conditions and not directly due to coldness itself.

Ferns have provided a considerable amount of food for thought in explaining the very genuine differences in polyploidy in different parts of the world. No single explanation covers all the cases. Many factors, such as climatic and geological disturbances (or stability), geographical isolation, or proximity have all played a part in differing degrees.

In addition to considerations of the incidence of polyploidy in whole floras, investigations into the geographical distribution of ploidy within genera can be very informative. Lovis showed that in *Asplenium* there was an irregular cline in the frequency of diploids along a NW/SE axis from Europe (53%) via the Himalayas (21%),[73,77] Sri Lanka (13%) to New Zealand (0%), coupled with a much lower mean level of polyploidy in the north temperate zone (2.9 to 3.8) than in the south temperate zone (4.7 to 5.5).[9,53] He suggested that *Asplenium* originally diversified in the Southern Hemisphere and that the present great diversity of diploid European species represents a secondary and more recent radiation. In *Adiantum* the contrast between its representatives in the Old and New Worlds is most striking. Bidin demonstrated that in the Old World only 26% of the species were diploid,[50] contrasting sharply with the New World figure of 62%. Much of the speciation in this genus in Central and South America is proceeding by genetic means and by gradual loss of chromosomal homologies at the diploid level.[12] Elsewhere in the world, polyploid complexes in this genus tend to be either ancient relicts such as the *A. reniforme* group with its very disjunct distribution, or as plants of somewhat specialized habitats as in Australia.

V. APOSPORY, APOGAMY, AND APOMIXIS

The normal life cycle of pteridophytes which involves both a morphological and a cytological alternation of generations may be modified in a variety of ways, some being of a purely temporary nature, such as apospory and sporadic apogamy, while apomictic processes are repetitive. Those of a temporary nature appear to have been of relatively little significance in evolution while apomixis has played a very significant role in speciation.

A. Apospory

In apospory the prothallial tissue develops from the sporophyte directly instead of via the intermediacy of spores. For the most part these prothalli develop normal sex organs with fully functional gametes. The latter have the same chromosome number as the parental sporophyte; hence self-fertilization results in a new sporophyte generation with double the chromosome number of the preceding one. Such a doubling imposes its own limits on apospory and confines it to being a more-or-less single event in the life of a plant.

Apospory rarely may occur naturally, but it may be induced experimentally. To understand the process in the hopes of its throwing light on the nature of the gametophyte-sporophyte relationship, a variety of organs have been used to produce prothalli.[99] However, for the cytologist, apospory provides a means of inducing polyploidy without recourse to various chemical treatments, although success has been reported in creating an autotetraploid in *Adiantum capillus-veneris* by the use of colchicine.[100] The autopolyploid series in *Osmunda regalis* originated from depauperate leaves which had reverted to a juvenile condition following starvation and neglect. These were pegged out on damp soil and prothalli developed from the margins.[4] Good results have been obtained by pinning down normally produced juvenile leaves onto damp soil or onto a nutrient medium as in diploid *Asplenium ruta-mararia* subsp. *dolomiticum*.[101] Such autopolyploids have enabled a study of meiosis involving chromosomal pairing and multivalent formation to be made. In *A. ruta-mararia* the newly created autotetraploids were used to compare meiosis found in them and in natural populations of the tetraploid *A. ruta-mararia* subsp. *ruta-mararia*. As a consequence,

Bouharmont[101] concluded that the wild tetraploid had been formed essentially in an autoploid manner from slightly differing diploids.

B. Apogamy

The term apogamy has unfortunately been used almost universally in pteridophyte cytology to involve repetitive apogamy as one of the essential ingredients. Strictly, apogamy refers to the production of sporophytes on gametophytic tissue without involving sexual reproduction and fertilization. The cause of the phenomenon may be genetic as in the case of obligately apogamous species, but sporadic apogamy may be induced by the prevention of fertilization over a more-or-less prolonged period. The most common factor preventing fertilization is the absence of free water in which the spermatozoids can swim to fertilize the egg cell; hence apogamy can be simply induced by watering the prothalli from below to prevent such a film of water forming on the undersurface. After several months, apogamously produced sporophytes may appear. In a hybrid derivative in *Ceratopteris,* in which a population of individually isolated prothalli was regularly watered from above,[102] sporophytes were produced normally from every prothallus with one exception. In this instance the sporophyte which was produced at a later date was apogamous in origin, fertilization having been prevented by a mutation causing the production of nonfunctional spermatozoids. Other successful means of inducing apogamy include raising prothalli in the presence of ethylene and sugar.[99]

Sporophytes formed as a result of induced apogamy will possess the gametophytic chromosome number. Use has also been made of this phenomenon to investigate the polyploid nature of a species. In the wild tetraploid *Asplenium septentrionale* and *A. trichomanes* subsp. *quadrivalens* bivalents only are regularly formed at meiosis. However, diploid plants produced apogamously showed extensive pairing of the chromosomes. From this Bouharmont deduced that both tetraploid taxa were autopolyploid in origin but that the expected multivalent formation had been suppressed.[101] Conversely, in *Dryopteris dilatata* and *D. felixmas*, diploid plants which had been apogamously induced had the vast majority of their chromosomes unpaired at meiosis with only 2 to 5 bivalents being formed.[103] The genomes involved in each of the two species were nonhomologous and the plants were allotetraploid in nature.

As in the case of apospory, apogamy must be sporadic in its occurrence unless there is also a compensating mechanism which stabilizes the chromosome number and prevents it being halved at each event. Just such a mechanism occurs in apomixis.

C. Apomixis

In apomixis involving spore production, apogamy is repetitive and accompanied by one of two main devices whereby the chromosome number is maintained at a constant level. Although the morphological alternation of generations is maintained, there is no cytological alternation, with the result that both sporophyte and gametophyte have the same chromosome number. The sum total of these events has widely been given the name apogamy, but this is somewhat inaccurate. The entire process basically consists of diplospory coupled with apogamety. A satisfactory name was coined by Löve et al., the term "agamospory".[16] This phenomenon is very common in the ferns but much less so in other groups of pteridophytes.

During sporogenesis in a typical homosporous sexual fern the young sporangium contains a single archesporial cell at first which then divides four times by mitosis, resulting in 16 spore mother cells. Each spore mother cells than undergoes meiosis resulting in 64 spores, produced in tetrads, with the reduced chromosome number. This course of events is modified in two main ways in agamospory.

The commoner method was first described by Döpp in *Dryopteris remota* (= *Aspidium remotum*).[2] A more complete and photographically well-documented account based on a number of species belonging to several genera was given by Manton.[4] This method has been variously described as the "normal" type or the Döpp-Manton type.[9,104] The sporangia follow one of two main lines of events, in the first of which the archesporium divides as for a sexually reproducing form, producing 16 spore mother cells by mitosis and then undergoing meiosis. However, in all cases investigated meiosis is highly irregular due to the formation of variable numbers of chromosomal associations such as univalents, bivalents, and multivalents and the spores are misshapen and abortive. Although this line of events is of little significance to the plant it is of considerable interest to the cytologist, since here at meiosis the chromosomes are believed to show their true homologies. Use has been made of this in determining the interrelationships of sexual and apomictic members of the *Cheilanthes farinosa* complex in Africa.[105] The second sequence of events results in the production of viable spores. The archesporial cell divides only three times by normal mitosis, resulting in 8 spore mother cells, followed by a fourth mitotic division which is greatly modified. The early stage of this fourth division is perfectly normal with the chromosomes undergoing prophase and moving to the equator and splitting. However, there is no further movement of the chromosomes to the poles and a restitution nucleus is formed in each cell without cytoplasmic cleavage. As a consequence the number of cells is retained at 8 but the chromosome number has been doubled. These now undergo regular meiosis due to the pairing of sister chromosomes only. The 32 spores (= diplospores) which result are well-filled, and on their germination prothalli are produced which have the same chromosome number as the parent sporophyte. Manton also described two other courses of events, one producing 8 spore mother cells with partial cytoplasmic cleavage and the other resulting in giant spore mother cells with double the original chromosome number, which may form the basis of some tetraploid apomicts although other explanations are possible.

The contrast between sexual species normally producing 64 spores as against the apomicts producing 32 spores has unfortunately sometimes been taken by some workers to be diagnostic of the breeding behavior. However, although this is the more normal pattern, variations do occur. Knobloch surveyed the reproductive systems of *Cheilanthes* and *Notholaena*, based on the number of spores per sporangium produced by each species.[106] Although the conclusions in the main may be correct, this is a particularly unfortunate group of species to have used since Vida et al. showed that *Cheilanthes catenensis* produced sporangia containing only either 32 or 16 spores thought to be characteristic of an apomict despite its sexual nature.[107]

In the second scheme of apomixis or agamospory, first described by Braithwaite in *Asplenium aethiopicum*, known as the "Asplenium-aethiopicum" or Braithwaite type, all the sporangia follow the same course of development.[108] The initial archesporial cell divides by 4 mitotic divisions to give rise to 16 spore mother cells. Meiosis starts with normal first prophase stages but there is complete failure of pairing of the chromosomes, resulting in the production of univalents only. These move irregularly toward the equator and restitution nuclei are formed without any division or polar distribution of the chromosomes. A typical second meiotic division then follows, resulting in approximately 32 well-filled viable diplospores arranged in diads. Numbers of these spores most commonly vary somewhat between 23 and 32. Again, a reduction in chromosome number is circumvented.

A supposedly third type of sporangial development, termed ameiotic, has been described by Evans for *Polypodium dispersum*.[109] However, recent reviews of the evidence suggest that this is only another example of the Braithwaite scheme.

Approximately 10% of all ferns of which the type of reproduction is known are agamosporous,[104] a condition recorded in a wide variety of families, although concentrated in only 18 genera. These are Schizaeaceae *(Anemia)*; Hymenophyllaceae *(Trichomanes s.l.);* Adiantaceae *(Actiniopteris, Adiantum, Bommeria, Cheilanthes, Hemionitis, Notholaena, Pellaea, and Pteris)*; Polypodiaceae *(Polypodium)*; Aspleniaceae *(Asplenium)*; Dryopteridaceae *(Cyr-*

tomium, Diplazium, Dryopteris, Polystichum); Thelypteridaceae *(Phagopteris)*; and Grammitidaceae *(Xiphopteris)*. Some of the more primitive families such as Marattiaceae, Osmundaceae, Gleicheniaceae, and Cyatheaceae have not had agamosporous species recorded in them. The vast majority of such species occurs in the two families Adiantaceae and Dryopteridaceae. In *Asplenium* both the Braithwaite and the Döpp-Manton systems occur, strongly suggesting that these systems have arisen independently on several occasions and are not indicative of a common ancestry among those genera possessing one or other systems.

Apart from the ferns, agamospory is not prevalent in pteridophytes, although to some extent this may reflect a less complete knowledge of meiosis and prothallial characters in the non-fern groups, especially in the case of *Selaginella*. Prothalli also tend to be more difficult to raise than in ferns; those of *Lycopodium*, for example, take an inordinate length of time to germinate, often many years, and hence there are few direct observations on the gametophyte generation. Tryon considers that there is clear evidence for agamospory in a race of *Selaginella rupestris* in which microsporangia are absent from the strobili, coupled with a reduction in the number of megaspores per megasporangium to 1 or 2 as compared with other races of this species.[110] In a few instances in *Selaginella* triploid numbers have been reported, and probably these also are apomictic.[111]

The only other group apart from the ferns in which agamospory has been closely investigated is in some species of *Isoetes*.[112-115] The following account of agamospory is a somewhat simplified and composite picture of the course of events as described in *I. coramandelina*. Numerous specimens of this plant in India bear megasporangia only and are present in many localities as both diploid and triploid cytotypes. In the course of megasporogenesis only a few megaspore mother cells are formed.[113] During meiosis the chromosomes are mainly present as univalents although in some cells rings of 4, etc., may be observed suggesting translocation heterozygotes. In a proportion of cells with complete asynapsis, regular anaphase separation occurs and the somatic number is thus found at each pole. Verma reported that there was no indication of a second meiotic division but that cytokinesis occurred,[114] resulting in 2 nucleate and 2 enucleate megaspores. Parthenogenetic (or agamosporous) development of the megaspore is inferred by the subsequent production of the sporophytes in the absence of microspores. This method shows similarities to the Braithwaite system except that in *Isoetes* the first meiotic division is fully functional and the second one is abortive. In addition, there are many different forms of behavior in other cells which result in nonviable spores. The impression gained is of a system more variable in its details than those operating in the ferns and less "efficient". Microspore development shows very irregular meiosis and the mature microspores are for the most part inviable. (See also Figure 1).

Agamospory appears to be completely absent from *Equisetum* despite the presence of asynapsis in a number of populations of hybrid origin.

By comparison, relatively little is known concerning agamosporously produced prothalli in pteridophytes other than in ferns. Much of the evidence for agamospory here depends on the production of sporophytes from megaspores in the absence of microsporangia in heterosporous groups such as *Isoetes* and *Selaginella*. In the ferns, prothalli are more-or-less normal in gross morphology but lack functional archegonia. However, in those produced as a result of the Döpp-Manton system, antheridia are present which produce fully functional spermatozoids capable of fertilizing the egg cell of a suitable sexual species. In every case examined, such hybrids resulting from the crossing of an agamosporous species and a sexual one have inherited the complete system of agamospory as a dominant character and are therefore fertile. There is always an increase in ploidy level following an apomictic × sexual species cross because of the unreduced chromosome number in the apomict (e.g., a diploid apomict × a diploid sexual species gives rise to a triploid apomict).

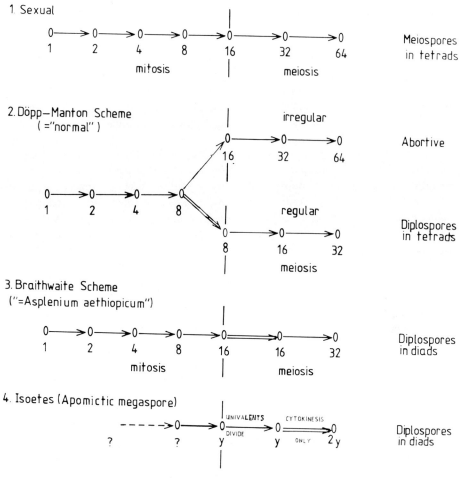

FIGURE 1. Sporogenesis in sexual and apomictic pteridophytes.

In ferns with the Braithwaite system neither type of sex organ is formed and hence they are unable to form hybrids in combination with sexual species. It clearly demonstrates that the formation of antheridia and archegonia is under separate genetic control.

Since functional archegonia are not produced, agamospory in ferns is obligate and is inherited as a single dominant character, although the whole process involves several different genetic factors. This situation contrasts on both counts with that in flowering plants. Only one case of facultative apomixis has been reported in the ferns, but this needs confirmation and further investigation of the cytological features shown in this particular line of *Matteucia orientalis*.[116]

Several lines of evidence suggest a hybrid origin for the vast majority of apomictic pteridophyte species. These include the irregular typically hybrid type of meiosis seen in the 16-mother-celled sporangia in plants having the Döpp-Manton system. Braithwaite concluded that the failure of pairing in *Asplenium aethiopicum* is genetically imposed and that the proposed relationships between the various members of the complex suggest that a proportion of pairing or possibly multivalent formation would be present in the absence of such a genetic restraint.[108] The preponderance of triploids — 89 out of 125 apomictic cytotypes from many parts of the world — suggests that these could only have arisen by hybridization. It may be argued that such triploids could have resulted from hybridization occurring between a diploid apomict and a diploid sexual species. However, there are many cases in which no likely diploid apomictic parent has been forthcoming.[12] Yet another line of evidence for the hybrid nature is to be seen in *Polypodium dispersum*, where the chromosome lengths show three peaks, suggesting the presence of three genomes at least partially different.[12]

Since each type of apomictic system depends upon the bringing together of a number of separate genetically controlled features in both the sporophyte generation — such as the modification of meiosis or mitosis and other cytological events — and in the gametophyte — such as the suppression of functional archegonia and the spontaneous production of baby sporophytes — it is difficult to conceive of these having been acquired one by one in an individual plant. Hybridization suggests a means whereby such features may be assembled rapidly from several different sources. The only apomictic plant experimentally produced from sexual forebears was derived from an interspecific hybrid of *Ceratopteris* which showed sticky chromosomes.[117] This plant would, however, have been doomed to failure in the wild in that less than 5% of the spores produced were viable and the apogamously produced sporophytes were dwarfed, lacked well-developed roots, and showed several severe abnormalities of the frond.

Wagner and Wagner suggest that apomictic ferns are hybrids of the ABC type in triploids and imply that meiosis is irregular because of univalent formation.[29] They pose the question as to why apomictic ferns do not simply double their sporophytic chromosome number and have normal meiosis, normal sexual gametophytes, and normal syngamy. However, this suggestion appears to be based on one or several misconceptions. In the Döpp-Manton system nearly all the species investigated do not show all univalents in the 16-spore-mother-celled sporangia but a truly irregular meiosis with univalents, bivalents and frequently, multivalents. The premeiotic doubling of the chromosomes gives every individual chromosome an exact and equal partner with which to pair in very close juxtaposition. Doubling of the type suggested by Wagner and Wagner would only result in the more irregular formation of multivalents, etc. In ferns of the Braithwaite system, where univalents only form and superficially conform to the type proposed by the Wagners—the ABC type—in fact some are not of this type. In *Asplenium aethiopicum* these univalents are suggested to be brought about by genetic suppression of synapsis.[108] It follows, therefore, that a doubling of the chromosome number should still result in the formation of univalents and not cause a reversion to a polyploid sexual condition.

Many apomicts are highly successful in that they frequently have a more extended geographical range than their near sexual relatives, e.g., *Polypodium dispersum*,[118] *Bommeria*,[119] and in *Pellaea*.[120] However, this is by no means a universal rule and does not apply in *Pteris* when the facts are critically examined.[76] Here, supposedly very widespread apomictic species such as *P. cretica* and *P. biaurita* consist of several different cytotypes which have been given the same specific name. The geographical range of each of these components tends to be limited when compared with some sexual species such as *P. vittata* and *P. tripartita*.

The rate of germination of fern spores is independent of the sexual or apomictic nature of the species but in general the apomicts mature and produce sporophytes more rapidly than do sexual species, there often being a difference of several weeks. This early maturity coupled with the freedom from the requirement for a free film of water to effect fertilization in apomicts has led some workers to consider this to be an adaptation to dry habitats.[29,121,122] While a few, predominantly American, genera such as *Pellaea, Cheilanthes*, and *Notholaena* which have appreciable numbers of apomictic species are predominantly xeric, this is not true for the bulk of the remaining genera with large numbers of apomicts such as *Dryopteris, Polystichum,* and *Trichomanes.* Indeed many of them favor very damp habitats. This situation is further highlighted by comparing the distribution of apomictic species in various floras of the world. Jamaica with both very wet and very dry habitats shows only some 3% apomicts and these do not demonstrate any preference for the dry areas. Japan has the remarkably high figure of 18% apomicts and cannot be considered outstandingly xeric.[76]

D. Vegetative Apomixis

Vegetative reproduction occurs in a wide variety of pteridophytes as a means of accessory increase, reinforcing the normal reproduction by spores. In those cases where no viable spores are produced, it must be regarded as a form of vegetative apomixis. Here there ceases to be an alternation of generations in both morphology and cytology. A number of species are known to reproduce normally by spores over most of their range, failing to do so on reaching their climatic limits. *Polypodium dispersum* in much of Florida perpetuates itself by root buds,[123] while some species of *Grammitis, Vittaria,* and *Trichomanes* are present in some parts of the U.S. only as prothalli persisting by means of gemmae.[124,125] Apart from such cases of environmental stress, almost all other examples known of exclusively vegetative reproduction refer to hybrids which fail to produce viable spores. These are numerous, and a wide range of forms of vegetative apomixis have been recorded.[104] Mention has been made of gametophytic gemmae in ferns, but these small platelets of cells are also found on the rhizomes of *Psilotum* in abundance.[125] These gemmae are probably responsible for the maintenance in garden areas in the Finisterre Mountains of Papua New Guinea of large populations of *Psilotum* which show irregular pairing and numerous lagging chromosomes at meiosis.[12]

Stolon production has led to the formation of large clones of the triploid *Blechnum* × *caudatum* (*B.* × *antillanum*) in various West Indian Islands, and profusely branched creeping rhizomes have built up large thickets of a sterile triploid *Gleichenia* hybrid in Trinidad.[12] *Asplenium* × *fawcettii* is abundant on the summit area of Blue Mountain Peak in Jamaica where it perpetuates by plantlets produced on the fronds.[76,104]

Possibly the most famous case in temperate regions is that of the clubmoss *Lycopodium selago* which in Europe, at least, reproduces by bulbils formed in the axils of its leaves. It is a high polyploid with c. 264 chromosomes which pair irregularly and show a high degree of lagging at meiosis.[4] Similarly Manton has demonstrated the hybrid origin of *Equisetum* × *littorale, E.* × *trachyodon* and *E.* × *moorei.*[4] The first two have built up large populations in Ireland by vegetative means.

One of the most spectacular cases of very efficient vegetative reproduction is that of the floating aquatic fern *Salvinia molesta.* Here the delicate thin rhizomes branch, fragment, and continue growing at a rapid rate. Large tracts of water have sometimes been covered over completely in an almost explosive fashion where this plant has been introduced and has caused serious problems, such as the initial infestation of Lake Kariba in Africa. Figure 2 summarizes the various forms of departures from the normal life cycle of pteridophytes considered above.

It has been suggested that not only may vegetative apomixis ensure the immediate survival of otherwise sterile hybrids but that in a number of such instances the hybrids may be maintained until fertility is improved either by selection or by a doubling of the chromosome number giving rise to allopolyploids.[104] Lovis has suggested that a genotype may be allowed to persist by such means until genetic factors are acquired by mutation, enabling sporangial apomixis to become established.[9] Yet another suggestion is that vegetative reproduction allows the rapid increase of especially favorable genotypes of particular importance in the invasion and colonization of new habitats which may become suddenly available.

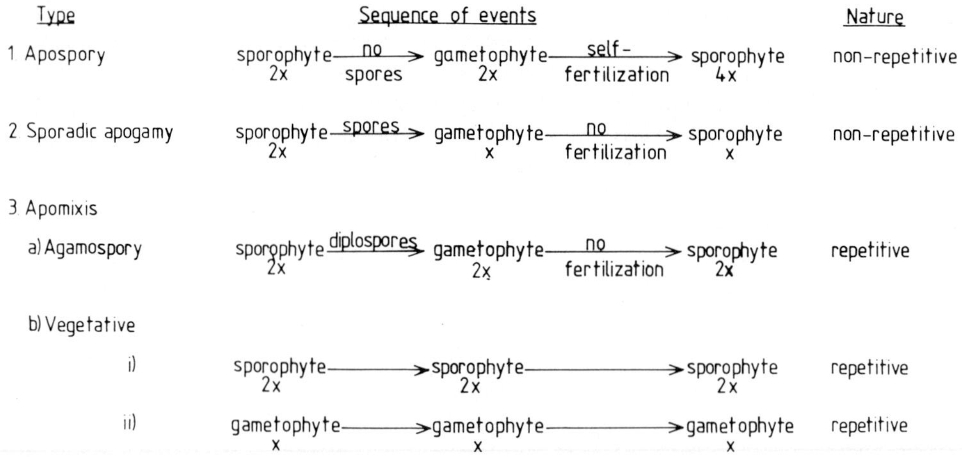

Type	Sequence of events			Nature
1 Apospory	sporophyte ──no spores──► gametophyte 2x	──self-fertilization──►	sporophyte 4x	non-repetitive
2 Sporadic apogamy	sporophyte ──spores──► gametophyte 2x / x	──no fertilization──►	sporophyte x	non-repetitive
3 Apomixis				
a) Agamospory	sporophyte ──diplospores──► gametophyte 2x	──no fertilization──►	sporophyte 2x	repetitive
b) Vegetative				
i)	sporophyte 2x ──────►	sporophyte 2x ──────►	sporophyte 2x	repetitive
ii)	gametophyte x ──────►	gametophyte x ──────►	gametophyte x	repetitive

FIGURE 2.   Departure from the normal life cycle in pteridophytes. Comparative levels of chromosome number for each stage are shown.

VI. GENERAL CYTOLOGICAL FEATURES

The pteridophytes vary greatly in a number of general cytological features seen at meiosis and concerning which only isolated remarks are to be found in the literature. One of the most prominent features which varies from group to group is the size or volume of the cell relative to that of the number of chromosomes. As a consequence, plants with very high numbers are not necessarily the most difficult to analyze cytologically if there is a large amount of cytoplasm in which the chromosomes may be spread, while conversely those with low numbers and little cytoplasm may be more troublesome. Thus, some members of Gleicheniaceae with comparatively low numbers such as $n = 34$ are usually rather more difficult than *Ophioglossum* species with $n = 480$ or higher.

In *Psilotum* the cells are very large, as also are the chromosomes which show a large range in size. Lovis has pointed out that *Psilotum* and *Tmesipteris* have a characteristic laxity of spiralization of the chromosomes at metaphase I.[9] At the end of meiosis the members of the tetrad are very elongated, virtually banana-shaped.

Equisetum also tends to have a large volume of cytoplasm and its 108 pairs of chromosomes (or 216 univalents in the hybrids) spread very easily and uniformly in it. In the heterosporous members both *Isoetes* and *Selaginella* have relatively massive megaspores and in the megaspore mother cells the chromosomes only occupy a very small amount of the total volume as seen in the illustrations of Ninan for *Isoetes* and of Manton for *Selaginella*.[4,113]

The meiotic cells of *Lycopodium* s.l. are the most variable and among the most difficult to handle owing to their somewhat peculiar properties. In general, the cells have a remarkable opacity and often additional contents of a nonchromosomal nature. Each spore mother cell is invested in a sheath or pellicle and, unlike the situation in the other groups of pteridophytes, very little can be seen of the chromosomes when the cells are just lying in stain on the microscope slide before heating and squashing. After squashing, the chromosomes may prove to be very compact and well-defined as in *Lycopodium clavatum*, but in a proportion of the species the chromosomes assume some very peculiar, rather diffuse configurations which make them very difficult to analyze, especially if they are not all bivalents but show irregular pairing as in *Lycopodium selago*.

A. Mitosis

The study of mitotic chromosomes in the pteridophytes has until very recently been inhibited by a number of factors with the consequence that most chromosome studies have been concerned with meiotic divisions. In a number of instances mitotic divisions are difficult to obtain, for example, in *Psilotum* which produces neither roots nor croziers and in *Equisetum*, in which the chromosomes do not usually spread very satisfactorily. Even freely produced root-tips have properties which are somewhat restrictive. Thus the percentage of dividing nuclei, especially of those at metaphase, tends to be very low. For example, in a root tip dividing very actively by pteridophyte standards, out of 6497 nuclei examined only 22 were in metaphase ($= 0.34\%$) and of these only a single one ($= 0.015\%$) showed the chromosomes spread suitably.[12] The very tough nature of the intercellular cement now no longer presents a practical problem following enzymic treatment. The normally very high chromosome numbers can also be a handicap in that what are otherwise very good spreads may be spoiled by only a small clump of chromosomes failing to separate properly. In addition many of the best chemical agents used in flowering plant cytology have been very disappointing when applied to pteridophyte material. Thus, colchicine is not an effective pre-treatment agent and Feulgen staining has so far proved to be ineffective. As a result, the centromeres are not always as clearly defined as they might otherwise be, especially when they are in the terminal or near-terminal position.

These difficulties have tended to greatly inhibit karyotype analyses in the pteridophytes and those that have been carried out have been confined to one or, at most, only a few species in a genus, an exception to this being Bidin's work on *Adiantum*.[50]

Mitui published the mean chromosome lengths for many species of ferns covering a wide spread of genera and families.[88] These ranged from the lowest of 3.0 μm up to 7.5 μm. However, including figures for other pteridophytes, particularly those of *Selaginella*, the lower end is extended down to the 0.5 μm in *S. angustiramea*. The range in mean chromosome lengths in flowering plants according to Stebbins is from 0.8 μm in *Drosera rotundifolia* to 20 μm in *Lilium pardalinum*.[127] The pteridophytes thus span approximately the lower one third of this range. Although these are the mean values Table 4 shows that the range in chromosome length in an individual species is also normally relatively small, the longest chromosome rarely being more than twice that of the shortest. An obvious exception is *Asplenium auritum* which ranges in chromosome length from 1.5 to 6.0 μm but here it is clearly an allopolyploid of a special type (see below). Even more prominent is the difference found in *Psilotum nudum* where the chromosomes range from 4.5 μm to 18 μm, although the cause of this is less certain than in the preceding example. Mitui concluded that closely related genera have a similar chromosome length but as more species are being analyzed this generalization does not seem to be particularly valid except possibly in a few cases such as Osmundaceae. He commented on the large size of the chromosomes in *Bolbitis subcordata* as compared with those of other members of the family and figures for *B. portoricensis* from the West Indies (see Table 4) confirm this exceptional size. Again, in the very ancient family Marattiaceae the more typical members of *Angiopteris* and *Macroglossum* have chromosomes about half the size of those in the very different-looking *Christensenia*.[12]

If the chromosome lengths of a species are plotted graphically they usually will be found to form a normal distribution curve, with a prominent peak. However, in a number of cases the graph may show a double peak as in *Asplenium auritum* or even a treble peak in other species.[12] Such cases are possibly allopolyploids which combine different ancestral chromosome sets, each with its own characteristic and different peak. The presence of a double peak may therefore prove to be a relatively easy way of determining the alloploid nature of a plant. However the converse is not necessarily true and a single peak is not peculiar to any particular form of polyploidy.

Although the total number of pteridophytes analyzed in this way is not extensive, nevertheless a sufficient range of types has been sampled and quite a large body of unpublished

Table 4
LENGTHS OF MITOTIC CHROMOSOMES IN PTERIDOPHYTES

Species	Chromosome lengths Minimum — maximum (μm)	Authority
Psilotaceae		
Psilotum nudum	4.5—18.0	51
Selaginellaceae		
Selaginella schlecteri	1.7—3.7	From 111[a]
S. opaca	1.7—2.7	From 111
S. kerstingii	1.0—1.3	From 111
S. densa	0.7—1.7	From 111
S. angustiramea	0.5—0.7	From 111
S. polystachya	1.0—1.7	From 111
Isoetaceae		
Isoetes asiatica	4.8—6.5	128
I. sampathkumarani	2.0—4.0	113,51
I. coromandelina (2x)	2.0—5.3	113,(51)
	(2.0—4.0)	
I. coromandelina (3x)	3.3—6.0	113,51
Equisetaceae		
Equisetum hyemale	3.7—9.3	12
Ophioglossaceae		
Ophioglossum reticulatum	1.5—4.5	51
Marattiaceae		
Angiopteris evecta	2.7—5.3	12
Marattia fraxinea	1.4—4.4	51
Macroglossum alidae	2.3—4.7	12
Christensenia aesculifolia	4.7—11.7	12
Osmundaceae		
Osmunda japonica	5.0—8.0	23
O. claytoniana	5.4—7.7	24
O. cinnamomea var. *fokiense*	5.0—8.0	24
O. bankseifolia	5.0—6.7	24
O. regalis	6.6—10.0	51
Schizaeaceae		
Lygodium venustum	3.3—6.0	12
L. micans	4.7—9.7	12
Parkeriaceae		
Ceratopteris thalictroides	2.6—7.3	51
Adiantaceae		
Doryopteris concolour	2.0—4.0	51
D. ludens	1.3—3.3	51
Adiantum polyphyllum	2.0—6.0	50[b]
Pteris biaurita	3.3—4.0	51
Hymenophyllaceae		
Hymenophyllum barbatum	2.9—4.4	27
Polypodiaceae		
Photinopteris speciosa	2.7—6.7	12
Pyrrosia adnascens	2.6—6.0	51
Microsorium alternifolium	3.3—6.0	51
Campyloneurum phyllitidis	2.7—6.0	12
Phlebodium aureum (2x)	2.0—4.0	12
Polypodium decumanum	3.0—5.7	12
P. loriceum	3.7—6.3	12
P. triseriale	2.7—4.7	12
P. chnoodes	3.0—5.7	12

Table 4 (continued)
LENGTHS OF MITOTIC CHROMOSOMES IN
PTERIDOPHYTES

Species	Chromosome lengths Minimum — maximum (μm)	Authority
Thelypteridaceae		
Macrothelypteris torresiana	1.0—4.5	51
(as *Thelypteris uliginosa*)		
Aspleniaceae		
Asplenium varians	2.5—4.8	21
A. incisum	1.6—3.5	19
A. auritum	1.5—6.0	12
A. unilaterale	2.6—6.0	51
Camptosorus sibiricus	1.4—3.0	21
Dryopteridaceae		
Diplazium esculentum	2.4—4.7	25
D. sylvaticum	1.3—6.6	51
Tectaria wightii	2.0—4.6	51
Lomariopsidaceae		
Bolbitis portoricensis	4.7—11.0	12
Elaphoglossum pteropus	3.3—4.3	12
Davalliaceae		
Nephrolepis cordifolia	2.6—6.0	51
Blechnaceae		
Blechnum occidentale	2.7—4.7	12
B. unilaterale	2.0—3.3	12
B. fraxineum (2x)	3.7—5.7	12
B. fraxineum (4x)	3.0—5.7	12
B. orientale	1.0—3.3	51
Salpichlaena volubile	4.3—8.3	12
Marsileaceae		
Marsilea quadrifolia	1.7—2.8	51
Regnellidium diphyllum	2.3—4.3	51

[a]   The measurements presented here were made on the illustrations of Jermy, Jones, and Colden.[111]

[b]   This species is a representative of the genus *Adiantum*; the other 29 species investigated by Bidin and 3 species by Abraham, Ninan, and Matthew are not included.[50,51]

evidence points to deviations from the normal distribution being caused by chromosomal rearrangements such as translocations. Thus species of *Lomariopsis* with the more usual chromosome number based on $x = 41$ show a normal distribution pattern, while if the lengths of the chromosomes of the highly derivative *Lomariopsis rossii* with $2n = 62$ and *L. hederacea* with $2n = 32$ are plotted they no longer form a single peak but show an extended range in size in which not more than any 4 chromosomes are of the same length. The graph or histogram is thus very flat with numerous very small peaks. This represents an extreme example but is a very useful test case, and other deviants from the well-marked peak patterns are worthy of investigation to see if they also represent derived or modified karyotypes.

Isoetes is exceptional among the pteridophytes in that 8 out of the 22 somatic chromosomes in *I. asiatica* are truly median or very nearly so.[128] In virtually all other pteridophytes that have been karyotyped median centromeres are rather infrequent and the majority of the chromosomes have their centromeres in either the terminal or subterminal portion. This is also found to hold true when analysis of species in a single genus is carried out, although

Table 5
CENTROMERE POSITIONS IN *BLECHNUM*
***OCCIDENTALE*, *B. FRAXINEUM*, AND THEIR HYBRID**
B.* × *CAUDATUM

	B. occidentale (2n = 124)	*B. fraxineum* (2n = 62)	*B.* × *caudatum*[12] (2n = 93)
Median point, M	8 (4)	2 (1)	5
Median region, m	20 (10)	12 (6)	16
Submedian region, sm	12 (6)	4 (2)	8
Subterminal region, st	10 (5)	12 (6)	11
Terminal region, t	—	4 (2)	2
Terminal point, T	72 (36)	28 (14)	50
Satellited	2 (1)	0	1
Total	124	62	93

Note: Figures in parentheses are the gametic complements of the parental species. Designations of centromere positions are those of Levan et al.[145]

the exact proportions in each particular position varies somewhat from species to species as shown by Bidin in the karyotypes of c. 30 taxa of *Adiantum*.[50]

An analysis of karyotypes in some plants of *Blechnum* from the West Indies demonstrated that in favorable cases this type of analysis could be very effective and useful.[12] In this instance, the chromosome complements of *Blechnum occidentale* with 2n = 124 and the diploid cytotype of *B. fraxineum* with 2n = 62 were analysed together with that of the triploid hybrid formed between these two species, *B.* × *caudatum* (= *B. antillanum.*) It will be seen from Table 5 that the centromere positions of *B. occidentale* differ from those of *B. fraxineum* and that the hybrid is indeed formed by the addition of one set of gametic chromosomes to the other.

VII. HYBRIDIZATION

Hybridization is of frequent occurrence in pteridophytes and has been recorded in every one of the major groups, encompassing *Psilotum, Lycopodium, Isoetes, Selaginella, Equisetum*, and the ferns. Knobloch has listed some 620 pteridophyte hybrids,[129] although included in this figure is a small number of apomictic species such as *Cyrtomium caryotideum, C. falcatum*, and *C. fortunei*. Of these hybrids, a little over half have been confirmed cytologically, the remainder being diagnosed on purely morphological characters. Nearly half (48%) are triploids, a high proportion being backcrosses of allotetraploids to one or other of their parents.

In every pteridophyte flora that has been well-sampled cytologically a wide range of hybrids has been found.[76] It may be suspected that the particular percentage of hybrids found bears some relationship as to how well-known a flora is or how well-sampled, rather than to any other intrinsic factor. Thus, the pteridophyte floras of Britain and North America have had 34% and 27% hybrids reported and these figures fall off sharply in other countries where, although sampling has been carried out, there has probably not been the same interest shown by numerous workers in particular genera such as shown in *Asplenium* and *Dryopteris* in both Europe and North America.

If attention is concentrated primarily on diploid hybrids in the first instance it will be seen that their cytological behavior ranges from a complete absence of chromosome pairing to complete synapsis with every intermediate stage and from completely sterile to fully fertile.

Numerous cases are reported of complete failure of chromosome pairing in the diploid hybrids, for example in *Asplenium* and *Dryopteris*. These may form the raw material of allopolyploidy which is one of the major evolutionary forces in homosporous pteridophytes, and this process is continuing at the present time. Sometimes asynapsis is not total in all cells, a small number of bivalents being formed, although not in all cells. In the wild hybrid *Adiantum villosum* × *A. lucidum* found in Trinidad, pairing is very variable — a small proportion of cells showing all bivalents ($x = 30$) while others show various stages down to 15 bivalents and 30 univalents.[76] Here the chromosomes appear to be losing their homologies, the process not having gone far enough to impose a regular pattern of asynapsis. The possibility at least exists that some spores formed as a consequence of complete chromosome pairing in their mother cells may be viable and capable of producing F_2 or introgressive forms, although this has not yet been tested.

Introgression has been reported in a few cases, although the evidence has been somewhat variable. Hickok and Klekowski, and Hickok synthesized hybrids between three diploid species of *Ceratopteris* and reported that the majority of the chromosomes paired but that some of the configurations suggested that chromosomal restructuring by translocations and inversions separated the species.[130,131] The hybrids were semisterile in the F_1 generation but fully fertile in the F_2.

In California the gold-back fern *Pityrogramma* may show introgression in certain situations.[132] Two of the species, *P. viscosa* and *P. triangularis*, tend to be separated by their ecological preferences, *P. viscosa* being a plant of sunny situations on sandy soils while *P. triangularis* is a plant of shady scrub woodland. Where the habitats are distinct and separated from one another the species are clear-cut, but in other areas where the habitats meet or intergrade a full range of morphological intermediates between the two species is found. A small sample of such intermediates showed 30 bivalents at meiosis but the spore fertility in 8 specimens tested varied from only 14% to 97%.

The best-documented case of introgressive hybridization occurs in *Pteris* in Sri Lanka. Here the simply pinnate *P. multiaurita* is connected by a very large range of forms to the completely pinnatifid *P. quadriaurita*, each form differing from the next by the addition of only one or two pinna segments or pinnules. Such hybrids are very common, and among the most common ferns of the island. They show 29 bivalents at meiosis and are fully fertile.[133] Spores of these intermediates give rise to a completely segregating population of which the members are also fully fertile. This interpretation has been confirmed by synthesizing the hybrid between *P. multiaurita* and *P. quadriaurita*. The characters of *P. quadriaurita* are completely dominant in the F_1. On allowing the prothalli obtained by sowing the F_1 spores to self-fertilize, an F_2 generation was obtained which exactly duplicated the range of forms found in the wild. Evidence suggests that these two species are separated by rather weak ecological requirements. *P. multiaurita* requires a higher light intensity for good prothallial development and is therefore a plant of open habitats, whereas *P. quadriaurita* appears to need freedom from competition with other herbaceous plants and as a consequence is restricted to forest in the drier zones with slight undergrowth. The prothalli of the latter species appear to be indifferent to light intensities. This example shows that the ecological requirements of both generations have to be taken into account when dealing with pteridophytes. The great plantation development of the last century in Sri Lanka, coupled partly with the crash of the coffee industry, led to many intermediate habitats being created in which the hybrids could establish themselves and thrive. The process has continued to the present with the road-building program, the embankments and verges forming ideal situations. A third species, *P. confusa*, may also be involved. *P. confusa* is a diploid apomict, having the Döpp-Manton system. This crosses with any of the sexual *P. multiaurita-quadriaurita* range of forms and gives rise to plants which also range in morphology from about the mid-part of the range to the *P. quadriaurita* end of the scale. Because the apomixis is inherited these triploid hybrids are also apomictic and breed true. Thus there is a diploid segregating range of forms partly duplicated in morphology by an apomictic triploid range of forms, each of which is now stabilized and true-breeding. Although this remains a unique example, it highlights the fact that triploid apomicts may gain a great deal of genetic diversity when one is compared with another by repeated formation from the crossing of the parental diploid apomict with genetically diverse elements of a sexual species.

While triploids form the numerically strongest category of hybrids, they, apart from the apomicts, are highly sterile. Nevertheless, they may provide a source of genetic variation as suggested by Vida and Reichstein in *Polystichum* × *illyricum*,[77] formed by the crossing of the tetraploid *P. aculeatum* with one of its diploid parents, *P. lonchitis*, (the other parent being *P. setiferum*). This hybrid has 41 bivalents and 41 univalents at meiosis but in a few cells the 41 univalents may fail to segregate and all pass to one pole, together with a further 41 chromosomes derived from the segregation of the bivalents. This mechanism may have been the cause of the introgression recorded between *P. aculeatum*, *P. lonchitis*, and *P. setiferum*. Genetic variation could be increased by this means even in an originally homozygous tetraploid, introducing new elements by introgression.

On the whole, pteridophyte genera are reasonably well circumscribed, and although a number of cases of intergeneric hybrids are known in the ferns, such as *Tectaria* × *Quercifilix*,[44,134] *Hemionitis* × *Gymnopteris*,[135] *Eriosorus* × *Jamesonia*,[136] and *Pleopeltis* × *Polypodium*,[137] these have usually suggested that the original generic boundaries were too stringently drawn taxonomically and that these really represent crosses between very different species of the same genus rather than between two genera. In the monotypic *Fadenya hookeri* the plants are sterile triploids but populations may build up by means of the plantlets produced at the frond-tips. Its basic chromosome number of $x = 40$ helps to confirm its tectaroid affinities, but its parentage is obscure although a simple-fronded species of *Tectaria* may be involved.[46] In another monotypic plant, *Pleuroderris michlerana*, the parents have been shown to be *Tectaria incisa* and *Dictyoxiphium panamense*.[138] These two species are remarkably different from one another in appearance both as regards frond shape and soral features. Despite appearances the closeness of *Dicytoxiphium* to *Tectaria* is demonstrated by the high degree of chromosome pairing which ranges from 15 to 35 bivalents in the hybrid. Lovis has compared the behavior of the Aspleniaceae, where intergeneric hybrids are known involving *Asplenium*, *Ceterach*, *Phyllitis*, and *Camptosorus*,[53] with such angiosperm families as the grasses and orchids.

That hybridization has played an even more profound role in the evolution of the ferns than just speciation or the introduction of new genetic variation in existing species is suggested by the unpublished work of C. N. Page on the predominantly Australasian genus *Doodia*.[139] This genus consists of a number of very ill-defined species. Page, as a result of extensive field studies and hybridization experiments estimates the actual number to be at least 17. Of 122 wild plants cytologically examined 41 were diploid, 60 were tetraploid, and 13 were hexaploid, all with a regular meiosis, while the remaining 8 consisted of 3 triploid, 3 tetraploid, and 2 pentaploid hybrids. New Zealand showed only a few diploids and many pentaploids, while Australia had 47% of its 80 plants sampled at the diploid and only about 2% at the hexaploid levels. Page suggests that the genus is actively evolving, particularly in the changing, specialized sclerophyll forests which are being invaded and exploited by *Doodia*. That the genus is of recent origin is suggested by its ill-defined species. Its ancestry is thought to lie among members of the predominantly Southern Hemisphere genus *Blechnum* and to involve hybridization between several taxa and subsequent selection among the offspring in response to the Australian environment. This has been followed by a spread to New Zealand, mainly in the form of polyploids.[139]

REFERENCES

1. **Friebel, H.,** Untersuchungen zur Cytologie der Farne, *Beit. Biol. Pfl.,* 21, 167, 1933.
2. **Döpp, W.,** Die Apogamie bei *Aspidium remotum* A. Br., *Planta,* 17, 86, 1932.
3. **Döpp, W.,** Cytologische und genetische Untersuchungen innerhalb der Gattung *Dryopteris, Planta,* 29, 481, 1939.
4. **Manton, I.,** *Problems of Cytology and Evolution in the Pteridophyta,* Cambridge University Press, London, 1950.
5. **Meyer, D. E.,** Untersuchungen über Bastardierung in der Gattung *Asplenium, Bibliotheca Bot.,* 123, 1, 1952.
6. **Chambers, T. C.,** Use of snail cytase in plant cytology, *Nature, Lond.,* 175, 215, 1955.
7. **Tindale, M. D. and Roy, S. K.,** Personal communication, 1981.
8. **Walker, T. G.,** Evidence from cytology in the classification of ferns, in *The Phylogeny and Classification of the Ferns,* Jermy, A. C., Crabbe, J. A., and Thomas, B. A., Eds., *Bot. J. Linn. Soc.,* 67, 91, 1973.
9. **Lovis, J. D.,** Evolutionary patterns and processes in ferns, *Adv. Bot. Res.,* 4, 229, 1977.
10. **Wagner, F. S.,** New basic chromosome numbers for genera of neotropical ferns, *Am. J. Bot.,* 67, 1980.
11. **Tryon, A. F., Bautista, H. P. and Araujo, I. S.,** Chromosome studies of Brazilian ferns, *Acta Amazonica,* 5, 35, 1975.
12. **Walker, T. G.,** unpublished data, 1981.
13. **Chiarugi, A.,** Tavole chromosomiche delle pteridophyta, *Caryologia,* 13, 27, 1960.
14. **Fabbri, F.,** Primo supplemento alle ''Tavole Chromosomiche delle Pteridophyta'' di Alberto Chiarugi, *Caryologia,* 16, 237, 1963.
15. **Fabbri, F.,** Secondo supplemento alle ''Tavole Chromosomiche delle Pteridophyta'' di Alberto Chiarugi, *Caryologia,* 18, 675, 1965.
16. **Löve, A., Löve, D. and Pichi Sermolli, R. E. G.,** *Cytotaxonomical Atlas of the Pteridophyta,* Cramer, Vaduz, 1977.
17. **Brownlie, G.,** Chromosome numbers in New Zealand ferns, *Trans. R. Soc. N.Z.,* 85, 213, 1955.
18. **Klekowski, E. J.,** Populational and genetic studies of a homosporous fern - *Osmunda regalis, Am. J. Bot.,* 57, 1122, 1970.
19. **Tatuno, S. and Kawakami, S.,** Karyological studies on Aspleniaceae, 1. Karyotypes of three species in *Asplenium, Bot. Mag., Tokyo,* 82, 436, 1969.
20. **Kawakami, S.,** Karyological studies in Aspleniaceae. I. Karyotypes of three species of *Asplenium, Bot. Mag., Tokyo,* 82, 436, 1969.
21. **Kawakami, S.,** Karyological studies on Aspleniaceae. II. Chromosomes of seven species in *Aspleniaceae, Bot Mag., Tokyo,* 83, 74, 1970.
22. **Takei, M.,** Karyological studies in Polypodiaceae. I. Karyotypes of a few species of the genus *Lemmaphyllum* and *Pyrrosia* in Japan, *Bot. Mag., Tokyo,* 82, 482, 1969.
23. **Tatuno, S. and Yoshida, H.,** Karyologische Untersuchungen über Osmundaceae. I. Chromosomen der Gattung *Osmunda* aus Japan, *Bot. Mag., Tokyo,* 79, 244, 1966.
24. **Tatuno, S. and Toshida, H.,** Karyological studies on Osmundaceae. II. Chromosomes of the genus *Osmundastrum* and *Plenasium* in Japan, *Bot. Mag., Tokyo,* 80, 130, 1967.
25. **Tatuno, S. and Okado, H.,** Karyological studies in Aspidiaceae. I. *Bot. Mag., Tokyo,* 83, 202, 1970.
26. **Kawakami, S.,** Karyological studies in Pteridaceae. I. Karyotypes of three species in *Pteris, Bot. Mag., Tokyo,* 84, 1971.
27. **Tatuno, S. and Takei, M.,** Karyological studies in Hymenophyllaceae. I. Chromosomes of the genus *Hymenophyllum* and *Mecodium* in Japan, *Bot. Mag., Tokyo,* 82, 121, 1969.
28. **Duncan, D. and Smith, A. R.,** Primary basic chromosome numbers in ferns: facts or fantasies?, *Systematic Botany,* 3, 105, 1980.
29. **Wagner, W. H. and Wagner, F. S.,** Polyploidy in pteridophytes, in *Polyploidy: Biological Relevance,* Lewis, W. H., Ed., Plenum Press, New York, 1980, 199.
30. **Pichi Sermolli, R. E. G.,** Pteridophyta, in *Vistas in Botany,* W. B. Turrill, Ed., Pergamon Press, London, 1959, 421.
31. **Pichi Sermolli, R. E. G.,** Tentamen Pteridophytorum genera in taxonomico ordine redigendi, *Webbia,* 31, 313, 1977.
32. **Nayar, B. K.,** A phylogenetic classification of the homosporous ferns, *Taxon,* 19, 229, 1970.
33. **Crabbe, J. A., Jermy, A. C. and Mickel, J. T.,** A new generic sequence for the pteridophyte herbarium, *Fern Gaz.,* 11, 141, 1975.
34. **Holttum, R. E.,** Studies in the family Thelypteridaceae. III. A new system of genera in the Old World, *Blumea,* 19, 17, 1971.
35. **Copeland, E. B.,** *General Filicum,* Chronica Botanica, Waltham, Mass., 1947.
36. **Manton, I.,** Chromosomes and fern phylogeny with special reference to ''Pteridaceae'', *J. Linn. Soc. (Bot.),* 56, 73, 1958.

37. **Brownsey, P. J.,** A chromosome count in *Loxosoma, N.Z. Jl. Bot.,* 13, 355, 1975.

38. **Mehra, P. N.,** Chromosome numbers in Himalayan ferns, *Res. Bull. Panjab Univ. N.S.,* 12, 139, 1961.

39. **Sorsa, V.,** Chromosome studies on Puerto Rican ferns (Gleicheniaceae), *Caryologia,* 21, 97, 1968.

40. **Holttum, R. E.,** Pteridophyta. (Ferns and fern allies), *Flora Malesiana,* 11, 1, 65.

41. **Bir, S. S.,** Cytological observations on the East Himalayan members of *Asplenium* Linn., *Curr. Sci.,* 29, 445, 1960.

42. **Bir, S. S.,** Cytological observations on some ferns from Simla (Western Himalayas), *Curr. Sci.,* 31, 248, 1962.

43. **Kuriachin, P. I. and Fabbri, F.,** Secondo supplemento alle Tavole cromosomiiche delle Pteridophyta di Alberto Chiarugi, *Caryologia,* 18, 675, 1965.

44. **Manton, I. and Sledge, W. A.,** Observations on the cytology and taxonomy of the pteridophyte flora of Ceylon, *Phil. Trans. R. Soc. Ser. B.,* 238, 127, 1954.

45. **Smith, A. R. and Mickel, J. T.,** Chromosome counts for Mexican ferns, *Brittonia,* 29, 391, 1977.

46. **Walker, T. G.,** A cytotaxonomic survey of the pteridophytes of Jamaica, *Trans. R. Soc. Edinb.,* 66, 169, 1966.

47. **Roy, S. K. and Manton, I.,** The cytological characteristics of the fern sub-family Lomariopsidoideae sensu Holttum, *J. Linn. Soc. (Bot.),* 59, 343, 1966.

48. **Tsai, J.-L. and Shieh, W.-C.,** Chromosome numbers of the fern family Aspidiaceae (sensu Copeland) in Taiwan (1), *J. Sci. and Eng.,* 12, 321, 1975.

49. **Lorence, D. H., in Löve, A.,** IOPB chromosome number reports. LXII, *Taxon,* 30, 706, 1981.

50. **Bidin, A. A. Bin Hj.,** Studies in the fern genus *Adiantum* L., Ph.D. Thesis, University of Newcastle upon Tyne, 1980.

51. **Abraham, A., Ninan, C. A. and Mathew, P. M.,** Studies on the cytology and phylogeny of the pteridophytes. VII. Observations on one hundred species of South Indian ferns, *J. Indian Bot. Soc.,* 41, 339, 1962.

52. **Brownlie, G.,** Chromosome numbers in some Pacific Pteridophyta, *Pacif. Sci.,* 19, 493, 1965.

53. **Lovis, J. D.,** A biosystematic approach to phylogenetic problems and its application to the Aspleniaceae, in *The Phylogeny and Classification of the Ferns,* Jermy, A. C., Crabbe, J. A., and Thomas, B. A., Eds., *Bot. J. Linn. Soc.,* 67(Suppl. 1), 210, 1973.

54. **Brownsey, P. J.,** *Asplenium* hybrids in the New Zealand flora, *N.Z. J. Bot.,* 15, 601, 1977.

55. **Manton, I.,** Cytological information on the ferns of west tropical Africa, in *The Ferns and Fern Allies of West Tropical Africa,* 2nd ed., Alston, A. H. G., London, 1959.

56. **Wagner, W. H.,** A natural hybrid, × *Adiantum tracyi* C. C. Hall, *Madroño,* 13, 195, 1956.

57. **Wagner, W. H.,** Cytological observations on *Adiantum* × *tracyi* C. C. Hall, *Madroño,* 16, 158, 1962.

58. **Smith, A. R.,** Taxonomic and cytological notes on ferns from California and Arizona, *Madroño,* 22, 376, 1974.

59. **Klekowski, E. J. and Baker, H. G.,** Evolutionary significance of polyploidy in the Pteridophyta, *Science,* 153, 305, 1966.

60. **Klekowski, E. J.,** Sexual and subsexual systems in homosporous pteridophytes: a new hypothesis, *Am. J. Bot.,* 60, 535, 1973.

61. **deWet, J. M. J.,** Origins of polyploids, in *Polyploidy: Biological Relevance,* Lewis, W. H., Ed., Plenum Press, New York, 1980, 3.

62. **Wagner, W. H. and Whitmire, R. S.,** Spontaneous production of a morphologically distinct fertile allopolyploid by a sterile diploid of *Asplenium ebenoides, Bull. Torrey Bot. Club,* 84, 79, 1957.

63. **Rasbach, von H., Rasbach, K., Reichstein, T., Schneller, J. J. and Vida, G.,** *Asplenium* × *lessinense* Vida et Reichst. in den Bayerischen Alpen und seine Fahigkeit zur spontanen Chromosomenverdoppelung, *Ber. Bayer. Bot. Ges.,* 50, 23, 1979.

64. **Lovis, J. D.,** Artificial reconstruction of a species of fern, *Asplenium adulterinum, Nature, Lond.,* 217, 1163, 1968.

65. **Lovis, J. D., and Reichstein, T.,** Die zwei diploiden *Asplenium trichomanes* × *viride* Bastarde und ihre Fahigkeit zur spontanen Chromosomenverdoppelung, *Bauhinia,* 4, 53, 1968.

66. **Lovis, J. D.,** The synthesis of a new *Asplenium, Br. Fern Gaz.,* 10, 153, 1970.

67. **Moran, R. C.,** × *Asplenosorus shawneensis,* a new natural fern hybrid between *Asplenium trichomanes* and *Camptosorus rhizophyllus, Amer. Fern J.,* 71, 85, 1981.

68. **Klekowski, E. J.,** Genetical features of ferns as contrasted to seed plants, *Ann. Mo. Bot. Gdn.,* 59, 138, 1970.

69. **Klekowski, E. J.,** Homoeologous pairing in ferns, in *Current Chromosome Research,* Jones, K. and Brandham, P. E., Eds., Elsevier/North Holland Biomedical Press, Amsterdam, 175, 1976.

70. **Wagner, W. H., Wagner, F. S., Miller, C. N., and Wagner, D. H.,** New observations on the royal fern hybrid *Osmunda* × *rugii, Rhodora,* 80, 92, 1978.

71. **Stebbins, G. L.,** *Variation and Evolution in Plants,* Oxford University Press, London, 1950.

72. **Vida, G.,** The nature of polyploidy in *Asplenium ruta-muraria* L. and *A. lepidum* C. Presl, *Caryologia,* 23, 525, 1970.

73. **Manton, I. and Ghatak, J.,** Cytotaxonomic studies in the *Adiantum caudatum* complex of Africa and Asia. I. Parentage of *A. indicum* Ghatak, *J. Linn. Soc. (Bot.),* 60, 223, 1967.

74. **Manton, I. and Sinha, B. M. B.,** Cytotaxonomic studies in the *Adiantum caudatum* complex of Africa and Asia. II. Autoploidy and alloploidy in African representatives of *A. incisum. Bot. J. Linn. Soc.,* 63, 1, 1970.

75. **Sinha, B. M. B. and Manton, I.,** Cytotaxonomic studies in the *Adiantum caudatum* complex of Africa and Asia. III. Segmental allopolyploid origin of *A. malesianum* Ghatak, *Bot. J. Linn. Soc.,* 63, 247, 1970.

76. **Walker, T. G.,** The cytogenetics of ferns, in *The Experimental Biology of Ferns,* Dyer, A. F., Ed., Academic Press, London, 87, 1979.

77. **Vida, G. and Reichstein, T.,** Taxonomic problems in the fern genus *Polystichum* caused by hybridization, in *European Floristic and Taxonomic Studies,* Walters, S. M., Ed., published for The Botanical Society of the British Isles by E. W. Classey Ltd., Faringdon, 126, 1975.

78. **Shivas, M. G.,** Contributions to the cytology and taxonomy of species of *Polypodium* in Europe and America. I. Cytology, *J. Linn. Soc. (Bot.),* 58, 13, 1961.

79. **Shivas, M. G.,** Contributions to the cytology and taxonomy of species of *Polypodium* in Europe and America. II. Taxonomy, *J. Linn. Soc. (Bot.),* 58, 27, 1961.

80. **Shivas, M. G.,** The *Polypodium vulgare* complex, *Br. Fern Gaz.,* 9, 65, 1962.

81. **Walker, S.,** Cytogenetic studies in the *Dryopteris spinulosa* complex. II. *Amer. J. Bot.,* 48, 607, 1961.

82. **Walker, S.,** Identification of a diploid ancestral genome in the *Dryopteris spinulosa* complex, *Br. Fern Gaz.,* 10, 97, 1969.

83. **Gibby, M.,** A cytogenetic and taxonomic study of the *Dryopteris carthusiana* complex, Ph.D. Thesis University of Liverpool, 1977.

84. **Gibby, M., Jermy, A. C., Rasbach, H., Rasbach, K., Reichstein, T., and Vida G.,** The genus *Dryopteris* in the Canary Islands and Azores and the description of two new tetraploid species, *Bot. J. Linn. Soc.,* 74, 251, 1977.

85. **Brownsey, P. J.,** A biosystematic investigation of the *Asplenium lepidum* complex, *J. Linn. Soc. (Bot.),* 72, 235, 1976.

86. **Walker, T. G.,** The *Anemia adiantifolia* complex in Jamaica, *New Phytol.,* 61, 291, 1962.

87. **Tryon, A. F.,** Comparisons of sexual and apogamous races in the fern genus *Pellaea, Rhodora,* 70, 1, 1968.

88. **Mitui, K.,** Chromosomes and speciation in ferns, *Sci. Rep. Tokyo Kyoiku Daig.,* B 13, 285, 1968.

89. **Bir, S. S.,** Correlation between spore size and polyploid-level in the Himalayan asplenioid and athyrioid ferns, *J. Palynology,* 2, 3, 41, 1966-67.

90. **Baquar, S. R.,** Polyploidy in the flora of Pakistan in relation to latitude, life form, and taxonomic groups, *Taxon,* 25, 621, 1976.

91. **Walker, T. G.,** Additional cytotaxonomic notes on the pteridophytes of Jamaica, *Trans. R. Soc. Edinb.,* 69, 109, 1973.

92. **Manton, I. and Vida, G.,** Cytology of the fern Flora of Tristan da Cunha, *Proc. R. Soc. B.,* 170, 361, 1968.

93. **Page, C. N.,** Cytotaxonomic and anatomical studies in certain Pteridophyta, Ph.D. Thesis, University of Newcastle upon Tyne, 1968.

94. **Page, C. N.,** Ferns, polyploids, and their bearing on the evolution of the Canarian flora, *Monographiae Biologicae Canarienses,* 4, 1973.

95. **Larsen, K.,** Cytological and experimental studies on the flowering plants of the Canary Islands, *Biol. Skrift Dan. Vid. Selsk.,* 11, 1, 1960.

96. **Larsen, K.,** Contribution to the ecology of the endemic Canarian element. II, *Bot. Notiser,* 116, 409, 1963.

97. **Manton, I.,** Evolutionary mechanisms in tropical ferns, *Biol. J. Linn. Soc.,* 1, 219, 1969.

98. **Tarling, D. H.,** *Continental Drift and Biological Evolution,* Carolina Biological Supply Co., Burlington, North Carolina, 1980.

99. **White, R. A.,** Experimental investigations of fern sporophyte development, in *The Experimental Biology of Ferns,* Dyer, A. F., Ed., Academic Press, London, 1979, 505.

100. **Verma, S. C. and Loyal, D. S.,** Colchi-autotetraploidy in *Adiantum capillus-veneris, Nature,* 188, 1210, 1960.

101. **Bouharmont, J.,** Meiosis and fertility in apogamously produced diploid plants of *Asplenium trichomanes, Chromosomes Today,* 3, 253, 1972.

102. **Hickok, L. G.,** An apomictic mutant for sticky chromosomes in the fern *Ceratopteris, Can. J. Bot.,* 55, 2186, 1977.

103. **Manton, I. and Walker, S.,** Induced apogamy in *Dryopteris dilatata* (Hoffm.) A. Gray and *D. filix-mas* (L.) Schott. emend. and its significance for the interpretation of the two species, *Ann. Bot.,* 18, 377, 1954.

104. **Walker, T. G.,** Apomixis and vegetative reproduction in ferns, in *Reproductive Biology and Taxonomy of Vascular Plants,* Bot. Soc. Br. Isles Conf. Rep., 9, 152, 1966.
105. **Manton, I., Roy, S. K. and Jarrett, F. M.,** The cytotaxonomy of some members of the *Cheilanthes farinosa* complex in Africa and India, *Kew Bull.,* 18, 553, 1966.
106. **Knobloch, I. W.,** A preliminary review of spore number and apogamy within the genus *Cheilanthes, Am. Fern J.,* 56, 163, 1966.
107. **Vida, G., Page, C. N., Walker, T. G., and Reichstein, T.,** Cytologie der Farn-Gattung *Cheilanthes* in Europa und auf den Canarischen Inseln, *Bauhinia,* 4, 223, 1970.
108. **Braithwaite, A. F.,** A new type of apogamy in ferns, *New Phytol.,* 63, 293, 1964.
109. **Evans, A. M.,** Ameiotic alternation of generations: a new life cycle in the ferns, *Science,* 143, 261, 1964.
110. **Tryon, R. M.,** The process of evolutionary migration in species of *Selaginella, Brittonia,* 23, 89, 1971.
111. **Jermy, A. C., Jones, K. and Colden, C.,** Cytomorphological variation in *Selaginella, J. Linn. Soc. (Bot.),* 60, 147, 1967.
112. **Abraham, A. and Ninan, C. A.,** Cytology of *Isoetes, Curr. Sci.,* 27, 60, 1958.
113. **Ninan, C. A.,** Studies on the cytology and phylogeny of the pteridophytes. V. Observations on the Isoetaceae, *J. Indian Bot. Soc.,* 37, 93, 1958.
114. **Verma, S. C.,** Cytology of *Isoetes coromandelina, Amer. Fern J.,* 51, 99, 1961.
115. **Pant, D. D. and Srivastava, G. K.,** Cytology and reproduction of some Indian species of *Isoetes, Cytologia,* 30, 239, 1965.
116. **Lloyd, R. M.,** Facultative apomixis and polyploidy in *Matteucia orientalis, Amer. Fern J.,* 63, 43, 1973.
117. **Hickok, L. G.,** The cytology and derivation of a temperature-sensitive meiotic mutant in the fern *Ceratopteris, Am. J. Bot.,* 64, 552, 1977.
118. **Evans, A. M.,** Interspecific relationships in the *Polypodium pectinatum-plumula* complex, *Ann. Mo. Bot. Gdn.,* 55, 193, 1969.
119. **Gastony, G. J. and Haufler, C. H.,** Chromosome numbers and apomixis in the fern genus *Bommeria* (Gymnogrammaceae), *Biotropica,* 8, 1, 1976.
120. **Tryon, A. F.,** Spores, chromosomes and relations of the fern *Pellaea atropurpurea, Rhodora,* 74, 220, 1972.
121. **Tryon, A. F. and Britton, D. M.,** Cytotaxonomic studies on the fern genus *Pellaea, Evolution,* 12, 137, 1958.
122. **Wagner, W. H.,** Structure of spores in relation to fern phylogeny, *Ann. Mo. Bot. Gdn.,* 61, 332, 1974.
123. **Evans, A. M.,** The *Polypodium pectinatum-plumula* complex in Florida, *Amer. Fern J.,* 58, 169, 1968.
124. **Farrar, D. F. and Wagner, W. H.,** The gametophyte of *Trichomanes holopterum* Kunze, *Bot. Gaz.,* 129, 210, 1968.
125. **Wagner, W. H. and Wagner, F. S.,** Pteridophytes of the Mountain Lake Area, Giles Co., Virginia: biosystematic studies 1964-65, *Castanea,* 31, 121, 1966.
126. **Holloway, J. E.,** The gametophyte, embryo, and young rhizomes of *Psilotum triquetrum* Swartz, *Ann. Bot., Lond.,* N.S., 3, 313, 1939.
127. **Stebbins, G. L.,** *Chromosomal Evolution in Higher Plants,* Edward Arnold, London, 1971.
128. **Tatuno, S.,** Zytologische Untersuchungen der Pteridophyten. I. Chromosomen von *Isoetes asiatica* Makino, *Cytologia,* 28, 293, 1963.
129. **Knobloch, I. W.,** Pteridophyte hybrids, *Pub. Mus. Michigan State Univ.,* 5, 277, 1976.
130. **Hickok, L. G. and Klekowski, E. J.,** Inchoate speciation in *Ceratopteris:* an analysis of the synthesised hybrid *C. richardii* × *C. pteridoides, Evolution,* 28, 439, 1974.
131. **Hickok, L. G.,** Cytological relationships between three diploid species of the fern genus *Ceratopteris, Can. J. Bot.,* 55, 1660, 1977.
132. **Alt, K. S. and Grant, V.,** Cytotaxonomic observations on the gold-back fern, *Brittonia,* 12, 153, 1960.
133. **Walker, T. G.,** Hybridization in some species of *Pteris* L., *Evolution,* 12, 82, 1958.
134. **Manton, I.,** Evolution in the Pteridophyta, in *A Darwin Centenary,* Wanstall, P. J., Ed., *Bot. Soc. Br. Isles, Conf. Rep.,* 6, 105, 1961.
135. **Mickel, J. T.,** A redefinition of the genus *Hemionitis, Am. Fern J.,* 64, 3, 1974.
136. **Tryon, A. F.,** A monograph of the fern genus *Eriosorus, Contr. Gray Herb. Harv.,* 200, 54, 1970.
137. **Wagner, W. H. and Wagner, F. S.,** A hybrid polypody from the New World tropics, *Fern Gaz.,* 11, 125, 1975.
138. **Wagner, W. H., Wagner, F. S., and Gomez, P. L. D.,** The singular origin of a Central American fern, *Pleuroderris michleriana, Biotropica,* 10, 254, 1978.
139. **Page, C. N.,** personal communication, 1981.
140. **Mitui, K.,** Chromosome numbers of Japanese pteridophytes, *Bull. Nippon Dental Coll., Gen. Education,* 4, 22, 1975.
141. **Brownlie, G.,** Introductory note to cytotaxonomic studies of New Zealand ferns, *Trans. R. Soc. N.Z.,* 82, 665, 1954.

142. **Brownlie, G.,** Cyto-taxonomic studies on New Zealand Pteridaceae, *New Phytol.,* 56, 207, 1957.
143. **Brownlie, G.,** Additional chromosome numbers of New Zealand ferns, *Trans. R. Soc. N.Z., Bot.,* 1, 1, 1961.
144. **Roy, R. P., Sinha, B. M. B. and Sakya, A. R.,** Cytology of some ferns of Kathmandu Valley, *Br. Fern Gaz.,* 10, 193, 1971.
145. **Levan, A., Fredga, K., and Sandberg, A. A.,** Nomenclature for centromeric position on chromosomes, *Hereditas,* 52, 201, 1965.

Chapter 6

MECHANISMS OF CHROMOSOME CHANGE IN THE EVOLUTION OF THE TRIBE TRADESCANTIEAE (COMMELINACEAE)

Keith Jones and Ann Kenton

TABLE OF CONTENTS

I. INTRODUCTION

The Commelinaceae is a family of herbs distributed in tropical, subtropical, and warm temperate regions. The bulk of the species are found in Central and South America, Africa, India, and S.E. Asia, and some in the U.S., China, Japan, and Australia. The family has been subdivided in various ways, on the basis of different characters and organs,[1] but here we will accept the two tribes, the Tradescantieae and the Commelineae as recognized by Rohweder.[2] The Tradescantieae, is confined to the New World, with the highest concentration of species in Central America but some in the U.S., South America, and the Caribbean. Rohweder saw it as consisting of 13 genera resulting from the gradual taxonomic dismemberment of *Tradescantia* and *Callisia;* the addition of genera by other taxonomists has brought this number up to 17. Hunt has concluded that this number of genera is excessive and has made a start on reconstituting *Tradescantia*.[3,4] Although partly accepting his view, we find it more useful to adopt the philosophy of Rohweder.

The species of the Tradescantieae are mostly low-growing perennial herbs reproducing sexually by outbreeding systems,[6] and vegetatively by rhizomes, tubers, and nodal rooting. A small number of species are annual inbreeders and an uncertain number are perennials with some capacity for self-fertilization. Phenotypes vary considerably from species to species but some similarities in flora and inflorescence characters pose problems for the taxonomists. Those who are unfamiliar with the range of variation in the tribe will probably know it from *Tradescantia virginiana* and *T. paludosa*, both widely used for demonstration and experiment in the chromosome field, and from *T. fluminensis* and *Rhoeo spathacea*, widely used as ornamentals.

A substantial living collection of the Tradescantieae, mostly from Mexico, has been built up at Kew and has been the subject of chromosome study for more than 10 years. Some of this work has been directed towards improving its taxonomy and some to elucidating problems of chromosome differentiation. In the present paper we will concentrate attention on the latter interest, but taxonomic matters will be included.

II. THE RANGE AND DISPOSITION OF KARYOTYPE VARIATION

The extremely extensive range of karyotype diversity in the entire family has been described,[7] when it was pointed out that all of the larger chromosome species are New World and most of these fall in the Tradescantieae, whose genera are entirely confined to this hemisphere. Within this tribe, however, there are considerable differences in chromosome size, karyotype pattern, and basic and absolute chromosome numbers. A few examples are shown in Figures 1 to 12.

As judged by the number of species which possess it, the commonest and possibly the most successful karyotype in the tribe is that with $x = 6$ large metacentrics occurring at diploid, tetraploid and more rarely hexaploid levels (Figure 1). This is present in all species of *Tradescantia* s.s. in the U. S. and the northern parts of Mexico, and in the disjunct *T. ambigua* with an Andean distribution in South America. All these species can be considered as temperate ones, those in Mexico not generally occurring below 1500 m altitude. The same karyotype pattern is present in species of *Setcreasea* at much lower altitudes but in the most northern regions of Mexico. In all of the 40 or so species of these genera there is little distinction between any of the karyotypes in overall pattern apart from the frequency and placement of terminal small satellites, but it is clear from the unpublished data of the late Arnold Sparrow and his colleagues that there are substantial differences in genome volume, and Martinez showed large differences in DNA amount between some Mexican diploids.[8] Tests of relative chromosome homologies between the sets of distinct species have given varying results. Hybrids between the closely related species of the *T. virginiana* alliance

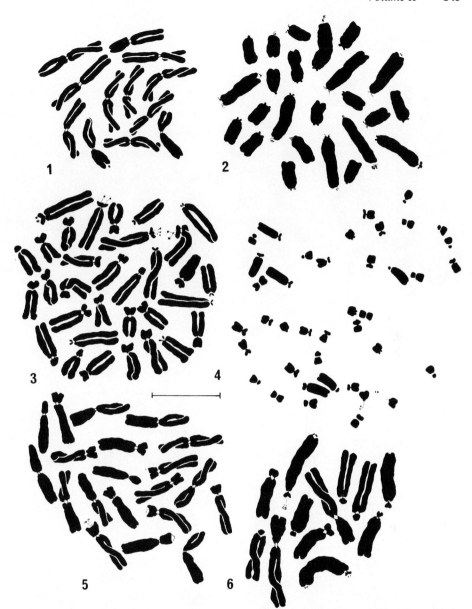

FIGURES 1—6   Karyotypes of some genera and species of the Tradescantieae. 1: *Tradescantia crassifolia* (2*n* = 12); 2: *T. micrantha* (2*n* = 24); 3: *Phyodina navicularis* (2*n* = 32); 4: *T. fluminensis* (2*n* = 40); 5: *Gibasis karwinskyana* (2*n* = 20); 6: *Callisia elegans* (2*n* = 12). Bar represents 10 μm.

of the U.S. show complete pairing with no major structural differentiation, though fertility was reduced.[9] Hybrids of well-differentiated species in Mexico were more irregular in behavior at meiosis with some heterozygosity for small interchanges and inversions.[8] The present results indicate that genomes show increasing degrees of difference as species diverge but there is no evidence for the involvement of major types of chromosome change, and nothing to suggest how this stable and highly symmetric karyotype originated. Some would doubtless see it as a primitive, as opposed to a derived, pattern but there is as yet no evidence that this is so, particularly as it is associated with a very advanced inflorescence type.

In absolute contrast with this symmetry is the bimodal complement of generally small

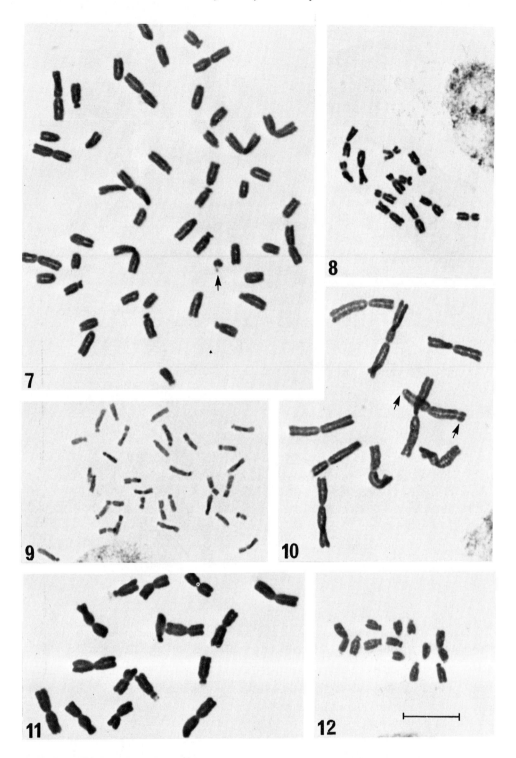

FIGURES 7—12   *Zebrina* sp. $2n = 46 +$ centric fragment (arrow); 8: *Rhoeo spathacea* ($2n = 12$); 9: *Thyrsanthemum floribundum* ($2n = 32$); 10: *Gibasis pulchella* ($2n = 10$); 11: *Gibasis schiedeana* ($2n = 16$); 12: *Cymbispatha plusiantha* ($2n = 12$)

and acrocentric chromosomes of *Tradescantia fluminensis* and its allies of South America (Figure 4) which share a basic number of $x = 10$. These are subtropical plants with a different habit from those in the north and have been suggested for removal to a separate genus. There is certainly connection between the two groups in their cytology.

High degrees of asymmetry in complements of large chromosomes can be seen in three species of northern Mexico and the U.S. which have been suggested for placement in the unsatisfactory genus *Phyodina*. *P. graminea* of the southeastern U.S. has a basic number of $x = 6$ but only three of these large chromosomes are metacentric and the rest acrocentric.[10] *P. micrantha*, whose cytology was described under *Tradescantia*,[11] is also $x = 6$ and all its chromosomes are telocentric (Figure 2). *P. navicularis* with $2n = 32$ is probably tetraploid and has a completely acrocentric complement (Figure 3). There is no obvious relationship between any of these karyotypes. The species are isolated from each other and the *Tradescantia* species; their only possible common characteristic is that they are relics of an earlier and tropical flora.

The species of *Cymbispatha* are distributed widely in Mexico and extend into the northern parts of South America. Their cytology has been described in detail.[5,12] Although their basic number and karyotype morphology vary from species to species, the chromosomes are either metacentric or acutely acrocentric (Figure 12). In this case each species, whether diploid or polyploid, has a total number of major chromosome arms which is a precise multiple of 7. There is ample additional evidence to show that evolution within the genus has been a combination of Robertsonian fusion and periods of chromosome doubling. It is a clear example of karyotype orthoselection in the tribe, although species of *Zebrina* (Figure 7) can also be interpreted in the same way.

The genus *Gibasis*, relatively primitive in its inflorescence and deserving to be classified separately from other members of the Tradescantieae,[2] is cytologically heterogenous and its species can be divided into four groups on this basis:[13]

1. Those with small asymmetric chromosomes and $x = 8$ (*G. geniculata* and allies)
2. Those with medium chromosomes and asymmetric complements with $x = 5$ ($\rightarrow 4$) (*G. karwinskyana* and allies; Figures 5,11).
3. Those with medium chromosomes and asymmetric complements with $x = 6$ ($\rightarrow 5$) (*G. linearis* and allies)
4. Those with large chromosomes and symmetrical complements and $x = 5$ (*G. pulchella* and allies; Figure 10).

In groups 2 and 3 the standard basic number is reduced by one in some cases by Robertsonian fusion and this is the only modification to karyotype pattern in the groups. In this respect it resembles *Cymbispatha*, where with this exception the karyotype is highly conservative and apparently resistant to changes in centromere position and relative chromosome lengths. Large differences in DNA amount in these *Gibasis* species is another permissible change.

The small chromosomes of species in group 1. cannot be related to any of the others and this feature in conjunction with morphological differences makes it likely that they may eventually find a place in another genus.

The species in group 4. have metacentric chromosomes much larger than any others in the genus and it is here that polymorphisms for interchanges occur, particularly in *G. pulchella*, culminating in complex interchange heterozygosity in some populations. These karyotypes again show no clear evolutionary connection with any other group in the genus, but since they exhibit unsolved features of karyotype change and pairing behavior, it is premature to believe that they are as unrelated as they seem.

Callisia shows wide variation in phenotype and inflorescence characters, but all species

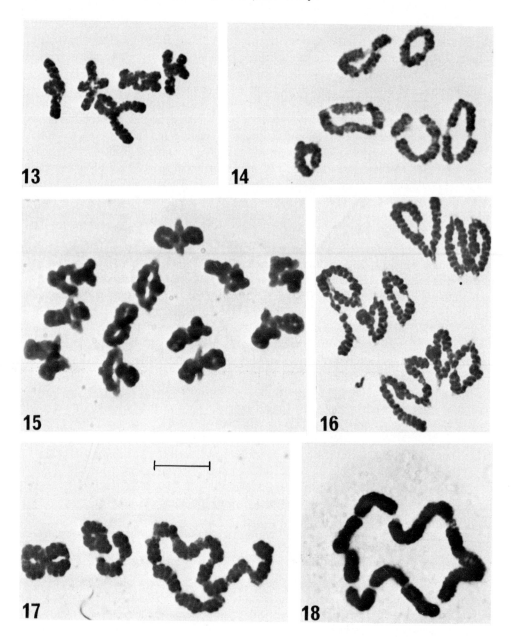

FIGURES 13—16   Chiasma localization and ploidy level. Figure 13: *G. speciosa* (2*n* = 10); 14: *G. rhodantha* (2*n* = 20); 15: *T. ambigua* (2*n* = 24), 12 bivalents, with proximal, interstitial and terminal chiasmata; Figure 16: *T. pinetorum* (2*n* = 24), 5 IV + 2 II, with terminal chiasmata only. FIGURES 17—18   Interchange heterozygosity. Figure 17: *G. consobrina* (2*n* = 20), 1 XII + 1 IV + 2 II; Figure 18: *G. pulchella* (2*n* = 10), a complex interchange heterozygote forming a complete ring of 10 chromosomes.

have relatively large chromosomes, a basic number of *x* = 6 and karyotypes with a preponderance of acrocentrics (Figure 6) but with clear variation in constitution from species to species. *Tripogandra*, with rather smaller chromosomes, has basic numbers of *x* = 7 and *x* = 8 and highly asymmetric and variable karyotypes.

The last group of genera comprises *Thyrsanthemum*, *Gibasoides*, and *Matudanthus*. These are considered in their inflorescence structure to be the most primitive in the Tradescantieae, or as Hunt puts it,[14] they may be relics of an evolutionary stage which gave rise to the Tradescantieae. They have very similar karyotypes of small asymmetric chromosomes and uncertain basic numbers.[15] Their chromosome numbers of $2n = 28$, 30, and 32 (Figure 9) may be polyploid or polyploid derivatives but could equally well represent high basic numbers of $x = 14$, 15, and 16.

The description of karyotypes given above should suffice to show that chromosome evolution in the tribe has involved major changes of chromosome dimension, shape, and number. Karyotypes may be conservative, allowing either no change in morphology or only clearly permissible types of modification. In some genera there is no evidence for the imposition of karyotype conformity. The largest and most symmetrical karyotypes are associated with advanced inflorescence types; small and asymmetric complements occur in those with the most primitive inflorescence features.

Although there are these general features of karyotype variation in the tribe, the location of large and interesting differences in species groups so differentiated makes it most unlikely that we shall be able to determine the mechanisms which produced the different situations or the pathways of evolution involved in the evolution of the species. The cytological method is most effective when dealing with close relatives which leave clearer indications of their past, and it is to these that we must turn for examples of the types of change which may have occurred elsewhere.

III. POLYPLOIDY

It has been estimated that the Commelinaceae comes second only to the grasses in its content of weedy species and shows a very high frequency of polyploids. In the Tradescantieae a survey of more than 80 species spread across all its genera shows that at least 76% are either polyploid or include polyploid populations. The great bulk of these are tetraploid, but hexaploids, octoploids, and decaploids are present in some species. These polyploids represent an important type of chromosome change which has initiated, accompanied, or followed evolutionary discontinuities.

Meiotic behavior in the *Tradescantia* species of the north, in *Cymbispatha* and *Gibasis* is most frequently of the autopolyploid type in tetraploids (Figure 16), hexaploids or the rare octoploid. Their high frequencies of multivalents contrast with the smaller number of polyploids which are either purely bivalent-forming or have low numbers of multivalents. In an intermediate position are those which show multivalents of a higher order than the level of polyploidy due to interchange heterozygosity (Figure 17) which may reflect initial differences between genomes or subsequent changes. There is some evidence as in the case of the Andean *Tradescantia ambigua* that bivalent pairing (Figure 15) is controlled by autosyndesis or specific genes. Although chromosome doubling within the species appears to be a common form of polyploid change throughout the tribe, hybridity has also played its part.

An interesting comparison can be made between chiasma frequency and disposition in diploids and the cytological autopolyploids in the same species or genus. Martinez studied this in Mexican *Tradescantia* species,[8] and we have made the same sort of analysis in populations of *Gibasis karwinskyana* from different regions. In both genera chiasma frequency per paired arm is markedly higher in diploids and the terminalization coefficient lower. In the tetraploids, chiasmata are usually limited to one per arm pair at the terminal position as shown for *Gibasis* (Figure 19). While there is a general correlation between chiasma frequency and terminalization ($r = -0.69$ with 17 d.f.; $p = 0.001$) the data fall into distinct groups determined by regional distribution and ploidy level. The most obvious clustering is that of the polyploids of the eastern race with low chiasma frequency and high terminalization. The diploids are quite distinct, but within their limits the eastern diploids

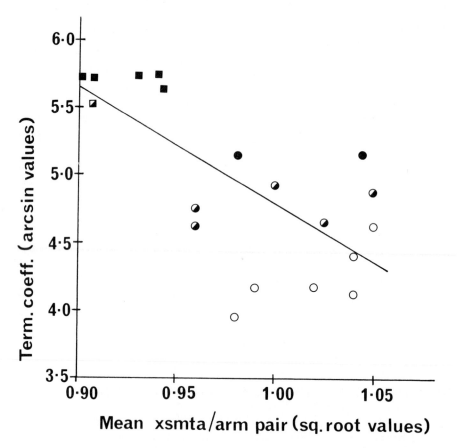

FIGURE 19   Regression of terminalization on chiasma frequency in populations of *G. karwinskyana* from different localities. ○ western diploids; ◕ intermediate diploids; ● eastern diploids; ■ eastern polyploids (4*x* and 6*x*); ◩ intermediate tetraploids. The regression line is based on projected values for known chiasma frequencies and is represented by $y = 13.1 - 8.31x$; $r = -0.69$, $p = 0.001$.

show the highest terminalization and the western the lowest. This difference may be due to the distinctive distribution of C-band heterochromatin in the two races.[16]

In both the *G. karwinskyana* and *G. linearis* alliances, polyploids generally have well-defined geographical limits. In *G. karwinskyana* they are limited to the north and east of the species distribution, and are typical cytological autopolyploids, apart from minor C-band polymorphisms. Diploids, tetraploids, and hexaploids of the eastern race are similar in all respects, and the F_1 between them shows regular autopolyploid meiotic behavior. In this case, the higher ploidy levels are almost certainly derived directly from the sympatric diploids, their visible relationship with these reflecting either their recent formation or their immutability over long periods of time. The latter would be consistent with Ehrendorfer's idea that autotetraploids arise in stable environments where they have a developmental advantage, and remain unchanged while related diploids continue to mutate and evolve,[17] as in forming the western race of *G. karwinskyana*.

Tetraploids of the *G. linearis* group tend to be confined to the Sierra Madre Occidental, on the west of the range in Mexico. Although they exhibit frequent quadrivalents, the maximum number is formed only rarely or not at all. Hybrids between them exhibit a high degree of autosyndesis with expected high fertility, but also show some multivalent formation which may be indicative either of parental interchanges or of residual homology between the chromosome sets.

In this group, tetraploids seem most likely to be segmental allopolyploids whose selective

Table 1
DNA AMOUNT IN TRADESCANTIEAE

Species	Altitude (m)	x	2n	No. of replicates (plants)	x̄ 2C DNA (pg)	DNA/ genome
Gibasis karwinskyana	1365—2000	5	10	12	17.01—19.82	8.5—9.9[a]
	1350	5	20	6	34.06	8.51
G. consobrina	1350	5	10	2	22.00	11.00
	1800—1900	5	20	4	40.66	10.16
G. schiedeana	1500—1600	5	10	4	17.78	8.89
	1300	5	16	4	32.23	8.05
G. pulchella	2200—2850	5	10	8	42.10	21.05
G. matudae		5	10	4	35.55	17.78
G. speciosa	2650	5	10	2	20.95	10.46
	2150	5 + 6	22	4	40.94	10.23
G. rhodantha	2100	6	12	1	22.82	11.41
	2300	6	20	2	38.15	9.54
G. linearis	2500	6	12	3	20.38	10.19
	2500	6	18	1	31.64	10.55
G. heterophylla		6	12	1	17.16	8.58
G. venustula ssp. robusta	1900	6	12	5	17.31	8.66
	1274—2123	6	12	31	10.87—17.30	5.44[a]
						8.65
G. aff. rhodantha		5—6?	32	1	54.75	?
Tradescantia llamasii[c]	700	6	12	4[b]	23.40	11.7
T. crassifolia[c]	2600	6	12	2[b]	21.50	10.8
T. pallida[c]	600	6	12	2[b]	16.50	8.30
T. rozynskii[c]	1820	6	12	3[b]	12.20	6.60
T. tepoxtlana[c]	1600	6	12	4[b]	11.80	5.90
T. subaspera		6	24	1	79.86	19.97
T. fluminensis		10	c. 67	1	11.99	2.00
Phyodina navicularis	2062	8	32	1	51.43	12.93
P. rosea		6	24	1	77.31	19.33
Thrysanthemum floribundum		8	32	1	14.47	3.62

[a] See Table 2.
[b] Replicates = roots of one individual.
[c] Data of Martinez[8]

advantage rests in conferring fertility on hybrids between chromosomally or genetically differentiated genomes. This may also be the case in *G. consobrina* where the dominant tetraploid cytotype, distributed over a wide range in southern Mexico, is always heterozygous for one or more interchanges, whereas in *G. karwinskyana*, autopolyploids probably owe their success more to their increased gene dosage, affecting their size and competitive ability.

IV. VARIATION IN DNA AMOUNT

Karyotypic studies alone indicate that there must be large differences in DNA amount per genome. Microdensitometry measurements have so far been limited to *Gibasis* and a few species of *Tradescantia*, *Phyodina*, and *Thrysanthemum*.

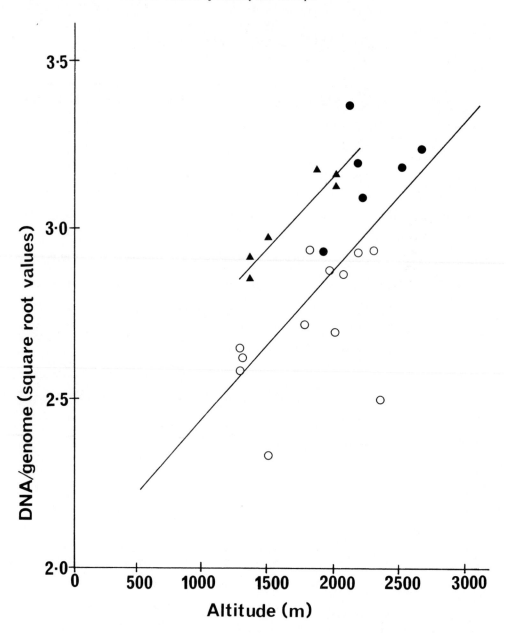

FIGURE 20 Linear regression of DNA content on altitude in *Gibasis*. Species of the *G. linearis* complex (●) and populations of *G. venustula* ssp. *venustula* (○), belonging to the same group, have been amalgamated in the regression line $y = 2.23 + (3.19 \times 10^{-4}) \times$ $(r = 0.66)$. The regression equation given by *G. karwinskyana* (▲) is $y = 2.32 + (4.28 \times 10^{-4}) \times$ $(r = 0.94)$. Both are highly significant $(0.01 < p > 0.001)$.

The lowest amount has been found in *T. fluminensis* (all small chromosomes and a bimodal karyotype) and *Thyrsanthemum*, seen by some as the most primitive genus in the tribe. The highest values are found within the $x = 6$ tradescantias with large metacentrics, in the *Gibasis pulchella* group, also with metacentrics, and in *Phyodina rosea*. However, within the $x = 6$ tradescantias from Mexico there are very large differences in DNA amount ranging from 5.9 to 11.7 pg per genome.[8]

Gibasis has been examined in some detail, measurements being made of each of the species available. In Table 1 the species are grouped together according to shared karyotype pattern, in harmony with the taxonomist's view of their relationships.

Table 2
DNA CONTENT IN *G. KARWINSKYANA*

Sample	Altitude (m)	2n	No. of replicates (plants)	$\overline{x}$ 2C DNA (pg)	DNA/genome
G	1500	10	2	17.68	8.83
H	1365	10	2	16.34	8.17
B	2000	10	5	19.35	9.68
D	1850	10	2	20.17	10.09
A	2000	10	4	20.01	10.00
J	1350	20	6	34.06	8.52

Table 3
DNA CONTENT IN *G. VENUSTULA* SSP. *VENUSTULA*

Sample	Altitude (m)	No. of replicates (plants)	$\overline{x}$ 2C DNA (pg)	DNA/genome
A	2059—2077	3	16.47	8.24
B	2123	3	17.18	8.59
C	1800	2	17.30	8.65
D	2275	1		8.63
E		2	16.48	8.24
F	1941	2	16.60	8.30
G	1289	1	14.09	7.05
H	1274	2	13.27	6.64
J	1304	2	13.77	6.89
K	2000	6	14.62	7.31
L	1789	1	14.76	7.38
M	2366	2	12.70	6.35
N	1500	4	10.87	5.44

In the *G. karwinskyana* group, the lowest DNA content per genome is present in the tetraploid *G. schiedeana* with its basic number reduced to $x = 4$ by Robertsonian fusion, and the highest in diploid and tetraploid *G. consobrina*. In *G. karwinskyana*, a species known to be polytypic for C-heterochromatin,[16] consistent differences in DNA content were found between populations, when DNA amount was correlated with altitude (Figure 20). Samples from the lowest altitudes (G, H, and J, Table 2) constitute part of a race distinct in phenotypic and ecological characteristics.

In the *G. linearis* group, with both $x = 5$ and $x = 6$, and a probable combination of them at polyploid levels,[18] the overall picture is of DNA amounts comparable to those in the *karwinskyana* group, the range of values also being similar to those in the Mexican tradescantias based on $x = 6$. In the two closely related species, *G. speciosa* (Figure 13) and *G. rhodantha* (Figure 14), whose four cytotypes differ by Robertsonian fusion, the lowest DNA value is in *G. rhodantha*, $(2n = 20)$, consistent with the loss of four short arms.

In 13 populations of *G. venustula* ssp. *venustula*, DNA amount showed a wide variation between 5.44 and 8.24 pg per genome, (Table 3), representing an east-west cline, with higher DNA values concentrated in the west of the distribution, where they are equalled by that of ssp. *robusta*, a phenotypically distinct and allopatric subspecies distributed much further west. Eastern populations of ssp. *venustula* are distinct in their phenotype and ecology, as in *G. karwinskyana*, but these features are not directly coincident with DNA differences. Populations differing in DNA content did not readily hybridize, but the few hybrids obtained showed clearly heteromorphic bivalents (Figure 21) and almost complete sterility, in contrast

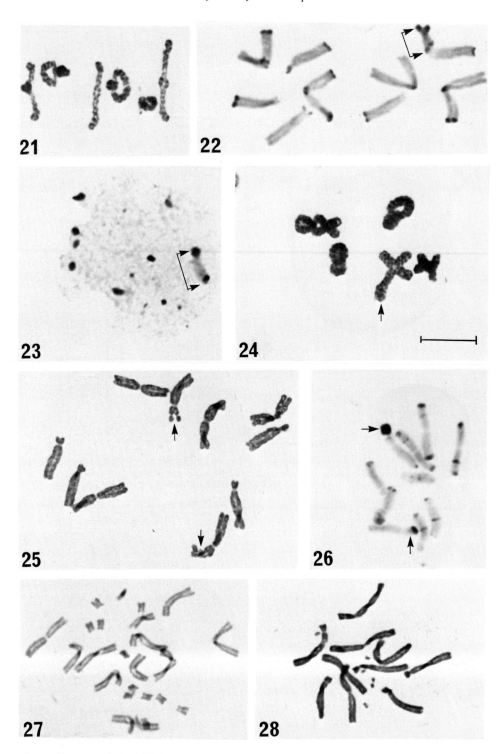

FIGURES 21—28   The accommodation of additional DNA. Figure 21: *G. venustula* F₁ hybrid, medium (K) ×
low (N) DNA, with 6 heteromorphic bivalents; Figures 22—24: *G. schiedeana* ($2n = 10$); Figure 22: Incomplete
cell showing large extra segment; Figure 23: Extra segment in C-banded interphase: Figure 24: Heteromorphic
bivalent at meiosis: Figures 25—26: *G. karwinskyana* ($2n = 10$), heterozygous for a supernumerary segment; Figure
27: *G. venustula* ssp. *robusta* ($2n = 12$) × *G. rhodantha* ($2n = 12 + 3B$) with 12 acrocentric B chromosomes; Figure
28: *G. karwinskyana* ($2n = 10 + 6B$).

to the F_1 between races of *G. karwinskyana,* where chiasma frequency was decreased but fertility only slightly reduced in most cases. This less disruptive effect may be due either to the relatively smaller amount of DNA discrepancy, or to the fact that in *G. karwinskyana,* the additional material is heterochromatic as shown by C-banding. On pooling populations of ssp. *venustula* with other species in the *G. linearis* complex, a highly significant and positive correlation emerges between DNA amount and altitude (Figure 20). The same situation applies in *G. karwinskyana,* plotted on the same graph. These results suggest that DNA increase is an adaptive change possibly having its effects by influencing cell cycle time and hence rate of development under different environmental conditions as suggested by Bennett.[19] Although plants which differ in DNA amount cannot interbreed successfully, it seems likely that it is a change which takes place following isolation of populations and not one which initiates disruption of interbreeding populations. More DNA measurements in the Tradescantieae may provide a more adequate understanding of the origin and influence of changes in amount.

V. INTERCHANGE HETEROZYGOSITY

A. Floating Interchanges

Heterozygosity for a single interchange occurs sporadically in the Tradescantieae as occasional mutants not of adaptive significance, though for the great majority of species there are insufficient individuals examined to allow any estimate of frequency of occurrence. Comment will therefore be confined to those few examples where interchange heterozygosity is relatively prominent and seems either to be adaptive or to reflect on evolutionary matters.

Gibasis, one of the most studied genera, also shows most evidence of interchange. This is particularly prominent in *G. pulchella* ($2n = 10$) whose populations are polymorphic. Some show complete rings of all ten metacentrics and are complex interchange heterozygotes (Figure 18). Its companion species with a similar karyotype, *G. matudeae,* has not been examined as extensively but small collections from two Mexican populations contained some plants heterozygous for one and in one case two interchanges. Both this species and *G. pulchella* show other features of chromosome variability which require closer study.

In the *G. karwinskyana* group of species, interchange heterozygosity is particularly prominent in the autotetraploid species *G. schiedeana* and *G. consobrina.* Four of six plants of the former have associations of more than four chromosomes while each of 11 of *G. consobrina* from three populations are of this type. Interchanges have also been found in tetraploid *G. karwinskyana* but on the present limited evidence they have not been as frequent. In none of these cases it is possible to know whether the interchanges have arisen after the origin of the species and have some adaptive significance in the control of recombination, or whether they reflect original and maintained differences between ancestral sets in the same way as for the grass *Anthoxanthum odoratum.*[20]

In the *G. linearis* group, floating interchanges have been found sporadically in diploids and polyploids in four of the six species, being commonest in the two subspecies of *G. venustula* where five of a total of 18 plants, all diploid, were heterozygous for one interchange.

Interchange heterozygosity is very frequent in *Tradescantia crassifolia.* Martinez found this condition in over two-thirds of his sample of the cytological autotetraploid cytotype gathered from nine sites.[8] Five from his much smaller sample of ten diploids were also heterozygous for a single interchange. Although in this species interchange is obviously of some significance, we cannot yet say what this is.

B. Complex Interchange Heterozygosity

Rhoeo spathacea (Swartz) Stearn ($= R. discolor$ [L'heritier] Hance) is the familiar example of complex interchange heterozygosity in the Commelinaceae. It forms rings and chains of all its twelve chromosomes and has been studied by many since first noted by Belling.[21] Purely bivalent-forming plants were later discovered by Wimber[22] in a mixed collection of *concolor* and *discolor* individuals, which differ in having either purely green leaves or ones with purple lower surfaces. As his bivalent-formers were only of the *concolor* types he suggested that this character could generally distinguish the two types of chromosome behavior. At Kew, however, of three plants collected by Adams in Belize, two, one *concolor* and one *discolor*, were unusual in having the inflorescence on an elongated peduncle. The third plant had the usual short-pedicelled inflorescence and was *concolor*. Despite these differences, all three were bivalent-forming and the possibility now emerges that the occurrence of complex interchange heterozygosity in this species is more a matter of location than phenotype.

Gibasis pulchella is another example of complex interchange heterozygosity in the family. Following the discovery of complete ring-forming plants in Mexico by Handlos,[23] more extensive collections have been examined at Kew. All plants came from central Mexico and could be divided into two types on the basis of their genetic systems. Plants from five populations widely separated in Guanajuato, Hidalgo, and Distrito Federal formed rings of all ten chromosomes (Figure 18), were self-fertile, and had small flowers with some tendency towards cleistogamy. In cultivation they had reduced pollen stainability and low seed-set per capsule. All the progeny produced on selfing two of these plants bred true for the ring-forming character. It seems that they are complex interchange heterozygotes whose condition is preserved by balanced lethal systems.

Plants from three other populations in Hidalgo state were bivalent-forming, self-sterile, and had larger flowers which opened normally. Pollen stainability was high.

Intermediate in chromosome behavior were plants which formed rings of four, a ring of four, and one of six and other multivalents of variable size. Triploid individuals showing signs of interchange heterozygosity were found at several sites.

Hybridization between plants with rings of ten and with bivalents only gave rise to progeny with either a ring of four or a ring of six and generally with many univalents. These parents are differentiated chromosomally in various ways and not solely by interchanges; further studies are now in hand.

C. Interspecific Differences

Distinct populations or species may be differentiated partly by smaller or larger interchanges. Martinez[8] recorded some multivalent formation in several diploid crosses within the *Tradescantia crassifolia* group of species. In *Gibasis*, interchange heterozygosity was frequent in interpopulation hybrids of diploid *G. karwinskyana* and in the diploid *G. venustula* spp. *venustula*. It was also present in hybrids of the two subspecies of *venustula* and in all but one of the remaining diploid and polyploid hybrids in the *G. linearis* complex. The chromosome relationships between the parents were, however, by no means simple since they could differ in DNA amount, by Robertsonian change and inversion in complex and interacting ways. The limited amount of hybridization accomplished in the tribe so far indicates that interchange does play a role in species differentiation but cannot anywhere be seen as a predominant mechanism.

VI. ROBERTSONIAN FUSION

The fusion of long arms of nonhomologous acrocentrics following breaks close to or within the centromere is accepted as a relatively common type of chromosome change during species evolution in animals. It has not been considered to be of such importance in plant evolution, although its involvement has been suggested for a number of species. The Tradescantieae provides clear evidence for the occurrence of Robertsonian fusion in the evolution of species karyotypes in several genera.

The first evidence of its presence in the family as a whole was in *Gibasis schiedeana*[24] where there exist diploid populations with a basic number $x = 5$ and a genome formula 2 M + 3 A and tetraploid populations with $x = 4$ and 3 M + A (Figure 11). The diploids are self-sterile and the tetraploids self-fertile. These plants show the Robertsonian relationship each having seven major chromosome arms in a set. Since the genome constitution of the diploid is common to both diploid and tetraploids in the allied species *G. karwinskyana* and *G. consobrina*, that of the $2n = 16$ *G. schiedeana* is considered to be derived; its self-fertile condition would also support this view. To test whether the additional metacentric of this tetraploid was produced by fusion of the missing acrocentrics it was crossed with the diploid to produce a $2n = 13$ triploid. Meiosis in the tetraploid was of the autotetraploid type. Several types of association between two metacentrics and two acrocentrics were expected in the hybrid if the anticipated Robertsonian homologies were present (Figure 31); all were found. Furthermore, all the remaining chromosomes of the two parents showed a high degree of trivalent pairing. Here then Robertsonian fusion was the main type of structural differentiation upon which polyploidy had been imposed. The diploid and tetraploid showed differences in leaf size and in the extent of rhizome formation but neither feature has been thought adequate to separate them as two species.

In the *G. linearis* alliance, found in more northerly parts of Mexico, there is further and more extensive evidence of Robertsonian change.[18,25] Most species have the basic number of $x = 6$ but others have $x = 5$, both occurring at the diploid and polyploid levels. One population with $2n = 22$ is considered to be a tetraploid combining both these basic numbers. In the $x = 6$ plants the genome consists of two metacentrics and four chromosomes with decidedly unequal arms which in a previous publication were described as three submeta-centrics and one subtelocentric.[25] The latter are here referred to as acrocentric, giving the formula 2 M + 4 A. This constitution, on comparison with that of the $x = 5$ plants, 3 M + 2 A shows the Robertsonian relationship. The constitution of the $2n = 22$ plants, 10 M + 12 A is seen to combine these.

The species with $x = 6$ are *G. linearis, G. venustula* ssp. *venustula* and ssp. *robusta, G. heterophylla* (all $2n = 12$), and *G. graminifolia* ($2n = 24$). Of the two other species, both of uncertain taxonomy, two populations of *G. rhodantha* were $2n = 12$ and $2n = 20$ (Figure 14), while two *G. speciosa* populations were $2n = 10$ (Figure 13) and $2n = 22$. In *G. rhodantha*, then, we find the standard $x = 6$ at the diploid level and the supposed Robert-sonian derivative, $x = 5$, at the tetraploid level; in *G. speciosa,* the diploid represents the fused state ($x = 5$), while the tetraploid has $x = 5 + x = 6$.

Apart from these karyotype comparisons a further demonstration of the reality of Robertsonian change comes from this last $2n = 22$ plant, for here an A-M-M-A quadrivalent is a feature of meiosis (Figure 30). Such pairing testifies to the origin of these metacentrics from acrocentric ancestors.

In addition, artificial F_1 hybrids have been produced between *G. speciosa* ($2n = 10$) and *G. rhodantha* ($2n = 12$) and the same *speciosa* with *G. venustula* spp. *robusta*.[25] In both cases the Robertsonian trivalent A-M-A was observed, although an A-M bivalent was more usual (Figure 39).

Triploid hybrids were produced by crossing $2n = 12$ diploids with *G. rhodantha* ($2n = 20$). Here again pairing of metacentric and acrocentric chromosomes took place as bivalents, trivalents, or quadrivalents similar to those in the *G. schiedeana* triploids.

Tetraploid hybrids produced from crossing $2n = 22 \times 24$, and $2n = 22 \times 20$ were simplex and triplex, respectively, for the Robertsonian metacentric, and the pairing of acrocentrics and metacentrics proceeded in ways to confirm the Robertsonian relationship (Figure 32).

Overall the meiotic analysis of these hybrids between quite distinct species revealed that the parents differed not only in the Robertsonian fusion but also in other structural ways,

Table 4
THE CHROMOSOME
CONSTITUTION OF
CYMBISPATHA **SPECIES**

2n	Meta	Acro	NF
14	—	14	14
12	2	10	14
16	12	4	28
14	14	—	28
36	6	30	42
22	20	2	42
30	26	4	56
28	28	—	56

including interchange and inversion, which could affect the Robertsonian group of chromosomes. This influenced the pairing relationships which proved to be complex and demonstrated increasing differentiation of the complement following what are probably long periods of evolutionary time. This contrasts with the phenotypically similar plants in the *G. schiedeana* example.

Cymbispatha has provided evidence for the occurrence of Robertsonian fusion on a more extensive scale.[5,12] In this genus, successive fusions in association with periods of chromosome doubling have been a major factor in the evolution of karyotypes and species, changing chromosome numbers in a way which confuses origins and relationships.

The somatic chromosome numbers found in *Cymbispatha* are $2n = 12, 14, 16, 22, 28$, 30, and 36. Karyotypes contain only metacentric and acutely acrocentric chromosomes either alone or in mixtures. This allows an easy calculation of the number of major chromosome arms (nombre fondamental) for each chromosome complement (Table 4). The only NF values which emerge are 14, 28, 42, and 56. This harmonious series which contrasts with the generally aneuploid series of somatic numbers suggests an ancestral basic number of $x = 7$ and that the present species are in genetical terms diploid, tetraploid, hexaploid, and octoploid. It also implies that Robertsonian fusions have produced the metacentric chromosomes and are responsible for the distortion of the expected numerical sequence consequent upon polyploidy. The beginning of the sequence of changes could be expected to be a constitution of $2n = 14$ acrocentrics as found in one of the diploid species, although we do not anticipate that this particular one represents the starting point for the evolution of the genus. The first step in the development of the Robertsonian sequence is found in *C. plusiantha* ($2n = 12$), homozygous for one fusion, and so far no further fusions have been found in diploids. At the tetraploid level, we find *C. standleyi* ($2n = 12$ M $+ 4$ T, NF $= 28$) with most chromosomes metacentric and a chromosome number divisible by four, giving a basic number of $x = 4$. In the *C. commelinoides* aggregate the cytotype with $2n = 14$ M is genetically tetraploid though it patently has a diploid chromosome number. Confirmation of its tetraploid constitution comes from its meiotic pairing characterized by quadrivalent formation, up to three quadrivalents being present in some cells of the many individuals examined. The frequency distribution of cells with from 0 to 3 quadrivalents shows that these are the consequences of random pairing of groups of four homologues and not of interchange heterozygosity. Thus, the $2n = 14$ M constitution disguises its state of auto-tetraploidy. Similarly, high degrees of multivalent formation confirm that the $2n = 22$ cytotype is autohexaploid and the $2n = 28$ and 30 are autooctoploids. The $2n = 36$ species shows bivalent pairing but its $2n = 6$ M $+ 30$ T constitution shows it to be hexaploid. Their detailed descriptions have been given in earlier publications.[5,12]

The type of chromosome evolution in *Cymbispatha* is one where a minimal development of Robertsonian fusion at the diploid level is accompanied by the establishment of successive fusions at several polyploid levels. There is no direct evidence that fusions proceeded to the maximum level, i.e., $2n = 6 M + 2 A$ in diploids prior to the formation of tetraploids from them, but for reasons given elsewhere,[5] we deem this as the most likely course of events. If, then, the tetraploids were of autopolyploid origin the four acrocentrics of a $2n = 12 M + 4 A$ plant would be homologous and their fusion to give the $2n = 14 M$ constitution would produce a pair of pseudoisochromosomes. Although this might seem unlikely, one individual of the $2n = 14 M$ cytotype of *C. commelinoides* occasionally formed two ring univalents at meiosis indicating that such chromosomes do in fact form part of the normal complement of the species.

A genus which may prove to be fairly closely related to *Cymbispatha* is *Zebrina*. Two karyotypes have been described, i.e., $2n = 6 M + 16 A/T$ for *Z. flocculosa* and $2n = 4 M + 20 A/T$ for *Z. pendula* and *Z. purpusii*.[26] These plants are essentially bivalent forming, but each can occasionally form a quadrivalent of the A-M-M-A type. NF value for all plants is 28 and *Z. flocculosa* shows a Robertsonian relationship with the other two species. It seems likely on all the available evidence that these species are allotetraploids originally based on $x = 7$, and with some structural homology between a few chromosomes of the ancestors. Robertsonian fusions taking place after tetraploidy could produce the present constitutions and the A-M-M-A quadrivalent as argued earlier.

Mattsson had noted a small satellited acrocentric in some plants and this was seen again in *Z. pendula* var. *quadricolor* by later authors,[27] who considered it to be the small product of centric fusion between a satellited and nonsatellited acrocentric. A chromosome of the same type was present in *Z. pendula* at Kew, studied by Karp.[27a] Two roots of this plant had the usual chromosome constitution, i.e., $2n = 4 M + 20 A/T$, one root had six metacentrics and the small satellited acrocentric and others had five metacentrics and a very small fragment. A tendency to spontaneous fusion between acrocentrics is seen, producing smaller or larger centric fragments depending on the precise nature of the fusing acrocentrics. This propensity in this popular cultivated material could lead to the establishment of various constitutions in cultivation.

VII. FOLD-BACK PAIRING

So-called 'fold-back' or intrachromosomal pairing is the result of chiasmate association between the two arms of a single chromosome which may occur when it is univalent or when engaged in bivalent or multivalent pairing. When present in reasonable frequency, it can be taken to show the existence of homologous segments in the two arms reflecting either recent and random changes or even unsuspected homologies which may characterize the species as a whole. Some hold different views of the significance of fold-backs, particularly in monoploids as discussed by Sadasivaiah.[28]

In the Tradescantieae, fold-back pairing has been seen in some individuals in both species and artificial hybrids. It was first noted in the genetically tetraploid cytotype of *Cymbispatha commelinoides* with $2n = 14$ metacentrics.[5] The presence of a pair of homologous pseudoisochromosomes had to be assumed as the consequence of its fusion origin, and a pair of ring univalents was formed occasionally in one individual to prove the point. If such chromosomes are able to form part of any normal complement, their pairing must be controlled to avoid the potentially disastrous consequences of internal pairing. It follows that the expression of intrachromosomal homologies will only rarely occur and be confined to the occasional individual as in *Cymbispatha* or in other plants where normal pairing controls are disturbed, such as hybrids.

A very clear case of fold-back pairing was found in three plants representing the entire

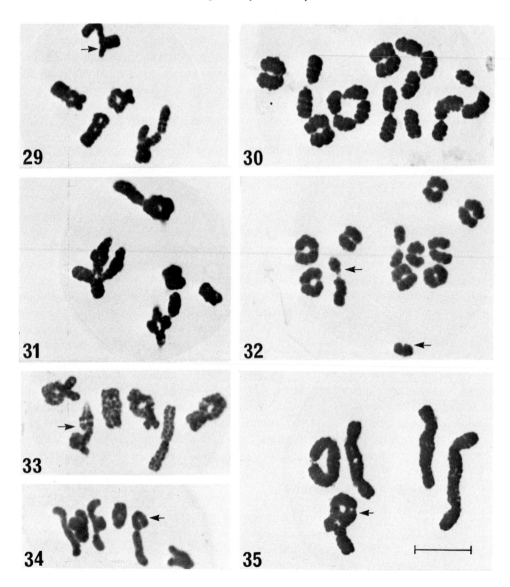

FIGURES 29—35   'Fold-back', fusion, and fission. Figure 29: *G. venustula* ssp. *robusta* ($2n = 12$) × *G. speciosa* ($2n = 10$), heteromorphic M-A bivalent (arrowed), and interchange trivalent; Figure 30: *G. speciosa* ($2n = 22$) A-M-M-A quadrivalent; Figure 31: *G. schiedeana* ($2n = 10$ × $2n = 16$) A-M-M-A quadrivalent; Figure 32: *G. speciosa* ($2n = 22$) × *G. graminifolia* ($2n = 24$), M-A bivalent and A univalent; Figure 33: *T. andreuxii* ($2n = 11$ M + 2 T), A-M-A trivalent; Figure 34: *G.* ssp. *robusta* × *G. heterophylla* ($2n = 12$), fold-back bivalent; Figure 35: *G. pulchella* ($2n = 10$), 'fold-back' bivalent.

sample of one bivalent-forming population of *Gibasis pulchella* ($2n = 10$ metacentrics). Here one chromosome of a bivalent frequently formed an internal chiasma terminally while at the same time forming a subterminal chiasma with its homologue (Figure 35). In addition, univalent frequency was substantial and the plants may be complex in their constitution. The simplest explanation of the fold-backs is the occurrence of an iso-chromosome, possibly as the consequence of misdivision of the centromere in a previous generation as described by Darlington.[29]

In artificial hybrids fold-backs were seen in the *Gibasis linearis* group of species. Here in several hybrids heterozygous for a Robertsonian fusion, it could be seen in one of the chromosomes taking part in the 'Robertsonian' association. It was seen again in hybrids of species of the same chromosome number, which did not differ by Robertsonian change (Figure 34).

In a hybrid of *Tradescantia crassifolia* ($2n = 12$) and *Tradescantia ambigua* ($2n = 24$) a fold-back was frequently observed in a univalent with distinctly unequal arms. These univalents were seen to undergo centromere misdivision to produce two telocentrics. Although no chiasmata could be seen, the fold-back was very likely the result of the formation of a proximal chiasma and its presence in some way stimulated the misdivision of the centromere. Since the univalent had unequal arms it was likely to be an isochromosome derivative.

Finally, ring univalents, up to two per cell, were sometimes found in hybrids of *Tradescantia tepoxtlana* and *Setcreasea pallida*.[8] In this case the plants were asynaptic, and apart from a very rare bivalent the only chiasmata formed were in these univalents.

More detailed consideration has been given to most of these events in an earlier paper,[12] and a full account will be produced on the findings in the *Gibasis* hybrids.

VIII. FISSION

The only clear case of centric fission in the Tradescantieae has been found in two individuals in one population of the diploid *Tradescantia andrieuxii*, with $2n = 12$ metacentrics. Both plants were heterozygous for the same fission through the centromere of one of a pair of satellited chromosomes, and both also possessed B chromosomes in some cells.

Chiasma frequency was high in both the fission heterozygotes with proximal, interstitial, and terminal chiasmata. The telocentrics paired with their metacentric counterpart producing T-M bivalents or the T-M-T trivalent (Figure 33), but the former were twice as common as the latter. Chiasmata in these heteromorphic associations were proximal or terminal with the proximals predominating. Although pollen grain mitosis showed that few departed from the balanced constitutions of 6 M and 5 M + 2 T, pollen stainability was reduced to 50%.

In the only $2n = 12$ plant from the same population, 40% of the pollen mother cells had two univalents and in most of the rest, one bivalent was conspicuous in being rod shaped with a single chiasma, contrasting with its multichiasmate companions. This observation, coupled with the pairing behavior in the fission heterozygotes, suggests that this bivalent results from the pairing of two chromosomes which differ substantially in a large terminal segment in one arm. The reduced pairing produces the high univalent frequency and ample opportunity for occasional misdivision. The two heterozygotes could result from this error.

IX. C-BANDING

Species of the Tradescantieae studied in some detail are limited to *Gibasis, Cymbispatha, Callisia*, and *Zebrina*. A few species of the *Tradescantia crassifolia* complex have been examined, but here the bands were very small and confined to secondary constrictions.

The most widespread pattern is one of centric and terminal bands, occurring in most species of *Cymbispatha*, in *Zebrina, Callisia, Gibasis schiedeana*, and some populations of *G. karwinskyana*, (Figures 36-37, and 40). An interesting feature of the centric bands in some species is their double nature, which may have important implications where fusion is involved, for example in *Cymbispatha plusiantha*, in which the double bands on the two metacentrics reflect large pericentric blocks on the acrocentrics.[5] Holmquist and Dancis suggested that fusion could lead to the formation of a dicentric in which the two centromeres were very close together and would therefore function as one,[30] and in the case of *C. plusiantha* the double centric bands may represent the two centromeres of the original acrocentrics. In *C. standleyi*, the bands on the telocentrics, which are close to, but not in

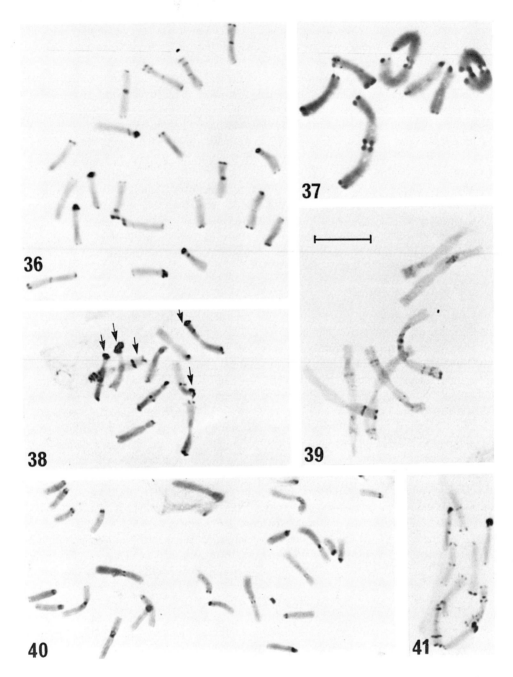

FIGURES 36—41   C-banding patterns in *Callisia* and *Gibasis* species. Figure 36: *G. karwinskyana* ($2n = 20$), an eastern race tetraploid with centric bands; Figure 37: *G. matudae* ($2n = 10$); Figure 38: *G. consobrina* ($2n = 10$) × *G. karwinskyana* (western race, $2n = 10$). *G. karwinskyana* chromosomes are arrowed; Figure 39: *G. pulchella* ($2n = 10$), 7 chromosomes only; Figure 40: *Callisia* sp. ($2n = 24$); Figure 41: *G. karwinskyana* (western race, $2n = 10$), 9 chromosomes only.

contact with, the centromere region, exactly match those of the metacentrics, reflecting their fusion relationship with the latter.

The Mexican species *G. karwinskyana* is polytypic for C-band patterns.[16] Populations in the states of Nuevo Leon and Tamaulipas have mainly centric and terminal bands, with only an occasional interstitial band (Figure 36). These plants are also characterized morphologically by vegetative characteristics, and cytologically by a high degree of polyploidy, tetraploids being the most frequent, with occasional diploids and hexaploids.

Tetraploids and hexaploids show some heteromorphy, both for a single interstitial band on one group of metacentrics and for some of the centric bands. Although the chromosomes are homologous with respect to their gene sequences, either C-heterochromatin has been generated during evolution, or the tetraploid is the result of doubling in a hybrid between two closely related, but not identical, diploids. Insufficient diploid plants have been obtained from the eastern race to enable us to test the validity of these proposals; however, the plants are all cytological autopolyploids and show no evidence of preferential pairing between identical homologues as found in grasshoppers.[31] In addition, the five diploids examined to date have lacked interstitial bands, whereas the polyploids all have at least one, making it likely that differentiation has occurred at the tetraploid level, rather than in ancestral diploids.

Populations on the west of the distribution, in the states of San Luis Potosi and Hidalgo, show mainly interstitial bands, usually with a complete absence of centrics (Figure 41). These plants are all diploids, which, combined with their unique C-banding, morphological, and distributional characteristics, and distinct localization of chiasmata, separates them as a discrete race. They are situated on the eastern cusp of the Sierra Madre Occidental, occupying regions of higher altitude than the eastern race, and show a corresponding increase in DNA content.

Between these two extremes are three populations in which C-bands are few in number and generally confined to terminal regions. Two populations are intermediately placed between the eastern and western races while the third is situated close to one of the western diploids, in Hidalgo. The terminally banded populations show various unusual cytological features and may be either evolutionary relics or remnants of interracial hybrids. The autopolyploid nature of the eastern tetraploids and hexaploids, and their close relationship to the diploids of the same race, can be demonstrated by the observation that, within the eastern race, cytological autotriploids and autopentaploids can be produced by hybridization, whereas western diploids will not easily hybridize with western polyploids and do not yield viable progeny.

Diploid hybrids between populations with the widest differences in C-banding, DNA content, and chiasma position showed decreased pollen stainability (46% and 72%) and a decrease in chiasma frequency effected by an increase in univalents and rod bivalents. F_2 progeny showed a wide range of pollen stainability, ranging from 44.4 to 93.7%. These results demonstrate genetic divergence of both eastern and western races and the intermediate form, although they are still partially interfertile.

In *Gibasis consobrina*, band patterns are almost exclusively terminal, the terminal blocks being much heavier than in the closely related *G. karwinskyana* (Figure 38). *G. consobrina* is confined to areas south of the trans-Mexican volcanic fault, one of the main features distinguishing it from *G. karwinskyana*, and shows an interesting comparison with *G. matudae*, which is also characterized by terminal bands (Figure 37) and a southern distribution. *G. pulchella*, closely related to *G. matudae*, has a more northerly distribution, covering a similar range to *G. karwinskyana* and like the latter species, is multibanded (Figure 39). The coincidence of C-band characteristics with distribution in both pairs of species may offer support for the same type of C-band differentiation having occurred more than once.

X. SUPERNUMERARY SEGMENTS

Supernumerary segments have been found in only two species of the Tradescantieae, namely, *Gibasis karwinskyana* and *G. schiedeana*, and are of some interest in themselves as possible examples of amplification. They are well demonstrated in a single population of *G. karwinskyana* ($2n = 10$) whose basic karyotype comprises 4 M = 6 A with one pair of acrocentrics having satellited short arms.

In a sample of 20 individuals from Santa Maria del Rio, Mexico, only 3 showed normal-sized satellites, the remainder being polymorphic for enlarged satellite regions of varying sizes, one of which was sufficiently large to transform the satellited acrocentric into a metacentric (Figure 25). The extra segment could be seen at prometaphase to consist of two tandemly repeated blocks, which were always C-heterochromatic and characteristically appeared as two fused chromocenters or as an attenuated, heavily-staining block. The supernumerary segments gave rise at meiosis to heteromorphic bivalents but never formed chiasmata and had no detectable effect on chiasma frequency. However, the population carrying them showed the lowest terminalization coefficient of 10 samples examined. This may be significant in the light of previous observations by John in grasshoppers, of the presence of terminal heterochromatic blocks resulting in fewer terminal chiasmata.

The low number of plants with normal satellites suggests that this polymorphism is maintained in the population either by clonal reproduction or by selection. In unpretreated mitoses, the extra segments tended to lag or form bridges, raising a possibility that different sized satellite regions could result from breakage and fusion events as suggested by Marchant and Brighton for *Ranunculus ficaria*.[33] This, coupled with intraplant variation for segment size, could be explained on a basis of clonal transfer through the population, which seems possible in a tuberous perennial such as *G. karwinskyana*. However, the preferred niches of the plants in rock clefts and the presence of B chromosomes in only one individual argue against the whole population representing a single clone.

Variation in the size of NORs is now well known and may be attributed to unequal crossover events. In *G. karwinskyana*, after initial amplification of the satellite regions by unequal SCEs, unequal crossing over at meiosis may have been responsible for the wholesale transfer of heterochromatic blocks, explaining the tandem arrangement at mitotic interphase and prometaphase.

Similar satellite polymorphism was present in two individuals of *G. schiedeana* ($2n = 10$) with a karyotype identical to that of *G. karwinskyana*, while in a third, from a different population, one of the nonsatellited acrocentrics was replaced by a submetacentric, identifiable by its unique C-band pattern (Figure 22). In this case, the supernumerary segment was not C-heterochromatic, but appeared in all C-banded interphases as a region of only slightly denser Giemsa staining flanked by two chromocenters (Figure 23). The possibility of the deviant chromosome being an isochromosome has been discarded on the grounds of its unequal arms, the presence of an interstitial band in one arm only, and the fact that the region proximal to the interstitial band is itself the size of a normal short arm. At meiosis, heteromorphic bivalents involving an acrocentric and the submetacentric were regularly formed (Figure 24). Chiasma frequency was higher than normal for *G. schiedeana* ($x = 2.17$ chiasmata/arm pair compared to 1.56/arm pair in 4 diploids from other populations). This is in contrast to *G. karwinskyana*, where plants with extra segments affect the distribution, but not the frequency, of chiasmata. The difference perhaps is in the nature of the extra material, which in *G. karwinskyana* may be supposed to contain tandem repeats and few genes, affecting chiasma distribution mechanically, whereas in *G. schiedeana*, its C-negative reaction suggests that it is euchromatic, and effects on pairing may be due to gene control.

XI. SUPERNUMERARY CHROMOSOMES

These are represented in only a few species of the Tradescantieae, mainly in *Gibasis, (G. karwinskyana, G. schiedeana, G. linearis,* and *G. rhodantha*) and *Tradescantia (T. cirrifera, T. andreuxii, T. crassifolia,* and *T. acaulis)*.

B chromosomes in *Gibasis* are of three types, all smaller than the A complement: small with indeterminate morphology, as in *G. karwinskyana* (Figure 28), medium-sized acrocentrics as in *G. rhodantha* and *G. linearis* (Figure 27), and larger telocentrics as in *G. schiedeana*. They may number from one, in a population of *G. schiedeana* to 17 in a single plant of *G. karwinskyana*.[13,18,34] In *G. karwinskyana*, and *G. linearis* where C-banding has been carried out, the B chromosomes appear to have very little C-heterochromatin.

Brandham and Bhattarai found that in *G. linearis*,[34] the number of B chromosomes was positively correlated with chiasma frequency both between and within individuals. Studies on interspecific hybrids have revealed that in these, the number of Bs is not correlated with chiasma frequency, but in some hybrids a very small B appears, as a result of deletion or centric misdivision of a B univalent, as demonstrated in *Allium schoenoprasum*.[35] Plants containing this diminutive B have considerably lower chiasma frequencies than those without it, and it is tentatively suggested that the loss of chromosome material may be responsible for the decrease, as seen in millet,[36] where deficient Bs had a suppressive effect on chiasma frequency.

The B chromosomes in *G. linearis* and *G. rhodantha* have a strong tendency for accumulation on the male side, as revealed by interspecific crosses with OB parents as female. In such hybrids up to 13 B chromosomes may be present (Figure 27), sometimes including a deficient, and often forming multivalents, with frequent nondisjunction.

In *G. schiedeana* ($2n = 16$), where there is evidence of Robertsonian fusion, the large telocentric B chromosome in one population was comparable in size to one of the metacentric arms.[13] In a second population, the Bs, up to 6 in number, were again telocentric and approximated in size to acrocentric short arms.[24] The presence of these supernumerary chromosomes in plants whose complements have been modified by fusion suggests that they may have been by-products of this event.[12] In the *G. linearis* group, the diploid, nonfusion complement ($2n = 12$) carries B chromosomes, and while fusion does occur in some diploids and tetraploids of the group and could have been represented in the ancestry of the $2n = 12$ cytotype, there is in this case no direct evidence to suggest that the B chromosomes are derived from fusion events.

XII. CONCLUDING REMARKS

The genera brought together in the Tradescantieae show such diversity in their karyotype characteristics that there is little to suggest what their evolutionary relationships may be. Even within some genera as presently conceived there are major chromosome discontinuities with no indications of common origin or clues to mechanisms responsible for the differences. Although by and large the dismemberment of *Tradescantia* has segregated chromosome differences, our attention is attracted most to those clusters of species sharing common karyotype patterns which indicate close evolutionary relationship. The more northerly *Tradescantia* species are segregated in this way from the *T. fluminensis* group of South America and they share their highly symmetrical karyotype with all the species of *Setcreasea*. This affinity is reflected in the recent reunion of this genus with *Tradescantia*.[3] The other species clusters occur within *Gibasis* though there are four distinct karyotype groups without any obvious cytological relationship to each other.

Similarity of karyotype pattern does not necessarily indicate high degrees of structural or genetic homologies and, indeed, can conceal a whole spectrum of differences which at one extreme may allow complete pairing in hybrids and at the other prevent pairing altogether. The preservation of karyotype pattern in the face of various types of differentiation of individual chromosomes implies that pattern is important, presumably for its influence on development and/or heredity or on the sequence of interphase chromosome associations.[37] Although pattern appears to be conserved within groups, it is clearly not inviolate, for in both the *Gibasis karwinskyana* and *G. linearis* groups it is modified, but only by Robertsonian fusion. This type of change presumably does not adversely affect development and conceivably would not alter sequences of interphase association. But it does have profound effects on heredity, severely restricting interchromosomal recombination and establishing new linkage relationships. In *Cymbispatha* Robertsonian fusions played a major role in repatterning of karyotypes though again it is possible to discern a degree of karyotype control which permits only acute acrocentrics and their fusion products.

Overlying karyotype patterns in the Tradescantieae are quantitative changes in total chromatin produced by increase (or decrease) of DNA amount per chromosome and by polyploid changes. DNA variation between species was demonstrated in both *Tradescantia*[8] and *Gibasis*. In *G. venustula*, significant differences were found between populations. The correlation in this last case between DNA amount and altitude suggests that modifications of DNA quantity are in some way adaptive. It is clear also that they can be disruptive, preventing successful interbreeding when parents differ in this way. For our present purposes, however, we need stress only the fact that chromosome size can change appreciably as a result of 'internal' adjustments and these can also affect relative chromosome size within complements depending on how the additional DNA is distributed.

The very high frequency of polyploids, most with high levels of multivalent pairing, adds significantly to the number of such plants known. Chromosome doubling is then a very common event in the tribe, producing chromosome polymorphism in some species and leading to speciation elsewhere. Allopolyploidy has its part to play in the speciation process as judged by low frequencies or absence of multivalents, but genetic or other controls to pairing seem to occur as in *T. ambigua*.

Polyploidy of course has many advantages associated both with the replication of sets *per se*, in the control of recombination and the restoration of fertility, but it also adds large amounts of DNA to complements and endows plants with the capacity for change of another type. In *Cymbispatha*[5,12] polyploidy in combination with Robertsonian fusion gave rise to plants with $2n = 14$ metacentrics as compared with $2n = 14$ smaller acrocentrics characteristic of the diploid ancestor. Because of the availability of favorable evidence we can understand the mechanism underlying this difference. Without it, however, we would be faced with another case of increase of DNA amount per genome. In looking at the obvious differences in chromosome size in the Tradescantieae there are then two potent mechanisms which can produce these results. They may occur together in the same species group or in the same complement and could be responsible for major differences between karyotypes of species which seem phenotypically related.

Only Rohweder[2] has attempted to assign genera to positions in an evolutionary tree of the Tradescantieae. It may be significant to the chromosome evolutionist to note that he places *Thyrsanthemum* and its allies in a primitive position close to that where the Tradescantieae and Commelineae may have diverged. *Tradescantia*, because of its highly advanced inflorescence, is placed at a high point. *Gibasis* occupies a position well below it and below all the remaining genera of the tribe. Thus primitiveness is associated with small acrocentric complements with high basic numbers and advancement with larger chromosomes, high degrees of symmetry, and low basic numbers in Rohweder's scheme. It would be premature to conclude that this is how chromosome evolution took place in the tribe but this possibility must be taken into account. In this connection the largest and most symmetrical complements in *Gibasis* are found in *G. pulchella* and its allies whose inflorescence is considered the most advanced in the genus.

We have singled out above only some aspects of chromosome variation in the tribe. These may add in a small way to a better appreciation of possible ways in which complements can evolve. What is evident throughout, however, is that only detailed studies of natural populations can give the answers that we are looking for.

While the assignment of taxonomic rank and groupings is based primarily on phenotypic characters and regional distribution, it is usual for the taxonomist to take chromosomes into account wherever possible. In the Tradescantieae there are many associations between chromosome and phenotypic distinctiveness and species have been place in alliances on both bases. When the taxonomic dismemberment of *Tradescantia* resulted in the unconscious segregation of major chromosome differences,[7] it could be considered that the chromosomes vindicated the taxonomists' judgement, but there can be many opinions on the definition of a genus and indeed on the desirability of using the genus as the most useful means for expressing species affinities. Hunt has preferred the larger genus and has partly reconstituted *Tradescantia*, grouping species into series and sections.[4] This treatment may cause some confusion to those not familiar with the processes of taxonomy and nomenclature, and there can be little doubt that chromosome cytologists would be happier to see species grouped as genera where the evidence of the karyotype would seem to indicate monophyletic relationships.

ACKNOWLEDGMENTS

We wish to thank Mrs. C. Colden, Mr. J. B. Smith, Mrs. C. Parry, and Dr. Angela Karp for their assistance in the chromosome analysis at various times during the last 15 years. We are grateful to Dr. A. Martinez for access to his extensive unpublished data on the cytology of Mexican species of *Tradescantia*. Information on breeding systems for many species has been continuously supplied by Dr. S. Owens; studies of the breeding system of *G. pulchella* were contributed by Dr. A. Cawood.

We are grateful to Mr. D. Hunt of the Kew Herbarium for dealing with all matters taxonomic and for building up a living collection whose early increase was due to a generous contribution of plants by the late Dr. H. E. Moore of the Bailey Hortorium, Cornell.

REFERENCES

1. **Brenan, J. P. M.,** The classification of Commelinaceae, *Bot. J. Linn. Soc.*, 59, 349, 1966.
2. **Rohweder, O.,** Die Farinosae in der Vegetation von El Salvador, *Abh. Gb. Auslandsk., Reihe C, Naturwiss.*, 18, 98, 1956.
3. **Hunt, D. R.,** The reunion of *Setcreasea* and *Separotheca* with *Tradescantia*, American Commelinaceae. I, *Kew Bull.*, 30, 443, 1975.
4. **Hunt, D. R.,** Sections and series in *Tradescantia*. American Commelinaceae. IX, *Kew Bull.*, 34, 437, 1980.
5. **Jones, K., Kenton, A. and Hunt, D. R.,** Contributions to the cytotaxonomy of the Commelinaceae. Chromosome evolution i *Tradescantia* section *Cymbispatha*, *Bot. J. Linn. Soc.*, 83, 157, 1981.
6. **Owens, S. J.,** Self incompatibility in the Commelinaceae, *Ann. Bot.*, 47, 567, 1981.
7. **Jones, K. and Jopling, C.,** Chromosomes and the classification of the Commelinaceae, *Bot. J. Linn. Soc.*, 65, 129, 1972.
8. **Martinez, A.,** Chromosome Relationships in the *Tradescantia crassifolia* Alliance, Ph.D. thesis, University of Reading, England, 1978.
9. **Anderson, E.and Sax, K.,** A cytological monograph of the American species of *Tradescantia*, *Bot. Gaz.*, 97, 433, 1936.
10. **Giles, N. H.,** Autopolyploidy and geographical distribution in *Cuthbertia graminea* Small, *Am. J. Bot.*, 29, 637, 1942.
11. **Jones, K. and Colden, C.,** The telocentric complement of *Tradescantia micrantha, Chromosoma (Berl.)*, 24, 135, 1968.
12. **Jones, K.,** Aspects of chromosome evolution in higher plants, *Adv. Bot. Res.*, 6, 119, 1978.
13. **Jones, K., Papes, D., and Hunt, D. R.,** Contributions to the cytotaxonomy of the Commelinaceae. II. Further observations on *Gibasis geniculata* and its allies, *Bot. J. Linn. Soc.*, 71, 145, 1975.
14. **Hunt, D. R.,** Three new genera in the Commelinaceae. American Commelinaceae. VI, *Kew Bull.*, 33, 331, 1978.
15. **Kenton, A.,** Cytology of *Thyrsanthemum, Gibasoides* and *Matudanthus* (Commelinaceae), *Kew Bull.*, 34, 187, 1979.

16. **Kenton, A.,** Giemsa C-banding in *Gibasis* (Commelinaceae), *Chromosoma (Berl.),* 65, 309, 1978.
17. **Ehrendorfer, F.,** Polyploidy and distribution, in *Polyploidy: Biological relevance. Basic Life Sciences,* Vol. 13, Part 1, Lewis, W. H., Ed., Plenum, New York, 1979.
18. **Jones, K., Bhattarai, S., and Hunt, D. R.,** Contributions to the cytotaxonomy of the Commelinaceae. *Gibasis linearis* and its allies, *Bot. J. Linn. Soc.,* 83, 141, 1981.
19. **Bennett, M. D.,** Nuclear DNA content and minimum generation time in herbaceous plants, *Proc. R. Soc. Lond. B.,* 181, 109, 1972.
20. **Jones, K.,** Chromosomes and the nature and origin of *Anthoxanthum odoratum, Chromosoma (Berl.),* 15, 248, 1964.
21. **Belling, J.,** The attachment of chromosomes at the reduction division in flowering plants, *J. Genet.,* 18, 177, 1927.
22. **Wimber, D. E.,** The nuclear cytology of bivalent and ring-forming Rhoeos and their hybrids, *Am. J. Bot.,* 55, 572, 1968.
23. **Handlos, W. L.,** Cytological investigations of some Commelinaceae from Mexico, *Baileya,* 17, 6, 1970.
24. **Jones, K.,** Chromosome evolution by Robertsonian translocation in *Gibasis* (Commelinaceae), *Chromosoma (Berl.),* 45, 353, 1974.
25. **Kenton, A.,** Chromosome evolution in the *Gibasis linearis* alliance (Commelinaceae). I. The Robertsonian differentiation of *G. venustula* and *G. speciosa, Chromosoma (Berl.),* 84, 291, 1981.
26. **Mattson, O.,** Cytological observations within the genus *Zebrina, Bot. Tidsskr.,* 66, 189, 1971.
27. **Lalithambika Bai, K. and Kuriachan, P. I.,** Evolution of *Zebrina pendula* var. *quadricolor* by centric fusions: evidence from karyotype, *Caryologia,* 34, 327, 1981.
27a. **Karp, A.,** unpublished.
28. **Sadasivaiah, R. S.,** Haploids in genetic and cytological research, in *Haploids in Higher Plants. Advances and Potential,* Part 5, K. J. Kasha, Ed., University of Guelph, 1974.
29. **Darlington, C. D.,** Misdivision and the genetics of the centromere, *J. Genet.,* 37, 341, 1939.
30. **Holmquist, G. and Dancis, B. M.,** A general model for karyotype evolution, *Genetica,* 52/53, 151, 1980.
31. **Giraldez, R. and Santos, J. L.,** Cytological evidence for preferences of identical over homologous but not- identical meiotic pairing, *Chromosoma (Berl.),* 82, 447, 1981.
32. **John B.,** Heterochromatin variation in natural populations, *Chromosomes Today,* 7, 128, 1981.
33. **Marchant, C. J. and Brighton, C. A.,** Mitotic instability in the short arm of a heteromorphic SAT-chromosome of a tetraploid *Ranunculus ficaria* L., *Chromosoma (Berl.),* 34, 1, 1971.
34. **Brandham, P. E. and Bhattarai, S.,** The effect of B chromosome numbers on frequency within and between individuals of *Gibasis linearis* (Commelinaceae), *Chromosoma (Berl.),* 64, 343, 1977.
35. **Bougourd, S. M. and Parker, J. S.,** The B-chromosome system of *Allium schoenoprasum.* XI. Stability, inheritance and phenotypic effects, *Chromosoma (Berl.),* 75, 369, 1979.
36. **Subba Rao, M. V. and Pantulu, J. V.,** The effects of derived B-chromosomes on meiosis in pearl millet *Pennisetum typhoides, Chromosoma (Berl.),* 69, 121, 1978.
37. **Bennett, M. D.,** Nucleotypic basis of spatial ordering of chromosomes in eukaryotes and the implications of the order for genome evolution and phenotypic variation, in *Genome evolution,* Academic Press, New York, 1982.

Chapter 7

CHROMOSOME EVOLUTION IN THE MONOCOTYLEDONS — AN OVERVIEW

Arun Kumar Sharma

TABLE OF CONTENTS

I. INTRODUCTION

A discourse on chromosome evolution in the plant system presents diverse problems. The evolutionary changes undergone in the morphology and number of chromosomes are no doubt undisputed, but the importance of such changes in the study of phylogeny is beset with an inherent limitation of their wide occurrence in almost all groups of angiosperms. Leaving aside karyotype orthoselection which is almost absent in the angiosperms, both numerical and structural changes associated with chromosome behavior have affected taxonomic groups at all levels of hierarchy.[1] However, despite these limitations, certain features of chromosomes at the structural and ultrastructural levels have a decided value in the assessment of evolutionary advance or regression. These features, when associated with numerical and structural alterations, are more often than not utilized in solving problems of taxonomic dispute.

A vast amount of chromosome number reports in several groups and detailed analyses of chromosomes in certain broad taxonomic units, both in dicotyledons and monocotyledons, are available, although most of the 44 families in Cronquist's system of classification[2] are still unexplored.[3]

The basic chromosome number in angiosperms has been suggested as $x = 7$, which is characteristic of major groups with slight deviation in certain orders,[3] and evolution has mostly been at the diploid level. Simultaneously, on the basis of high chromosome numbers in some of the perennial woody species of the Anoniflorae[4] and Hamamelidiflorae, it is claimed that polyploidy was present in some of the ancestral forms of the surviving families, though not so prevalent in the basal order of Dillenidae.[2,5,6]

The extent to which the similarity of basic chromosome number of the angiosperms to some of the gymnosperms (e.g., Gnetales) is to be taken as an index of phylogenetic relationships is still a matter of debate. Such an ancestral relationship must, however, await the reconciliation of data from varied lines of research.

The scope of dicotyledons is very wide and in view of claimed monophyletic and diphyletic origin of several groups, the present review has been restricted to monocotyledons. Limitations of space and time do not permit an analysis of angiospermic evolution as a whole which may hopefully be taken up later.

II. CHROMOSOME EVOLUTION IN MONOCOTYLEDONS

The analysis and interpretation of the considerable information available on chromosome evolution in monocotyledons need to be presented against a background of taxonomic classification in which assignment of status of taxa is based on their phylogeny. In the group as a whole, the system followed by Hutchinson[7] has, to a great extent, been vindicated on cytological grounds. The systems of Melchior,[8] Cronquist,[2] Thorne[9] and Takhtajan[10] have several features in common with Hutchinson's system and agree on broad principles. As such, for the sake of convenience, treatment of taxa, as assigned by the latter, will be followed in this article. For the sake of brevity, groups on which recent reviews are available have been summarized while others have been dealt with in slightly greater detail.

A. Helobiales and Commelinales

In Hutchinson's system of classification,[7] the Alismatales and Butomales form the basic stock for the origin of different orders of the monocotyledons. However, the Lilifloreae have also been regarded as secondary progenitors of several orders of monocotyledons. In the system followed by Cronquist,[2] the Alismatales form the first subclass, followed by Commelinoideae, Araceaedeae, and Lilioideae. Several of the orders originating from Lilioideae, such as the palms and aroids have all been included under Araceaedeae. The extent

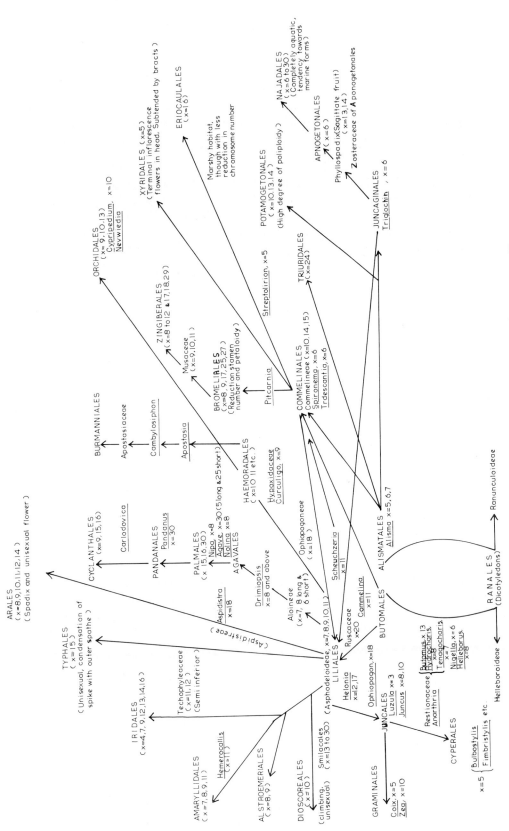

FIGURE 1. Evolution of Monocotyledons (Hutchinson[7]) and the chromosome profile.

to which $x = 7$ may be regarded as the basic chromosome number is not yet clear. Both Alismatales and Butomales show a series of chromosome numbers starting from $x = 5$ to 28 and in *Alisma* alone the numbers are $x = 5$, 6, 7, and 13. Other genera, like *Baldellia, Echinodorus,* and *Sagittaria* have $x = 7$, 8, 9, 10, and 11. It appears that $x = 7$ or 6 may be true basic chromosome number of the group as a whole with secondary derivatives. Intraspecific variations suggest an interrelationship of the various numbers and a monophyletic origin. In addition to low numbers, long chromosomes with symmetrical karyotypes are further evidences of primitiveness of the group, if taken in conjunction with other criteria like the aquatic habitat and flora morphology.[11]

The members of Butomales show a certain degree of asymmetry as especially evidenced in the family Hydrocharitaceae. A high level of polyploidy seen in genera like *Ottelia* ($2n = 22$, 44, 66), *Elodea* ($2n = 16$, 24, 48), *Lagarosiphon* ($2n = 14$, 22, 100) and others, indicates their comparatively advanced status. Butomales may have been derived from Alismatales through gradual evolution. In this respect Cronquist's system[2] appears to be more rational in including both *Alisma* and *Butomus* ($2n = 16$, 39) under Alismatales. The Hydrocharitaceae, with strong asymmetry in chromosome size and a high degree of polyploidy, may represent a separate order. The basic stock of the Alismatales has also been considered to be responsible for the origin of the Triuridales, Potamogetonales, Juncaginales, and Najadales. The last group represents the climax in the series with sexual differentiation and extreme reduction in size. However, *Najas* has a chromosome number of $x = 6$ as compared to *Potamogeton*, and in related genera multiples of $x = 13$ and 14 chromosomes have been noted. Similar multiples of 6 chromosomes are observed in *Triglochin* of the Juncaginales. It is preferable, therefore, to consider that all these orders represent diverse lines of evolution, not unrelated to each other and originating from a varying set of basic numbers. The manner or possibility of the origin of $x = 6$ from a base of $x = 7$ is difficult to ascertain at present.[12-16]

There is also no clear cytological evidence of a monophyletic origin of these orders, since it is difficult to visualize a single line of evolution due to limitations like the position of *Najas* with $n = 6$ at the culmination of the series, with *Potamogeton*, and *Aponogenton* located much lower down the ladder.

The Commelinales, as postulated by Hutchinson,[7] inhabiting relatively damp habitats in general, has a diphyletic ancestry from the Butomales and the Alismatales, respectively. Takhtajan,[10] however, considers Commelinadeae as a separate subclass which includes Commelinales, Eriocaulales, Restionales, and Poales. Chromosomal observations so far made, clearly indicate two distinct lines of evolution within the family — one originating with a basic number of 6 chromosomes and the other with 5 and its multiples. However, the number $x = 5$ of *Commelina* may be a derived one, since secondary association of bivalents shows four groups of bivalents. This line includes genera like *Pollia, Aneilema, Cyanotis* and others, in all of which structural alterations of chromosomes, duplication, and aneuploidy have played significant roles in evolution. A very high degree of polyploidy is recorded in *Commelina obliqua* with $2n = 150$ chromosomes. In *Commelinantia,* $x = 14$ and 28 became deep-seated after a certain stage of evolution. The Tradescantia line with $x = 6$ including *Tradescantia* ($x = 6$, 8, 12, 13, 15, 18), *Zebrina, Rhoeo,* and *Setcreasea* have comparatively much longer chromosomes. Polyploidy has been recorded, but the chromosomes, in general, are quite distinctive.

On the basis of chromosome studies, a separate generic status for the species *Cyanotis axillaris* ($2n = 20$) was adopted.[17-20] In another paper in this volume, Dr. Keith Jones and his colleagues have discussed the origin of the tribe, with special reference to the origin of metacentric chromosomes from telocentric ones through Robertsonian translocation. The extent to which chromosome size is related to the geographical distribution in *Gibasis* is still not certain.

B. Eriocaulales, Bromeliales, Zingiberales

Of the other orders, the Eriocaulales deserve special mention in that the two basic numbers reported are 8 and 9 with very high tetraploids and short chromosomes. A reduction in chromatin matter along with polyploidy has led to the evolution of Xyridales, which may represent a separate line from Commelinaceae.

The Bromeliales, due to their preference for a comparatively moist climate, has been considered by Hutchinson[7] as allied to Commelinales. The range of haploid number is quite high, starting from $x=8$ in *Tillandsia* to $x=25$, 28 in *Acanthostachys* and *Caraquata*, respectively. Polyploidy is rather common, as shown by $2n=50$, 75, and 100 chromosomes in *Ananas comosus* and $2n=54$, 72, and 108 in different species of *Billbergia*.[21] The very high number, coupled with small size of chromosomes, may indicate a higher evolutionary status of this family. The high numbers might have been derived throug aneuploidy and hybridization at the diploid level. A definite possibility exists of the origin of the Bromeliales from the *Commelina* stock.

The order Zingiberales, or in a wider sense, Scitamineae including Musaceae, Zingiberaceae,[7] Cannaceae, and Marantaceae, has been regarded to have originated from Bromeliales. Takhtajan,[10] however, placed it in the superorder Lilioineae of Lilideae, along with Liliales and Bromeliales. Irrespective of the mode of origin, the order appears to represent a homogeneous assemblage. The large chromosomes and asymmetrical karyotype of the Musaceae, including *Musa, Ensete, Rhodochlamys,* etc. have much in common with *Strelitzia, Heliconia,* and *Ravenala*. The basic numbers vary from $x=8$ to 22, and 11 is common for both the families indicating their close relationship. The comparatively primitive karyotype in Musaceae is quite different from the highly evolved karyotype of Bromeliaceae and as such there is no cytological evidence of any relationship between the two. Its origin from Commelinales, too, is yet to be established, in spite of the similarity in the chromosomes.

Zingiberales is a distinct natural assemblage with multiples of $x=8$, 9, 11, and 12 chromosomes. Intraspecific variations indicate relationships between the different genera. On the basis of cytological data, taken in conjunction with the progress of complexity in flora biology, *Costus, Zingiber,* and *Kaempferia* may represent a primitive line of evolution as compared to *Globba, Cautleya,* etc. It is not unlikely that the group is directly derived from the Musaceae or its immediate ancestor. Cannaceae has, however, $x=9$ and its multiples, and is very closely related to the Musaceae. Marantaceae, on the other hand, has chromosome numbers starting from $x=4$ in *Calathea*, 6 and 13 in *Maranta*, to 9 in *Ctenanthe*. The chromosomes in general are very small with marked size difference. The genera within the Marantaceae represent a homogeneous group but cytologically their relationship with the rest of the Zingiberales has yet to be clearly ascertained. But the short size of chromosomes, very low number, and asymmetrical karyotype make it difficult to place the Marantaceae either before or after Zingiberales. Chromosome characteristics may indicate a separate line of evolution for this family, parallel to the Zingiberales, perhaps originating from the Commelinales.[22,23]

C. Liliales

The family Liliaceae includes a vast number of genera distributed under different tribes. A large number of publications are available on the chromosomal evolution in this group.[24-38] It has been considered as the ancestral stock from which several petaloid families such as the Amaryllidaceae, Iridaceae, Palmeae, and Orchidaceae have been evolved.[7] A polyphyletic ancestry has been suggested, different parts of the family having been derived from the Butomales and the Juncaginales. Both in Juncaginales and *Tofieldia* of Heloniadeae, the chromosome number is $2n=12$. In the most primitive tribe Helonieae, the chromosome numbers so far noted are $n=12$ and 17. In species of *Chionographis* the somatic complement shows $2n=24$, with conspicuous difference in chromosome size.

Cytological data support the view that different orders of Liliaceae have originated from Hutchinson's[7] Asphodeleae, even though Scilleae[27] is regarded as a primitive tribe, having *Scilla sibirica* ($n = 6$) with long chromosomes representing an ancient status. The basic chromosome number in Asphodeloideae is 7 to 13. Karyotype and chromosome size show wide variation, from long symmetrical chromosomes of *Chlorophytum* to asymmetrical ones in *Eremurus*, whereas the genus *Asphodelus* shows reduction in chromosome size. This variation is also associated with diverse morphological variations.[32] There appear to be three lines of evolution in this tribe. One line shows increase in chromosome number along with structural changes but without any appreciable decrease in chromosome size, such as *Chlorophytum*. In the second line, a considerable reduction in the chromosome size has been associated with increase in chromosome number, as in *Alectorurus, Anemarrhena, Asphodelus,* etc. In the third line, without change in chromosome number, structural changes of chromosomes have led to an asymmetrical karyotype in an otherwise diploid complement, such as in *Eremurus* and *Arthropodium*.

Takhtajan[10] included Liliaceae and Amaryllidaceae under the Liliales and assumed origin of Amaryllidaceae and Agavaceae from ancestors of the Hemerocalleae ($x = 11$). The same order, however, includes Alliaceae ($n = 7$ and 8) and Agapantheae ($n = 6$ to 15) and Gillieseae.

The family Iridaceae is supposed to have originated from the Liliaceae in two distinct lines. The rhizomatous terminal-spiked Aphyllanthideae, as well as the nearly umbellate Johnsonieae, are considered to have led to the rhizomatous Aristeae of the Iridaceae. In the Johnsonieae the chromosome number is $n = 8$. In Aristeae, the lowest number so far reported is $n = 16$. The similarity in chromosome numbers may indicate some degree of affinity. The loosely panicled Dianelleae has been assumed to have given rise to the Tecophyllaceae in another line.

Takhtajan,[10] on the other hand, regarded Techophyllaceae as related to the Haemodoraceae, and the Iridales as being derived directly from the Melanthioideae of Liliaceae. The principal chromosome numbers in Dianelleae are $n = 8$ and 9; in Techophyllaceae $n = 10$ to 12, and in the Iridaceae, between 3 to 25. On the basis of chromosome number and morphology, the origin of the Techophyllaceae from the Dianelleae appears to be reasonable to a certain extent[37] though its further evolution towards the Iridaceae is debatable.[16,38]

Of the three genera in the Kniphofieae,[7] two are homogeneous, with $n = 6$ chromosomes. In the Aloineae, the size difference of chromosomes is marked, with a characteristic karyotype ($n = 7$) of 4 very long and 3 very short chromosomes. Hutchinson[7] suggested the origin of Aloineae from Asphodeleae through Kniphofieae. The chromosomes of some species like *Eremurus robustus* strongly resemble those of the Aloineae in their asymmetrical morphology and number. It is possible that *Eremurus* might have been a step in the evolution of Kniphofieae and ultimately led to Aloineae through continued development of asymmetry in chromosome size associated with aneuploidy. Structural alterations, hybridization, and to some extent polyploidy have been effective factors in the evolution of species in *Gasteria, Haworthia,* and *Aloe.*[41,42]

The bulbuous, subumbellate, noncoronate members of the Tulipeae are assumed to have given rise to the bulbous or cormuos, noncoronate umbellate Agapantheae,[43] now included under Amaryllidaceae by Hutchinson.[7] Hemerocallideae of the Liliaceae has been responsible for the evolution of Tulipeae.[24] The other line from the Tulipeae proceeds towards the Alstroemeriaceae. The chromosome numbers in the Tulipeae vary from 6 to 12, size ranging from long to medium, whereas in Agapantheae $n = 6$ to 15 and in the Alstroemeriaceae $n = 8$ and 9. The evolution of chromosomes of Alstroemeriaceae from Tulipeae may involve a genus or genera with short chromosomes.[31] The extreme size difference of chromosomes of the Haemantheae ($n = 9$)[44] precludes the possibility of the origin of this family through *Haemanthus* as suggested by Takhtajan.[10] In the Amaryllidaceae, the range of chromosome numbers is very wide, but the basic number has been suggested to be $x = 6$.[61] Takhtajan[10] assumed the origin of the Amaryllidaceae and Agavaceae from ancestors of the Hemerocalleae.

In the Tulipeae as well there have been extensive structural alterations, especially in the genus *Lilium*[39] which shows different karyotypes with identical chromosome number. Chromosome studies have indicated that though *Erythronium, Tulipa,*[40] and *Lilium* are related, yet structural alterations of chromosomes have played an important role in the comparatively primitive genus *Erythronium* in the diversification of its eastern and western species.

The origin of the Ophiopogoneae from the Asphodeloideae appears reasonable since series of chromosome numbers with variations in chromosome morphology are present. From the Ophiopogoneae, the Milliganieae having similar morphological characters have been assumed to have been evolved. Hypoxidaceae, with similar arrangement of leaves, may have evolved in turn from the Milliganieae. The chromosome number of the Ophiopogonaceae is $x = 18$[45] derived from $x = 9$; in the Milliganieae it is $x = 8$, whereas in the Hypoxidaceae, it is $x = 6$, 9, and 11. Reduction in chromosome number might have been involved in the process. Takhtajan[10] also placed the Hypoxidaceae very close to the Haemodoraceae and traced the relationship of the latter from the Ophiopogoneae of the Liliaceae. From the Hypoxidaceae two further lines of evolution can be traced. One has led to the Burmanniaceae through the Apostiaceae of Haemodorales. The other line proceeds from the bistaminate racemed *Neuwiedia* of the Haemodoraceae to a primitive tribe Cypripideae of the Orchidaceae with $x = 10$ and 11. This line of evolution has also been supported by Takhtajan.[10]

Ophiopogon is regarded as an intermediate step in the evolution from the Liliales to Juncales by Hutchinson,[7] notwithstanding the fact that Juncales, Bromeliales, and Commelinales have been derived directly from the Asphodeloideae by Takhtajan.[10] In *Juncus* there are different series of chromosome numbers, the predominant ones being 8 and 9, and in *Luzula* it is $x = 3$. If *Ophiopogon* is considered as the progenitor of the Juncales, it is difficult to visualize the evolution of the long chromosomes of *Luzula* with diffuse centromere from the medium-sized ones with localized centromere of *Ophiopogon*. Further studies on chromosome evolution may suggest that the Juncales, Cyperales, and Restionales may have a common ancestry as proposed by Cronquist.[2] In Hutchinson's system[7] Graminales is an offshoot from Juncales with basic chromosome sets of $n = 5$ and 10.

Trilliaceae of Hutchinson,[7] including *Trillium* and *Paris*, was placed under Paridae by Krause.[46] This family is characterized by whorled leaves, through which and *Medeola* the bisexual climbing Stenomeridaceae is supposed to have evolved. This line again led to unisexual, climbing Dioscoreaceae with $x = 10$ chromosomes. In fact, Stenomeridaceae has been merged in Dioscoreaceae by Cronquist.[2] This line of evolution involved increase in chromosome number with reduction in chromosome size which is rather frequent in angiosperms.[1] Species of Trilliaceae have $x = 5$, with long chromosomes, whereas in Dioscoreaceae the number is $x = 10$. *Medeola* has $x = 7$ with short chromosomes. Hutchinson[7] has derived Trilliaceae from Uvularieae. But in the distinctly homogeneous Trilliaceae, the chromosomes are much longer and fewer in number than those of *Uvularia*. In fact, in *Oakasia* of Uvularieae the size difference of chromosomes is much greater than that of *Trillium*. However, the chromosomal homogeneity of the family Trilliaceae justifies its inclusion within Liliaceae *sensu lato* as done by Takhtajan[10] and Cronquist.[2]

Another line of evolution leads from Asparageae to Ruscaceae. This family differs from Liliaceae in having united stamens, with highly specialized morphology. In *Asparagus*, though the root-stock is rhizomatous, vegetative structure is reduced to a spiny cladode. In Ruscaceae too, slightly woody nature is seen and the branchlets are very much modified and flattened. The basic chromosome numbers are also related, being $x = 10$ in Asparagaceae and $2n = 20$ in Ruscaceae. This family has been merged with in Liliaceae by Cronquist[2] and in Asparagaceae by Takhtajan.[10] The more primitive karyotypes of *Ruscus* justify its inclusion in a tribe more primitive than, but allied to the Asparageae.[28,31]

In the tribe Polygonateae all Himalayan species of *Disporum* form a complex with $n = 6$,[47] while the Asiatic and American species belong to two different categories. In *Polygonatum* the *"latifolium"* type ($n = 9$ or 10) represents a primitive state.[31] In *Smilacina*, extensive aneuploidy and structural heterozygosity have led to its wide distribution in the Himalayas. Of the three genera, *Disporum* of Pleistocene origin is regarded as having given rise to *Polygonatum* through diminution in chromosome size and asymmetry and to *Smilacina* through increased asymmetry.[31]

Hutchinson[7] considers the spadix and the unisexual flower of the Arales to have evolved from a floral structure, similar to *Aspidistra* of the tribe Aspidistreae. As against the wide range of chromosome numbers and graded karyotype in Arales in *Aspidistra*, the chromosomes are rather long and the number is only a multiple of $n = 18$. On the other hand, Takhtajan[10] visualized the origin of the Arales from the immediate ancestors of Liliales, including it in the Arecideae, a view supported by Cronquist[2] as well. Chromosome analysis does not provide any positive evidence of the origin of the Arales through the Aspidistreae of Hutchinson.

Similarly, the order Typhales has also been considered by Hutchinson[7] to have been derived from the Asphodeleae, — a view so far not substantiated on cytological grounds. This aquatic order has been placed under the Arecideae by Takhtajan[10] and under the Commelinideae by Cronquist.[2] In the Typhales, the chromosomes are very small and their number is uniformly $x = 15$ in both *Typha* and *Sparganium*.[48]

From the point of view of chromosome evolution, the origin of the Smilacaceae from the Asparageae is not difficult to visualize, the basic set being 10 chromosomes in both. The similarity of the herbaceous genus *Smilacina* in its leaf character ($n = 13$ to 16) is rather superficial.

The evolution of the Agavales from the Asphodeloideae is quite rational on cytological grounds, even though the homogeneity of the Agavales has been questioned.[43,45,49-52] The genus *Drimiopsis* of the Scilleae, with multiples of $x = 10$ chromosomes and a high degree of karyotype asymmetry, may form an intermediate link between the Asphodeloideae and Agavales. An ancestry of the Agavales from the Aloineae which has an asymmetrical chromosome evolution has also been claimed.[53] The extreme size difference in Aloineae[41,42] ($x = 7$, 4L + 3S) finds a parallel in *Agave* and related genera ($n = 30$, 5L + 25S). Notwithstanding such a similarity in karyotype morphology, the differences in chromosome number between the Aloineae and Agavales are distinct and any possible relationship between the two is yet to be substantiated.

The Amaryllidales, as separated by Hutchinson[10] and later Traub[61] from the Liliales and Agavales, present a relatively homogeneous complex, except for a few genera, with a wide range of chromosome numbers, of which $x = 11$ is rather frequent. The karyotypes are graded with long to medium-sized chromosomes. The different tribes present a general resemblance in the chromosome morphology. Aneuploid numbers are frequent and polyploidy is usually accompanied by structural alterations. The advanced status of the family as a whole is indicated by the high somatic numbers and a relatively larger frequency of submetacentric and acrocentric chromosomes.[44,54-60]

D. Graminales

The grasses have been regarded by Hutchinson as an offshoot of liliaceous stock parallel to the Cyperaceae. The family Bambuseae, in view of its tree-like habit and simple flower, is regarded as primitive. All tribes of the Gramineae and Festuceae are considered as ancient to all herbaceous temperate grasses. Takhtajan[10] placed the Gramineae as Poales under the subclass Commelinideae. Stebbins[62] visualized the evolutionary tree in a cross-section. The central star represented the hypothetical primitive stock, the distance from the star being directly proportional to the degree of specialization. The tree-like habit of the Bambuseae has been attributed to secondary specialization,[63] different from the growth habit of other trees. On the other hand, a majority of the families with which the grasses are related, such as the Restionaceae and Flagellariaceae are all herbaceous. The attributes of herbaceous nature with simple flower, like that of the bamboos, have been assigned to the primitive grasses. The grasses have been subjected to extensive investigation and different classifications have been proposed from time to time, based on varied lines of evidence.[64-72] Hilu and Wright,[73] summarizing the status, have shown that the present systematic classification of the Poaceae at the subfamily level is still sound for the most part.

The three distinct lines of evolution which characterize this family lead towards the festucoid, chloridoid, and panicoid complexes. The chromosomes of the Festuceae are undoubtedly very large, a condition which, however, according to Avdulov[74] developed later as an adaptation to cold climate. This contention has been questioned by Stebbins[6] since both genera with very large and short chromosomes of the Liliaceae and Cyperaceae, respectively, occur in the temerate climates. The modern day bamboos represent a less successful line of evolution, being restricted to moist tropical forests with exceptional specialized climbing forms. They exhibit a uniform series of $2n = 70$, 72 in Eubambuseae, while the more primitive Arundinarieae has $2n = 54$ chromosomes. Apparently the specialized habit has been an adaptation of this conservatism in karyotype.[75,76]

Avdulov[74] assumed that evolution in grasses was accompanied both by increase and decrease in chromosome number and size. He recognized three types of grass karyotypes, (1) $x = 9$ with small chromosomes, (2) $x = 12$ with small chromosomes, and (3) $x = 7$ with large chromosomes. The festucoid type includes 5, 6, or 7 large chromosomes in the haploid complement. In the Saccharoid group, including the Chlorideae, Andropogoneae, and Paniceae, $x = 10$ or 9 is most common.

According to Stebbins[62] the festucoid type maintains the primitive basic number of 6 to 12 and primitiveness in some of the floral characters. Stebbins[62] has, more cautiously, suggested that in delineating the grasses, such ideas should be reconsidered with other lines of evidence. Hubbard[77] suggested that the ancient primary grasses consisted of leafy, branched flowering shoots with many nodes and a sheathing base. The infrequent flowering periods in bamboos are held to lead to certain features regarded as primitive. Beetle[66] considered the Phareae as the most primitive of all tribes. Tateoka[67] included the Phareae and Oryreae in a primitive group, while Stebbins[62] suggested the two to have progressed about the same distance along the evolutionary path. He considered the Festuceae as advanced whereas others assumed the simplicity of leaf of the Festuceae to indicate primitiveness.[78]

The role of polyploidy in the evolution of grasses has been well established.[62] In the centers of origin, in general, the polyploids, because of their altered genetic setup, become less adapted as compared to diploids. This is especially true of a number of induced auto-tetraploids in grass and cereals. The adaptability of polyploids is principally restricted to new environments. Instead of assuming the adaptation of polyploidy in arctic region,[79,80] it is claimed that polyploidy of *Poa, Bouteloua, Calamogrostis*, etc. in the northern area is due to recent glaciation. In the millets, including *Triticum* and *Avena*, the chromosome numbers are 14, 28, and 42.[81-83] In the Paniceae they are $2n = 72$ whereas in the andropogons the somatic number ranges from 20 to 180, all in multiples of ten. The genus *Bromus*, which is one of the primitive taxa of the festucoid complex, has $2n = 14$ in some of the species.[84]

The oldest of the amphidiploids in the Triticineae, such as *Aegilops*, have been traced to the Pleistocene.[62] The final two sections of *Bromus, Centrophosum*, etc. are composed mainly of polyploids suggesting relics of once extensive polyploidy. *Bromus catharticus* belongs to the hexaploid South American species inhabiting temperate climates, maintaining livestock through its luxuriant growth in winter. In addition, several members of the family such as species of *Poa, Anthoxanthum*, etc. have shown B chromosomes. The adaptability conferred by B chromosomes to specialized habitats in several grass genera has been recorded, although negative evidences too, are known.

On the basis of observations so far made, multiples of $x = 5$, 6, 7, and 10 chromosomes are most prevalent in the grasses. The last number, no doubt, is the most deep-seated one. It is quite remarkable that Maydeae, which otherwise shows multiples of 5 chromosomes, is highly specialized in its vegetative and floral characters. Extreme specialization in chromosome characteristics, including heterochromatin knobs, has also been found in this group. If the attributes of the Bambuseae[75,76] as an advanced taxon are accepted, the primitiveness of the festucoid complex[78] can hardly be questioned. In fact, the Phalarideae, which on both chromosome count and other characters, has been assigned a primitive status, also belongs to the same festucoid line of evolution apart from the chloridoid and panicoid lines. The relative primitiveness of the different tribes and the relationship with the herbaceous grass-like simple-flowered progenitors is yet to be ascertained from cytological and other aspects. The base number of the family as a whole is, however, difficult to determine, especially in view of the fact that both increase and decrease in chromosome number have taken place in evolution. Assuming $x = 7$ as the base number of the monocotyledons as a whole,[3] considerable structural and numerical changes must have taken place during the derivation of secondary basic stock in this family.

In general chromosomal details, grasses present several interesting features. A high degree of polyploidy is common to many genera. Both structural and numerical alterations of chromosomes have been frequently observed. Interspecific hybridization is not uncommon. Large numbers of cytotypes, especially aneuploid complexes, have been recorded in several genera such *Cenchrus, Dicanthium, Bothriochloa,* and *Poa.*

The adaptability of wild grasses may, to a great extent, be attributed to the broad spectrum of genotypes. These features, in conjunction with the facultatively apomictic nature of the species, contribute to the survival of genotypes in wide areas of distribution. The partially vegetative mode of reproduction confers on this group an added advantage in evolution. All these factors have contributed to the success of the Gramineae in evolution.

E. Orchidales

The Orchidales have been derived from liliaceous stock through the Haemodorales, Hypoxidaceae, *Apostasia,* and *Neuwedia.*[7] It was included under the Lilineae along with the Zingiberales and Haemodorales under certain systems.[10] On the basis of cytological data it is difficult to state the origin of orchids. Schlechter[85] traced the phylogeny of this group from a *Curculigo*-type of ancestor due to the similarity of the chromosomes in *Curculigo recurvata* ($2n = 18$) with a mixture of long and short chromosomes, with that of *Cypripedium,* a primitive orchid ($n = 10$). There is no definite genetic evidence of such an origin. The taxonomic stand with regard to the phylogeny of orchids requires further information on both the chromosomes of orchids and Haemodorales and Hypoxidaceae.

Cytological data are available only on a fraction of the total number of living orchids.[86-90] Tanaka and Kamemoto[91] prepared a comprehensive list of chromosome numbers in orchids. Polyploidy in *Cymbidium, Paphiopedilum,* and *Cattleya* has been recorded by Mehlquist.[92] However, the most common numbers are 32 to 42. Chromosomes of the Diadrae are in general larger than those of Monandrae. The size difference between chromosomes is more marked in *Paphiopedilum* as compared to *Cypripedium.* Even within the Diandrae, there is a tendency towards reduction in chromatin content. In the Basitonae of the Monandrae, comparatively larger chromosomes are present in *Habenaria* and much shorter ones in *Orchis.* The intermediate stages are represented in *Gymnodenia and Pictanthera.*

In orchids, there has not only been reduction in chromatin matter but also disappearance of the size difference between chromosomes with all chromosomes becoming very small in size. In the Acrotonae of the Monandrae, reduction in size during chromosome evolution is quite conspicuous. Chromosomes of *Listea* and *Apipactis* are very long, with marked size differences, whereas in *Blettia* and *Goodyera* the size has been remarkably reduced. Similar evolution in chromosome size has also been recorded from *Plione* and *Dendrobium* through *Liparis* and *Tainia*. In *Pleuranthe* of the Acrotonae, Sympodiales is characterized by reduction in chromatin length with gradual increase in chromosome number. In *Odontoglossum* ($2n = 56$), the chromosome number is very high. On the other hand, in Monopodiales of the same group there is a uniformity in the complement, the chromosomes being short in size and high in number. In the family Orchidaceae in general, aneuploidy as well as increase and decrease in chromosome number are quite frequent even within a single genus. Jones[93] emphasized the variations in basic number in the Orchidaceae and brought forward evidences of their significance in working out evolution and affinities of genera like *Cychnoches* ($n = 32$, 34), *Catasetum, Mormodes* ($n = 27$)[94] (and others). Karyotype evolution has also been associated with reduction in size and decrease in chromatin content. On cytological grounds, therefore, the classification of Schlechter[85] appears to be viable.

However, in this group certain features of the chromosome complement are observed to be of special importance in relation to plant distribution. The orchids are widely distributed in the subtropical and temperate zones, where nearly 450 species occur in the Himalayas. The dense evergreen forests with heavy rainfall and high humidity in the eastern Himalayas have been responsible for the profusion of different taxa.[95] In the alpine zone, very few orchids have been recorded, a fact mainly attributed to the sturdy vegetation and stunted growth of plants.

In the family Orchidaceae, the general morphology of the chromosomes shows a general size range in most of the genera. The basic numbers are $n = 19$, 20, and 21, possibly derived from $x = 10$ and its derivatives through hybridization. A large number of cytotypes has also been recorded in the Himalayas. It is remarkable that the family, which is considered one of the most successful among the flowering plants, has a more-or-less constant basic set with certain exceptions. This is not only true for Himalayan orchids but holds good for other orchids as well. A possible association of such uniformity in chromosome number has been suggested with the epiphytic adaptation of the taxa. Such adaptations are quite characteristic, both from physiological and anatomical standpoints.[95] It is not unlikely that this deep-seated number may possibly represent genes or gene clusters in different chromosomes, all of which jointly contribute to the epiphytic characteristics. This suggestion is further confirmed by the fact that species with unusual or exceptional number for this group such as $n = 7$ in *Orchis drudei* M. Sch., $n = 9$ in *Pogonia ophioglossoides* (L.), $n = 14$ in several species of *Habenaria* Willd., and $n = 13$ in *Paphiopedilum bellatula* (Reichb. f.) Pfltz.[96] are not epiphytic but terrestrial in habit. However, much more data are necessary to substantiate the idea that the adaptive features of epiphytes are controlled by genes or gene clusters distributed on different chromosomes of the basic set. The deficiency of any member of the chromosome complement limits the survival of the individual. This group presents an excellent example of correlation of adaptive features with genetic architecture of the species. Another feature of the Orchidales, in common with the Liliales and Arales, is that interpopulation differences or cytotypes differing in chromosome number are not recorded in the Alpine zone. Possibly extreme condition of the environment in Arctic region exerts certain selection pressure acting against the viability of any variant.[52,95,97]

F. Palmales

Chromosome evolution in the Palmeae has followed a very distinct pattern. Hutchinson[7] has derived this group from the basic stock of the Liliales through the Agavales and leading to the Pandanales, based on their arborescent habit and inflorescence character. Palmeae is principally a tropical family, included along with the Areceae and Lemnaceae under the series Spadicifloreae by Rendle.[98] The similarity in sexual features, the relative size of embryo and endosperm, as well as the clustering of numerous small flowers in indefinite inflorescences are considered as valid criteria for the suggested relationship between palms and aroids. Agavales, as mentioned earlier, is characterized by extreme asymmetry in chromosome size and morphology. The genus *Funkia*, however, although a member of the Agavales, does not show size differences in at least one species *F. aromatica* ($2n = 20$), although in most other species a size difference ($x = 30 = 5L + 25S$) is present. It is difficult to ascertain the extent to which such size difference is an indication of relationship between the two groups.

In the Palmeae, the tribe Corypheae, otherwise regarded as primitive, has mostly medium-sized chromosomes as in *Rhapis, Corypha*, and *Livistona*. However, the genus *Thrinax* ($2n = 36$) has long chromosomes in the diploid set. Cytological data show that there has been a decrease in the chromatin content and chromosome size during the evolution of palms. On the basis of these observations, chromosomal evidences do not appear to contradict the derivation of the palms from the Agavales.[7] The high degree of polyploidy, often noted in the Agavales, might have also contributed to the evolution of the genus *Doryanthes*.

Several haploid numbers have been recorded in palms, including $x = 13, 14, 16$, and 18. The latter two are the most common ones. Cytological and other information indicate that this family is no doubt a primitive one. The tribe Corypheae is a rather natural assemblage in which *Thrinax* ($2n = 36$) may represent a primitive status as especially evidenced in its long chromosomes. Structural changes of chromosomes have been the major factor in the three possible lines of evolution in palms from a common ancestor; one line represented by *Livistona, Pichardia*, and *Sabal*, another by *Rhapis* and *Corypha*, and the third one by *Licuala*. In Borasseae, both $2n = 32$ and 36 chromosomes have been found in *Latania* and *Borassus*, respectively, though *Latania* has comparatively small chromosomes suggesting a diminution in chromatin matter. The tribe Calameae presents a homogeneous assemblage, most members having $2n = 28$ chromosomes.

In the Areceae, the genus *Caryota* ($2n = 32$) with large chromosomes may deserve a separate tribal status. In this tribe evolution may have proceeded in two lines, one branch including *Oreodoxa* with median to very short chromosomes. The other may be divided into two sublines, one containing *Dypsis* ($2n = 36$) and the other involving *Areca, Rentia*, and related genera with $2n = 32$ chromosomes. Secondary association of bivalents recorded in the genus *Caryota* suggests that the basic number is $x = 8$. The gross difference in the karyotypes of two species of *Areca* studied, represented in chromosome size, may suggest that they may not be included in the same genus. In the order Palmeae as a whole the basic numbers are evidently 7, 8, and 9 with characteristically smaller chromosomes in which polyploidy and structural changes associated with diminution in chromosome number have played very important roles in evolution.[99-101]

In Hutchinson's system,[7] Cyclanthales forms the culmination of the pandanus-palm line. But both the palms and the Pandanales are extremely specialized as compared to Cyclanthales, the former due to the establishment of sexuality[102] and the latter due to extreme specialization of the group. The genus *Carludovica* of the Cyclanthales has $2n = 18, 30$, and 32 in different species. The similarity of the small chromosomes of the Palmeae to the Cyclanthales is also remarkable. It may even be conjectured that both Palmeae and Pandanales ($n = 15$) might have been derived from the Cyclanthales in two different directions. However, there is no evidence yet of such a derivation, though the affinity of the three groups Pandanales,[48] Cyclanthales, and Palmeae (Arecales) has been supported by recent taxonomists. The evidences at present are too meager to derive one from the other.

G. Arales

The order Arales is quite complex and includes both herbaceous and climbing forms. Extensive variations in size and number of chromosomes in members belonging to different types have been revealed. Hutchinson[7] considers this order as a derivation from the broad liliaceous stock Aspidistreae. Takhtajan,[10] on the other hand, has included the Arales, along with Arecales, Cyclanthales, and Pandanales under a separate subclass Arisaedae. In view of the extensive range in number and size of chromosomes in the Arales, it is not possible either to confirm or reject the origin of the Arales through Aspidistreae. The range of the chromosome complement in the Arales can be well explained by the fact that in the tribe Richardieae, species of *Aglaonema* are characterized by somatic chromosome numbers ranging from 18 to 70; $2n = 26$ and 34 occur in species of *Homalonema*, whereas 28 and 32 chromosomes are found in species of *Zantedeschia*. *Aglaonema* possesses very large chromosomes, whereas very small chromosomes are observed in *Homalonema*, *Zantedeschia*, and *Schismatoglottis*, including different levels of aneuploidy and polyploidy.[103-111]

In this order, polyploidy has been found to be prevalent in several genera such as *Acorus*, *Symplocarpus*, and *Pothos*. In addition, extensive ranges in chromosome numbers, including both autopolyploid and aneuploid ones have been observed in the tribes Oronteae, Spathiphylleae, Anthurieae, Calleae, Monstereae, Stylochitoneae, Richardieae, Dieffenbachieae, Colocasieae, Philodendreae, Spathicarpeae, and Pythonieae. Similarly, structural alterations of chromosomes, both at intra- and interspecific levels, have been recorded in several genera. On the basis of chromosome studies, the homogeneity of majority of the tribes has been confirmed and the comparatively primitive status of *Aglaonema* and *Anthurium* has been suggested. Evolutionary advancement has been recorded in *Coelogyne*, *Pothos*, *Symplocarpus*, *Epipremnum*, *Monstera*, *Rhaphidophora*, *Scindapsus*, *Rhodospatha*, and *Pistia*.

The importance of intraspecific chromosomal variations has been recorded in several genera, indicating that diverse chromosome numbers might have been derived from a common basic set. It has further been shown that the majority of the taxa have involved a steady diminution in chromosome size from long chromosome complex with nearly symmetrical karyotype. This fact, taken together with the common basic number of $n = 14$ in most of the tribes, suggests that 14 or more likely 7 is the primary basic number for the family as a whole. A majority of the tribes, too, is characterized by long chromosomes. They have been regarded as representing primitive forms among the tribes, owing their origin to 7 or 14 chromosomes in the basic set. Divergent lines of evolution have been visualized emanating from this basic set in different directions, and as such the different genera do not possibly represent different levels from a single line of evolution. It has further been claimed[110] that the ancestral type has very long chromosomes like that of wild species of *Aglaonema*.

Chromosome evolution in relation to plant distribution has been well exemplified in this order. In the Himalayas, there are nearly 73 species distributed under 19 genera.[95] Of these again, species of *Alocasia* and *Colocasia* do not grow beyond the marshy, subtropical areas in foothills and their distribution merges into that of the Indo-Gangetic plain. The 8 species of *Rhaphidophora* occur as huge climbers in subtropical to temperate rain forests. The genus *Arisaema* is represented by 17 species,[112] starting from the subtemperate to alpine zones. Chromosome studies so far carried out in the Himalayan species of this family,[110,111,113] show a range from $n = 14$ to 17 chromosomes with aneuploidy, polyploidy, and structural changes of chromosomes even at the interspecific level. However, diploids and aneuploids near the diploid range are rather common as compared to polyploids with $n = 14$ as the basic number. Diploids and polyploids, no doubt, occur under the same ecological conditions but there are 3 species of *Arisaema*, namely, *A. nepenthoides* Mart., *A. sikkimense* Stapf., and *A. wallichiana* Hook. which occur beyond 10,000 ft in the alpine zone. They contain a variable number of Bs in a common chromosome number $2n = 26$. All the other species grow in the temperate zone and show either $2n = 26$ or 28 chromosomes. Moreover, the three species of *Arisaema* with B chromosomes have the normal chromosomes rather larger than the others. It has been assumed that in this genus the accessory chromosomes may have endowed them with some selective advantage in the alpine climate and may influence chromosome size through their control over chromosome coiling as well.

H. Cyperales and Juncales

The order Cyperales, too, forms one of the important families of monocotyledons, the distribution of which ranges from the tropical to the Alpine zone. In the Himalayas alone, of the family Cyperaceae nearly 175 species have been recorded. The range of chromosome number is quite high, starting from $x = 4$ and 5 to 208 as noted in *Cyperus*. This family shows the presence of certain chromosomal complements to restricted zones. In *Cyperus* alone, the high chromosome numbers are mostly restricted to the high altitudinal zones. The chromosomes of *Fimbristylis* show a low basic set of 5 with clear evidence of polyploidy, especially in the temperate areas. On the other hand, *Scirpus* is restricted to comparatively marshy areas and has a series of chromosomal cytotypes next to *Cyperus*. As in *Carex,* in *Eleocharis*, too, a diffuse centromere has been recorded.[114,115,116] However, localized centromeres have also been noted in the genus *Eleocharis*. In this family, evolution has been associated with a high degree of polyploidy, aneuploidy, and structural alterations of chromosomes.

The allied family Juncaceae shows not only a high degree of polyploidy, but also the occurrence of a diffuse centromere as a common phenomenon. In the genus *Luzula*, clear evidence has been brought forward to show that increase in chromosome number has principally been due to fragmentation of chromosomes with diffuse centromeres.[117] This has been confirmed from the study of total amount of chromatin matter, interspecific hybridization, and X-ray induced fragmentation.

I. Iridales

In the Iridales, Hutchinson[7] principally considered the Sisyrincheae to be followed by the Mariceae. In this family, extensive studies have been carried out in the genus *Iris* where a series of chromosome numbers including polyploidy and aneuploidy has been recorded. [118-122] In the tribe Mariceae, the species of *Marica*, especially those with a low chromosome number, have large chromosomes with nearly median primary constriction — a criterion for the primitive state. In several species of this genus, high chromosome numbers with gradation in chromosome size and asymmetry have been noted.[123]

The evolution of the Irideae from the Mariceae has principally been associated with increase or decrease in chromosome number involving structural alterations. In the Tigrideae, the culmination of the line starting from the Mariceae through Irideae, abrupt size differences of chromosomes and extreme asymmetry have been recorded in certain species of *Tigridia*. Remarkable heteromorphy in chromosome size indicating the role of structural hybridity has been noted in *Cipura* ($n = 6$). Elimination of chromatin matter and considerable reduction in size of chromosomes have principally affected the tribe Ixieae. In the Gladioleae, along with polyploidy, there has been considerable reduction in chromosome size. In general, chromosome evolution in the Iridales vis-à-vis Irideae has more or less run parallel with the evolutionary tendencies represented in Hutchinson's system of classification.[7]

III. GENERAL OBSERVATIONS

The chromosomes of monocotyledons indicate certain remarkable features in the structural behavior. One of the principal phenomena often noted in this group as a whole is the occurrence of a large number of cytotypes in nature. They involve both polyploidy and aneuploidy as well as genotypes with structurally altered chromosome complements. Undoubtedly, this feature has been responsible to a great extent for the success of evolution.

Prior to dealing with the other factors responsible for the success of this phenomenon, it may be recorded that the habitat under which the monocotyledons flourish may play a significant role for their successful existence. Excepting the alpine regions, restricted conditions do not allow the occurrence of variable genotypes. Monocotyledons as a whole to a great extent inhabit areas where nutrient supply is rather abundant. Humid tropical to temperate forests of the Himalayas with high rainfall provide an outstanding example of habitat where this group of plants grows in profusion. The same explanation also holds good for the epiphytes. Such abundance of nutrient supply in aquatic, marshy, or swampy regions has nullified the pressure of rigorous selection. This factor has contributed to the survival of a large number of genotypes mostly of herbaceous and shrubby nature. The comparative flexibility is also provided by the less differentiated nature of genotypes having practically no secondary growth excepting in palms and related genera.

The origin of such cytotypes, both polyploids and aneuploids, has been traced to the extensive occurrence of polysomaty in the somatic tissue.[124-127] Such variant nuclei principally arise through regulated nondisjunction, mitotic arrest, and endomitotic replication (partial or complete). Their role in speciation through entrance into the growing tip of the daughter shoots, i.e., bud mutation, is assured by the vegetative reproduction — an almost universal feature in a majority of the monocotyledons.

The occurrence of accessory chromosomes in these species is also very interesting. Such accessories or B chromosomes have been recorded in several families of plants, including the Gramineae and Liliaceae. The role of such accessories has not been fully established in assessing their positive or negative effect on the adaptability of species.[30,128,129,130]

It is remarkable that in one species of *Allium* definite evidence of accessories conferring an adaptive advantage has been recorded in *Allium stracheyii*.[129,130] It is yet to be noted to what extent the B chromosomes control crossovers and recombinations as indicated in certain genera.

The relationship of chromosome size to increase in chromosome number or polyploidy has been pointed out in several genera.[52] The diminution in size often noted along with this process has been attributed either to elimination of heterochromatin or differential coiling of chromosome segments. Similarly, there are genera where no such diminution is recorded. Possibly, an answer to this interesting behavior may have to await a complete analysis of the extent of repetitive DNA or redundant DNA occurring at the diploid level.

Leaving aside the behavioral pattern mentioned above, at the structural level of chromosomes considerable discussion has been held with regard to the evolutionary status of diffuse and localized centromeres on one hand, as well as other heterochromatic segments as noted in *Trillium, Fritillaria*, and *Paris*. In view of the fact that even at the intrageneric level both diffuse and localized centromeres have been found, it is difficult to consider these factors as progression or regression in evolution.

ACKNOWLEDGMENT

I would like to express my gratitude to Dr. Arati Raychaudhuri, Research Associate, Centre for Advanced Study, for her ungrudging assistance in preparing this manuscript.

REFERENCES

1. **Stebbins, G. L.,** *Chromosomal evolution in higher plants,* Addison-Wesley, Reading, Mass., 1971, 216 pp.
2. **Cronquist, A.,** *The Evolution and Classification of Flowering Plants,* Houghton Mifflin, Boston, 1968, 396 pp.
3. **Raven, P. H.,** The basis of angiosperm phylogeny: cytology, in *Ann. Mo. Bot. Gard.,* 62, 724, 1975.
4. **Raven, P. H. and Cave, M. S.,** Chromosome numbers and relationships in Annoniflorae, *Taxon,* 20, 479, 1981.
5. **Raven, P. H. and Axelrod, D. I.,** Angiosperm biogeography and past continental movements, *Ann. Mo. Bot. Gard.,* 61, 539, 1974.

6. **Stebbins, G. L.,** *Variation and Evolution in Plants,* Columbia University Press, New York, 1950, 643 pp.

7. **Hutchinson, J.,** *The Families of Flowering Plants,* Vol. 2, Clarendon Press, Oxford, 1959, 511.

8. **Melchior, H.,** Ed., A. Engler's *Syllabus der Pflanzenfamilien,* 12th ed., Vol. 2, Angiospermen, Gebrüder Borntraeger, Berlin, 1964, 666 pp.

9. **Thorne, R. F.,** Synopsis of a putatively phylogenetic classification of the flowering plants, *Aliso,* 6, 57, 1968.

10. **Takhtajan, A.,** *Flowering Plants: Origin and Dispersal,* Transl. by Jeffrey, C., Smithsonian Institution Press, Washington, D.C., 1969, 310 pp.

11. **Sharma, A. K.,** Cytology as an aid in taxonomy, *Bull. Bot. Soc. Bengal,* 18, 1, 1964.

12. **Sharma, A. K. and Mukherji, R. N.,** Cytology of two members of Alismaceae, *Bull. Bot. Soc. Bengal,* 9, 32, 1955.

13. **Sharma, A. K. and Bhattacharyya, B.,** A study of the cytology of four members of Hydrocharitaceae as an aid to trace the lines of evolution, *Øyton,* 6, 123, 1956.

14. **Sharma, A. K. and Chatterjee, T.,** Cytotaxonomy of Helobiae with special reference to the mode of evolution, *Cytologia,* 32, 286, 1967.

15. **Chaudhuri, J. B. and Sharma, A.,** Cytological studies on three members of Hydrocharitaceae in relation to their morphological and ecological characteristics, *Cytologia,* 43, 1, 1978.

16. **Sharma, A.,** Chromosome census of the plant kingdom. I. Monocotyledons, Part I. Butomales to Zingiberales, *Nucleus,* 15, 1, 1972.

17. **Sharma, A. K.,** Cytology of some members of Commelinaceae and its bearing on the interpretation of phylogeny, *Genetica,* 27, 323, 1955.

18. **Sharma, A. K. and Sharma, A.,** Further investigations on cytology of members of the family Commelinaceae with special reference to the role of polyploidy and the origin of ecotypes, *J. Genet.,* 56, 1, 1958.

19. **Sharma, A.,** Chromosomal evolution in Commelinaceae from the Eastern India, Proc. 1st All India Congr. Cytol. Genet., *J. Cytol. Genet. (Suppl.),* 1, 19, 1974.

20. **Bhattacharyya, B.,** Cytological studies on some Indian members of Commelinaceae, *Cytologia,* 40, 285, 1975.

21. **Sharma, A. K. and Ghosh, I.,** Cytotaxonomy of the family Bromeliaceae, *Cytologia,* 36, 237, 1971.

22. **Sharma, A. K. and Bhattacharyya, N. K.,** Inconstancy in chromosome complements in species of *Maranta* and *Calathea, Proc. Nat. Inst. Sci. Ind.,* 24, 101, 1958.

23. **Sharma, A. K. and Bhattacharyya, N. K.,** Cytology of several members of Zingiberaceae and a study of the inconstancy of their chromosome complements, *La Cellule,* 59, 299, 1959.

24. **Mookerjea, A.,** A cytological study of several members of the Liliaceae and their interrelationships, *Ann. Bot. Soc. 'Vanamo',* 29, 1, 1956.

25. **Sharma, A. K. and Chatterjee, A. K.,** Chromosome studies as a means of detecting the method of speciation in some members of Liliaceae, *Genet. Iber.,* 10, 149, 1958.

26. **Sharma, A. K. and Sharma, A.,** An investigation of the cytology of some species of Liliaceae, *Genet. Iber.,* 13, 25, 1961.

27. **Roy, S. C.,** Chromosome study in different species of *Scilla, Rev. Rom. Embryol. Cyt.,* 8, 29, 1971.

28. **Sen, S.,** Structural hybridity, intra- and interspecific level in Liliales, *Folia Biolog.,* 21, 183, 1973.

29. **Chatterjee, A.,** Cytological investigation on certain genera in Liliaceae, *Folia Biolog.,* 21, 199, 1973.

30. **Sen, S.,** Polysomaty and its significance in Liliales, *Cytologia,* 38, 737, 1973.

31. **Sen, S.,** Cytotaxonomy of Liliales, *Feddes. Repert. 2. Bot. Taxon. Geobot.,* 86, 255, 1975.

32. **Sharma, A. K. and Bhattacharyya, N. K.,** Cytology of *Asphodelus tenuifolius* Cav., *Caryologia,* 10, 330, 1957.

33. **Nakai, T.,** Referred in Sato, D., Karyotype alteration and phylogeny in Liliaceae and allied families, *Jpn. J. Bot.,* 12, 57, 1942.

34. **Riley, H. P.,** Chromosome studies in some South African monocotyledons, *Canad. J. Genet. Cytol.,* 4, 50, 1962.

35. **Noda, S.,** Cytogenetics of relationship between *Lilium leichtlinii* var. *leichtlinii* and *L. maximowiczii, Jpn. J. Breeding,* 17, 173, 1967.

36. **Riley, H. P. and Majumdar, S. K.,** *The Aloinea: a biosystematic survey,* University Press of Kentucky, Lexington, 1979, 177 pp.

37. **Cave, M. S.,** *Chromosomes of Californian Liliaceae,* Berkeley, 1970.

38. **Sharma, A.,** Chromosome census of the plant kingdom. I. Monocotyledons, Part II. Liliales, Liliaceae — Tribes I and II, *Nucleus,* 15, 1, 1972.

39. **Sen, S.,** Intraspecific differentiation in the karyotype of *Lilium, Cytologia,* 43, 305, 1978.

40. **Sen, S.,** Chromosomes and evolution in species of *Tulipa, Acta Botanica Indica,* 5, 128, 1977.

41. **Sharma, A. K. and Dutta, K. B.,** An investigation on the cytotypes of *Haworthia, Genet. Iber.,* 14, 131, 1962.

42. **Sharma, A. K. and Mallick, R.,** Interrelationships and evolution of the tribe Aloineae as reflected in its cytology, *J. Genet.,* 59, 116, 1965.

43. **Sharma, A. K. and Mukhopadhyay, S.,** Chromosome study in Agapanthus and the phylogeny of its species, *Caryologia,* 16, 127, 1963.
44. **Sharma, A. K. and Bal, A. K.,** A cytological study of a few genera of Amaryllidaceae with a view to find out the basis of their phylogeny, *Cytologia,* 21, 329, 1956.
45. **Sharma, A. K. and Chaudhuri, M.,** Cytological studies as an aid in assessing the status of *Sansevieria, Ophiopogon* and *Curculigo, Nucleus,* 7, 43, 1964.
46. **Krause, K.,** Liliaceae, in *Die natürlichen Pflanzenfamilien,* Engler, A. and Prantl, E., Eds., 2 Aufl., Bd. 15a, Wilhelm Engelmann, Leipzig, 1930, 227.
47. **Kumar, V.,** The cytogeography of the East Asiatic *Disporum, Proc. 12th Int. Cong. Genet.,* Science Council of Japan, Tokyo, 1968, 186.
48. **Mallick, R. and Sharma, A. K.,** Chromosome studies in Indian Pandanales, *Cytologia,* 31, 402, 1966.
49. **Sharma, A. K. and Bhattacharyya, U. C.,** A cytological study of the factors influencing evolution in *Agave, La Cellule,* 62, 259, 1962.
50. **Sharma, A. K. and Sarkar, A. K.,** A study on the structure and behavior of chromosomes in different species of *Yucca, Bot. Tiddskr.,* 60, 180, 1964.
51. **Bhattacharyya, G. N.,** Chromosomes in different species of *Agave, J. Cyt. Genet.,* 3, 1, 1968.
52. **Sharma, A. K.,** Evolution and taxonomy of monocotyledons, in *Chromosomes Today,* Vol. 2, Oliver and Boyd, London, 1967.
53. **Darlington, C. D.,** *Chromosome Botany and the Origin of Cultivated Plants,* New York, London, 1963.
54. **Sharma, A. K. and Ghosh, C.,** Further investigation on the cytology of the family Amaryllidaceae and its bearing on the interpretation of its phylogeny, *Genet. Iber.,* 6, 71, 1954.
55. **Mookerjea, A.,** Cytology of Amaryllids as an aid to the understanding of evolution, *Caryologia,* 7, 1, 1955.
56. **Sharma, A. K. and Bhattacharyya, N. K.,** An investigation on the karyotype of the genus *Crinum* and its phylogeny, *Genetica,* 28, 263, 1956.
57. **Sharma, A. K. and Ghosh, C.,** The cytology of two varieties of *Polyanthes tuberosa* with special reference to their interrelations and sterility, *Genetica,* 28, 99, 1956.
58. **Sharma, A. K. and Jash, M.,** Further investigation on the cytology of Amaryllidaceae, *Øyton,* 11, 103, 1958.
59. **Sharma, A. K. and Dutta, P. C.,** Chromosome study in species of *Dracaena* with special reference to their means of speciation, *J. Genet.,* 57, 43, 1960.
60. **Sharma, A. K. and Ghosh, I.,** Cytotaxonomy of Dracaena, *J. Biol. Sci.,* 11, 45, 1968.
61. **Traub, H. P.,** *The Genera of Amaryllidaceae,* 1st ed., American Plant Life Society, La Jolla, California, 1963, 1.
62. **Stebbins, G. L.,** Cytogenetics and evolution of the grass family, *Am. J. Bot.,* 43, 890, 1956.
63. **Prat, H.,** La Systématique des Graminées, *Ann. Sci. Nat. Bot. Ser-10,* 18, 165, 1936.
64. **Roshevits, R. Yu,** Sistema Zlakov v svyazi s ikh evolutsieyi, *Sb. Nauchn. Rab. Bot. Inst. in V. L. Komarova Akad. Nauk SSSR,* 25, 1946, 140.
65. **Pilger, R.,** Das System der Gramineae, *Bot. Jahrb.,* 76, 281, 1954.
66. **Beetle, A. A.,** The four subfamilies of the Gramineae, *Bull. Torrey Bot. Club,* 82, 196, 1965.
67. **Tateoka, T.,** Miscellaneous papers on the phylogeny of Poaceae 10. Proposition of a new phylogenetic system of Poaceae, *J. Jpn. Bot.,* 32, 275, 1957.
68. **Prat, H.,** Revere d' agrostologie; vers une classification naturelle des Gramineés, *Bull. Soc. Bot. France,* 107, 32, 1960.
69. **Stebbins, G. A. and Crampton, B.,** A suggested revision of the grass genera of temperate North America, *Rec. Adv. Bot.,* 1, 133, 1961.
70. **Gould, F. W.,** *Grass Systematics,* McGraw Hill, New York, 1968.
71. **Pohl, R. W.,** *How to Know the Grasses,* 2nd ed., Wm. C. Brown, Iowa, 1968.
72. **Clayton, W. D.,** Gramineae, in *Flowering Plants of the World,* Heywood, V. H., Ed., Mayflower Books, New York, 1978.
73. **Hilu, K. W. and Wright, K.,** Systematics of Gramineae: a cluster analysis study, *Taxon,* 3, 9, 1982.
74. **Avdulov, N.,** Karyo-systematische Untersuchung der Familie Gramineen, *Bull. Appl. Bot. Genet. Pl. Br.,* Supl. 44, 1, 1931.
75. **Ghorai, A. and Sharma, A.,** Bumbuseae — A Review, *Feddes Repertorium,* 91, 281, 1980.
76. **Ghorai, A. and Sharma, A.,** Cytotaxonomy of Indian Bambuseae. II. Dendrocalameae and Melocanneae, *Acta Botanica Indica,* 8, 134, 1980.
77. **Hubbard, C. E.,** Gramineae, in *British Flowering Plants,* Hutchinson, J., Ed., P. R. Gawthorn Ltd., London, 1948.
78. **Ghorai, A. and Sharma, A.,** Chromosome studies in some Festuceae, *J. Indian Bot. Soc.,* 60, 148, 1981.
79. **Tischler, G.,** Pflanzliche Chromosomen-Zahlen. IV, *Tabul. biol.,* 16, 162, 1938.
80. **Hagerup, O.,** On fertilization, polyploidy and haploidy in *Orchis smanlatus* L., *(sensu lato) Dansk. Bot. Arkiv.,* 11, 1, 1944.

81. **Sharma, A. K. and De, D. N.,** Cytology of some of the millets, *Caryologia,* 8, 294, 1956.

82. **Sharma, A. K.,** Chromosome studies in some Indian Barley. I, *Proc. Nat. Inst. Sci. Ind.,* 22, 246, 1956.

83. **Sharma, A. K. and Bhattacharjee, D.,** Chromosome studies in *Sorghum* I., *Cytologia,* 22, 287, 1957.

84. **Sharma, A. K. and Jhuri, L. K.,** Chromosome analysis of some grasses. I, *Genet. Iber.,* 11, 145, 1960.

85. **Schlechter, R.,** Planta sinenses, Orchidaceae, in *Meddle från Göteborgs,* Botaniska Taüdgard, 1, 125, 1934.

86. **Afzelius, K.,** Zytologische Beobacktungen an einigen Urchidacean, *Svensk. Bot. Tidskr.,* 37, 266, 1943.

87. **Duncan, R. E.,** Orchids and Cytology, in *The Orchids,* Withner, C. L., Ed., Ronald Press, New York, 1959, 189.

88. **Sharma, A. K. and Chatterjee, A. K.,** The chromosome numbers of a few more orchid genera, *Curr. Sci.,* 30, 75, 1961.

89. **Sharma, A. K. and Chatterjee, A. K.,** Cytological studies on orchids with respect to their evolution and affinities, *Nucleus,* 9, 177, 1966.

90. **Roy, S. C. and Sharma, A. K.,** Cytological studies of Indian orchids, *Proc. Nat. Acad. Sci. Ind.,* 388, 72, 1972.

91. **Tanaka, R. and Kamemoto, H.,** List of chromosome numbers in species or Orchidaceae, in *The Orchids: Scientific Studies,* Withner, C. L., Ed., John Wiley & Sons, New York, 1974, 411.

92. **Mehlquist, G. A. L.,** Some aspects of polyploidy in Orchids, with particular reference to *Cymbidium, Paphiopedilum* and the *Cattleya* alliance, in *The Orchids: Scientific Studies,* Withner, C. L., Ed., John Wiley & Sons, New York, 1974, 393.

93. **Jones, K.,** Cytology and the study of Orchids, in *The Orchids: Scientific Studies,* Withner, C. L., Ed., John Wiley & Sons, New York, 1973, 383.

94. **Jones, K. and Daker, M. G.,** The chromosomes of orchids. III. Catasetineae Schltr., *Kew Bull.,* 22, 421, 1968.

95. **Sharma, A. K.,** Chromosome and distribution of Monocotyledons in the Eastern Himalayas, in *Tropical Botany,* Larsen, K., Ed., Academic Press, New York, 1979, 327.

96. **Chaudhuri, M. and Sharma, A. K.,** unpublished.

97. **Sharma, A. K.,** Genetic diversity of some monocotyledonous crops and their wild relatives in the Eastern Himalayas, Region Workshop Cons. Trop. Pl. Res. SE Asia, New Delhi, March 8—9, 1982.

98. **Rendle, A. B.,** *The Classification of Flowering Plants,* Vol. 2, 2nd ed., Cambridge University Press, Cambridge, 1967.

99. **Sharma, A. K. and Sarkar, S. K.,** A new technique for the study of chromosomes of palms, *Nature,* 176, 261, 1955.

100. **Sharma, A. K. and Sarkar, S. K.,** Cytological basis of differentiation in palms, *Sci. Cult.,* 22, 175, 1966.

101. **Sharma, A. K. and Sarkar, S. K.,** Cytology of different species of palms and its bearing on the solution of the problems of phylogeny and speciation, *Genetica,* 28, 361, 1956.

102. **Sarkar, S. K.,** Male sterility in palms, *Agron. Lusit.,* 18, 257, 1956.

103. **Sharma, A. K. and Das, N. K.,** Study of karyotypes and their alterations in Aroids, *Agron. Lusit.,* 16, 23, 1954.

104. **Mookerjea, A.,** Cytology of different species of Aroids with a view to trace the basis of their evolution, *Caryologia,* 7, 221, 1955.

105. **Sharma, A. K. and Dutta, K. B.,** A cytological study to work out the trend of evolution in *Aglaonema richardia, Caryologia,* 14, 439, 1961.

106. **Sharma, A. K. and Bhattacharyya, U. C.,** Structure and behaviour of chromosomes in species of *Anthurium* with special reference to the accessory chromosomes, *Proc. Nat. Inst. Sci. Ind.,* 27B, 317, 1961.

107. **Sharma, A. K. and Sarkar, A. K.,** Cytological analysis of different cytotypes of *Colocasia antiquorum, Bull. Bot. Soc. Bengal,* 17, 16, 1963.

108. **Sharma, A. K. and Sarkar, A. K.,** Studies on the cytology of *Caladium bicolor* with special reference to the mode of speciation, *Genet. Iber.,* 16, 21, 1964.

109. **Sharma, A. K. and Mukhopadhyay, S. B.,** Chromosome studies in *Typhonium* and *Arisaema* with a view to find out the mode of origin and affinity of the two, *Cytologia,* 30, 58, 1965.

110. **Bhattacharyya, G. N. and Sharma, A. K.,** A cytotaxonomic study on some taxa of Araceae, *Genet. Iberica,* 18, 237, 1966.

111. **Chaudhuri, J. B. and Sharma, A.,** Chromosome studies in certain members of Araceae, *Genet. Iber.,* 30, 161, 1978.

112. **Hara, H.,** *The Flora of Eastern Himalayas,* (2nd rep.), University of Tokyo Press, Tokyo, 1971.

113. **Kurosawa, S.,** Cytological studies on some Eastern Himalayan Plants, in *The Flora of Eastern Himalayas,* Hara, H., Ed., University of Tokyo Press, Tokyo, 1966, 658.

114. **Sharma, A. K. and Bal, A. K.,** A cytological investigation of some members of the family Cyperaceae, *Øyton,* 6, 7, 1956.

115. **Sanyal, B. and Sharma, A.,** Cytological studies in Indian Cyperaceae. I. Tribe Scirpeae, *Cytologia,* 37, 13, 1972.
116. **Sanyal, B. and Sharma, A.,** Cytological studies in Indian Cyperaceae. II. Tribe Cypereae, *Cytologia,* 37, 33, 1972.
117. **Sharma, A. K. and Sharma, A.,** Recent advances in the study of chromosome structure, *Bot. Rev.,* 24, 511, 1958.
118. **Randolph, L. F.,** Chromosome numbers in native American and introduced species and cultivated varieties of *Iris, Bull. Amer. Iris Soc.,* 52, 61, 1934.
119. **Randolph, L. F. and Mitra, J.,** Chromosome number of *Iris* species, *Bull. Amer. Iris Soc.,* 140, 2, 1956.
120. **Sokolovskaya, A. P.,** Geografischeskoe rasprostranenie poliploidnykh. vidov rastenity (Issledovanie flory Primorskogo Kraya), *Vestn. Leningr. Univ. Ser. Biol.,* 23, 92, 1966.
121. **Banerjee, M. and Sharma, A. K.,** A cytotaxonomical analysis of several genera of the family Iridaceae, *Plant Sci.,* 3, 14, 1971.
122. **Sharma, A. K. and Sharma, A.,** Chromosome studies of some varieties of *Narcissus tazetta* L., *Caryologia,* 14, 97, 1961.
123. **Sharma, A. K. and Talukder, C.,** Chromosome studies in members of the Iridaceae and their mechanism of speciation, *Genetica,* 31, 340, 1960.
124. **Sharma, A. K.,** Plant Cytogenetics, in *The Cell Nucleus,* Vol. 2, Busch, H., Ed., Academic Press, New York, 1974, 264.
125. **Sharma, A. K. and Sharma (nee Mookerjea), A.,** Chromosomal alterations in relation to speciation, *Bot. Rev.,* 25, 414, 1959.
126. **Sharma, A. K.,** A new concept of a means of speciation in plants, *Caryologia,* 9, 93, 1956.
127. **Sharma, A. K. and Sharma (nee Mookerjea), A.,** Fixity in chromosome number of plants, *Nature (London),* 177, 335, 1956.
128. **Jones, R. N.,** B chromosome system in flowering plant and animal species, *Int. Rev. Cytol.,* 40, 1, 1975.
129. **Müntzing, A.,** Accessory Chromosomes, *Ann. Rev. Genet.,* 8, 243, 1974.
130. **Sharma, A. K. and Aiyangar, H. R.,** Occurrence of B-chromosomes in diploid *Allium stracheyii* Baker and their elimination in polyploids, *Chromosoma,* 12, 310, 1961.

Chapter 8

CHROMOSOMES IN EVOLUTION IN HETEROPTERA

G. K. Manna

TABLE OF CONTENTS

I. INTRODUCTION

Sex chromatin was first detected in *Pyrrhocoris apterus*[1] and subsequently in other Heteroptera,[2-5] grasshopper[6], human somatic cells[7] and other mammals.[8] The study of chromosomes in heteropteran species has generally been carried out in the testes and very rarely in the ovary[9] and in somatic cells of embryos.[10] The strong foundation of heteropteran chromosome cytology was laid by Montgomery,[2,3,11-14] Wilson,[4,5,15-25] Gross,[26-29] Foot and Strobell,[30-33] Payne,[34-36] Browne[37-39] and others (reviewed by Manna[40-42] and Ueshima[43]). The chromosome constitution and their behavior during spermatogenesis of nearly 1200 species belonging to 46 families of Heteroptera have been determined. Detailed lists were made by Ueshima,[43] Makino,[44,45] Takenouchi and Muramoto[46] and Manna et al.[47-51] of the materials studied by them and other workers. Unlike the Homoptera, in species of Heteroptera so far examined cytologically bisexuality has been distinct, as are their sex chromosomes, except in *Lethocerus* sp.[52] Evolutionary complexity is thus less than that of Homoptera because the latter contains the superfamily Coccoidea with various parthenogenetic species showing aberrant chromosome cycles and anomalous types of meiosis[53-56] while bisexuality and well-defined sex chromosome mechanisms are found in most other major groups.[57,58] Therefore, the evolutionary mechanism in Heteroptera is restricted solely to the bisexually reproducing species.

II. CENTROMERIC STRUCTURE AND EXPERIMENTAL STUDIES

The centromeric constitution of heteropteran chromosomes remained an enigmatic problem. Though the chromosome with a single centromere is of almost universal occurrence, yet the plant *Luzula*,[59,60] the heteropteran and homopteran insects, and related groups like lice,[61,62] earwigs,[63] etc. were claimed to have holocentric chromosomes.[8] The holokinetic chromosome has been interpreted as having diffuse kinetochore/diffuse centromeric activity or numerous but discrete and localized centromeres.

The holokinetic nature of the chromosomes in the Homoptera and Heteroptera was based mainly on their morphological behavior and experimental verification after breakage by ionizing radiations. Schrader[66] observed that the heteropteran species *Protenor belfragei* possesses a conspicuously large X chromosome which is oriented parallel to the equator and that the chromosomal fibers are organized along its entire length. The holocentric behavior of the chromosome in heteropteran bugs was supported further in his later publications,[64,67] as well as from experimental studies.[68] The concept of such chromosomes with diffuse centromeric activity received active support from many heteropteran cytologists.[43,69-74] Electron microscopic study of meiotic chromosomes of male *Oncopeltus fasciatus*[75] showed a correlation between the visual image obtained in the nuclei isolated on a Langmuir trough and the light microscopic observation. They noted that, in contrast to the somatic and gonial mitoses, the chromosomes of metaphase I showed a restriction of kinetic activity to a limited region of the chromatids. In *Oncopeltus*, Comings and Okada[76] observed that in organisms with holokinetic chromosomes the formation of the meiotic kinetochore apparatus might have to be suppressed to allow terminalization of chiasmata. In a reduviid species *Rhodnius prolixus,* Buck[77] also reported the presence of the kinetochore plate in mitotic chromosomes and its absence in meiotic chromosomes. The study on the determination of base composition, buoyant density, thermal stability and reassociation kinetics, renaturation of DNA, etc. in chromosomes in *Oncopeltus* revealed that the repeated sequences were primarily short and scattered throughout the genome in contrast to the tandem repeats found in DNAs of organisms with a localized centromere.[78] The holocentric nature of chromosomes received strong support from the parallel movement of X-ray-induced fragments observed in a homopteran coccid, *Steatococcus*, by Hughes-Schrader and Ris.[79] The X chromosome fragments during meiosis in the bearberry aphid *Tamallia* behaved in the same way.[80] Halkka[81] recorded the holokinetic behavior of the chromosomes of the leafhopper *Limotettix*. However, Seshachar et al.[82] claimed that the chromosomes of *Eurybrachis* possess localized centromere. Hughes-Schrader and Schrader,[68] in the X-irradiated male heteropteran bugs, *Euschistus servus, E. tristigmus* and *Solubea pygnax*, observed a high frequency of translocations and deletions besides free fragments and, at anaphase, laggards and bridges. Because of the free movement and perpetuation of the fragments, they[68] concluded that the chromosomes were holocentric. LaChance and his collaborators[83-86] observed the perpetuation of the fragments and translocations in the meiotic cells of F_1, F_2, and F_3 males of *Oncopeltus fasciatus* to support the holokinetic nature of chromosomes. However, other workers did not share this view of the holocentric chromosome structure.[87] From the morphological behavior, Mendes[88] in the pyrrhocorid bug *Dysdercus*; Rao[89] in the lygaeid bug *Oncopeltus*; and Parshad[90,91] in pyrrhocorid, lygaeid, and coreid bugs, opined that the chromosomes of Heteroptera were monocentric. Parshad,[91] from studies on *Cletus punctiger* and *Lanchnophorus singalensis*, concluded that in the chromosomes of Heteroptera, unlike those of homopteran coccids and aphids with diffused activity, the kinetic activity was localized and resembled the T-chromosome of rye.[92] Thus, the two independent properties of the centromere — "active mobility" and "the region of special cycle" — may not be evident in the chromosome itself which brought about the difference in chromosome behavior in Heteroptera. Desai,[93] from irradiating the chromosomes of *Ranatra* with high doses of X-rays, generalized that heteropteran chromosomes were monocentric. The monocentric nature of chromosomes claimed by Mendes[88] in *Dysdercus* could not be ignored because of some later supportive evidence. Ruthmann and Permantier[94] in *D. intermedius* found the chromosomal spindle microtubules to be inserted in discrete kinetochores in spermatogonial chromosomes. They stated that the holokinetic behavior during mitotic anaphase movement was due to chromosomal interconnections which developed during prometaphase and were absent in first meiotic stages. Later Ruthmann and Dahlberg[95] in *D. intermedius* explained the X_1X_2 pairing and orientation on the basis of telocentric organization of meiotic chromosomes through the 'telomere' — the special property of chromosome ends in the behavior of meiotic autosomes and sex chromosomes.

Many workers[68,76,96-98] agreed that the kinetic organization of mitotic chromosomes appeared different from that of meiosis. In *Dysdercus*, Ruthmann and Permantier[94] observed spindle microtubules at the end in single dense spots of kinetochore material. The bivalents were separate, apparently telocentric for their coorientation, and connected by their free telomeric ends to the nearest pole, and the remaining telomeres were joined to the sister chromatids to hold the half-bivalents together. The two telomeres of a chromatid were therefore capable of spindle attachment. The differential behavior of meiotic and mitotic chromosomes, with the former showing restricted centric activity to shift from holocentric to telocentric activity,[64,66] was given by Gassner and Klemetson[98] though opposite explanations were cited before.[94,95] They[68] admitted that the unsolved enigma was the disproportionate activity of chromosome ends in anaphase movement during meiosis of holocentric chromosomes. Ueshima[43] suggested that the telomere of chromosomes as envisaged by John and Lewis[99] has some properties resembling those of the centromere. Even a well-defined monocentric chromosome under a different situation is known to shift its activity, as seen in maize plant.[100] Piza[101] made a very unusual claim, that the chromosomes of Heteroptera possessed a dicentric structure, one centromere at each end, because he observed that the longer chromosomes bent at the middle while the two ends directed to the pole during anaphase movement. This claim did not receive any support.[8,43]

Manna with his collaborators[102-105] obtained various aberrations with X-rays in chromosomes of the testes of *Physopelta schlanbuschi* and *Lygaeus hospes*. The male gonial complements of *Physopelta* $(2n = 12 + 2m + X_1X_2Y)$[106] contained two conspicuously large but indistinguishable marker sex chromosomes, X_1 and Y, of the X_1X_2Y complex. The *Physopelta* nymphs showed that the free fragments induced by X-rays did not survive long. Further, in comparison to other type aberrations, the frequency of fragments was very low.

The study of the free fragments induced by 50 r—400 r in the testes of *Lygaeus hospes* also indicated a similar trend, although the frequency was higher. The data on X-ray-induced free fragments in *Physopelta* and *Lygaeus* clearly indicated that the free fragments did not survive for a sufficient time for the cell to undergo another divisional cycle. The data, therefore, did not support the holocentric structure of chromosomes in these two species of Heteroptera.

The different morphological and experimental studies on the chromosomes of Heteroptera did not render unchallenged support to the holocentric organization of chromosomes in this group. The shifting of the centromeric activity from a holocentric nature to a telocentric one during meiotic anaphase, the occurrence of regular heteroploidy in the harlequin lobe in some pentatomid bugs,[107-110] the X-ray-induced behavior of the chromosome aberrations and the free fragments in some species of Heteroptera, the radioprotective effect of penicillin in X-irradiated chromosomes in *Physopelta*,[103] and the findings of other workers[88,94,95] would require reconsideration of the centromeric organization of chromosomes in Heteroptera rather than taking for granted that they have been holocentric. In some species of *Graptostethus*, *Dieuches*, etc.,[40] a very striking change occurs in the length and and breadth in chromosomes from diplotene to metaphase I due to spiralization and condensation. For all these reasons, the chromosomes of Heteroptera may be regarded as basically monocentric and the variations in their behavior as due to their highly spiralized and condensed state. The centipeds (Chilopoda) are accepted to have localized centromeres, but in *Thereupoda clunifera* and *Thereuonema hilgendorfi* the X and the Y and in *Esastigmatobius longtarisis* a pair of autosomes were outstandingly large and have been considered as holocentric chromosomes.[111-114] Ogawa[112,113] from the study of the X-ray-induced fragments in mitotic division supported the holocentric nature of these giant chromosomes. It may be concluded that the chromosome organization in Heteroptera, though basically monocentric, shows uncommon behavior. It could be taken as the evolutionary diversification of the spindle elements, the spiralization process, etc. rather than the centromeric structure itself. The term 'holocentric' would, therefore, have a restricted meaning which is the monocentric chromosome with modified behavior different from the classical type.

III. MITOTIC CHROMOSOMES

The mitotic chromosomes from somatic cells have rarely been studied in Heteroptera except some in cleaving cells of the embryo.[10] In most cases, gonial chromosomes of males showed some degree of specialization. In many species of Heteroptera, spermatogonial metaphase chromosomes were found to be connected with thin, thread-like interchromosomal connections. DuPraw[115] interpreted this to mean that all the chromosomes of the complement represented merely segments of a continuous circular DNA molecule during the metaphase configuration of some coreid bugs. The interchromosomal connections seen at metaphase complements in some, but not in all species of Heteroptera were random, being both terminal and lateral,[40] thus giving no support to the hypothesis by DuPraw.[115] These lateral and terminal connections were regarded as the extension of DNA material of the chromosomes in the form of loose loops or else as fixation artifacts. In the mitotic complement, the longer chromosomes during anaphase moved towards the pole in a sheet-like manner. So far, there is no clear indication of their bend at the centric region as would be expected from monocentric chromosomes except some sporadic claims.[88,91,101] The metaphase chromosomes in the equator showed no tendency of radial orientation, in contrast to meiotic metaphases in most families of Heteroptera. The mitotic chromosomes characteristically did not show any longitudinal differentiation, including banding. However, Muramoto[220,221] obtained C-banded and G-banded chromosomes in testes of some species of Heteroptera, and stated that the chromosomes are generally acrocentric with diffuse centromeres. Some chromosomes were found with C bands at 2 ends, some at 1 end, while some had none.[220] The banding techniques have yet to be standardized, before any definite claim can be made. Manna[40] reported that the largest pair of autosomes in many spermatogonial metaphases showed a SAT-like structure in *Bracyplatys subaeneus* (Plataspidae), which could be traced in the meiotic bivalent to a knob-like structure at one end. Perhaps this represented the secondary constriction, although no trace of a primary constriction was observed. However, in spermatogonial chromosomes, the X-ray-induced breaks were localized more in the middle in some marker and longer chromosomes.[103,106]

Geitler[116-120] made a critical study of somatic polyploidy in some species of Heteroptera, especially in the pond skater *Gerris*, according to the number of heteropycnotic X chromosomes in *G. lateralis*. In *Lygaeus saxatilis* it was determined by the heteropycnotic Y because the X was normal. The significance of somatic polyploidy in tissues of Heteroptera has not been understood.

IV. MEIOTIC CHROMOSOMES AND SPECIALIZED BEHAVIOR

The evolution of the meiotic pattern seemed to have taken place in more orderly manner in Heteroptera for which major groups could be characterized on the basis of chromosome number and size, sex chromosome mechanism, arrangement of chromosomes at metaphase I and time of division of the sex chromosomes, the presence of m chromosomes, and so on,[40-43] and thus interrelationships between and within the groups could be derived. The patterns of meiosis in different groups of Heteroptera show uniformity in some basic behavior. The spermatogonial chromosomes were generally stumpy rods and oval structures when the diploid number was low while they were small spheres, or dots, in high numbers. They were evenly arranged at the metaphase spindle, showing no longitudinal differentiation. The m chromosome when present can be distinguished by minute size and somewhat negative heteropycnotic behavior, while the sex chromosomes can be identified by the size difference.

A. Sex Chromosomes

The sex chromosome types in male heteropteran species could broadly be generalized as XO, X_nO, XY, and X_nY. In the X_nO type, the number of X chromosomes could be determined only in the spermatogonial complement because during meiosis, except in some cases like *Dysdercus*,[41,122] the multiple Xs would generally fuse as a single element. In the case of the XY and X_nY type, the sex chromosomes were fused and heteropycnotic at the beginning and by diakinesis they were separate as univalents, and each of them might show a clear chromatid split. The first metaphase orientation of the sex chromosomes was somewhat characteristic as, in case of the XO type with m, the latter would invariably occupy the center of the hollow spindle while the X would lie at the peripheral ring formed of autosomal bivalents or even beyond the ring. In X_nO, the fused Xs mostly oriented outside the ring of autosomal bivalents. In *Dysdercus* the X_1 and X_2 would always remain side by side within the circle formed by the autosomal bivalents and the equationally separated chromatids fused during anaphase I. Therefore, the regular autosomes in most of the species of Heteroptera formed a ring-like orientation at both metaphases I and II while the positions of the sex chromosomes were variable. The m chromosome oriented regularly at metaphase I, but at metaphase II its position was variable.

Though postreductional behavior of the sex chromosome was the characteristic of Heteroptera, its change to the prereductional pattern has been witnessed in some species in a family, as in the Tingidae. It was uncertain if the prereductional meiosis represented the original mechanism[48] or if it was secondarily adapted.

Table 1
LIST OF HETEROPTERAN SPECIES SHOWING
PRE-REDUCTIONAL SEPARATION OF THE SEX
CHROMOSOMES

Taxonomy	Male sex mechanism	Ref.
Coreidae		
Coreinaeae		
Archimerus alternatus	X0	9, 123
A. calcarator	X0	4, 5
Pachylis gigas	X0	20, 124
P. lateralis	X0	125
Tingidae		
Agramminae		
Agramma nexile	$14 = 12 + XY$	126
Tinginae		
Acalypta parvula	$12 = 10 + XY$	128
Bredenbachius	$14 = 12 + XY$	127
consanguineus		
Cochlochila lewisi		
(as *C. conchata*)	$14 = 12 + XY$	144
Dasytingis rudis		
(as *Tingis bengalan*)	$14 = 12 + XY$	127
Dictyla humuli		
(as *Monathia humuli*)	$14 = 12 + XY$	128
Dictyonota fuliginosa	$14 = 12 + XY$	128
Leptobyrsa decora	$14 = 12 + XY$	129
Stephanitis nashi	$14 = 12 + XY$	130
S. takeyai (as *Menathis*	$14 = 12 + XY$	127
globulifera)		
Teleonemia elata	$14 = 12 + XY$	129
T. scrupulosa	$14 = 12 + XY$	129
Tingis ampliata	$14 = 12 + XY$	128
T. cardui	$14 = 12 + XY$	128
T. clavata	$14 = 12 + XY$	2, 12
T. lasiocera	$14 = 12 + XY$	126
Reduviidae		
Echrichodiinae		
Ectrychotes abbreviatus	$29 = 28 + X0$	48
E. dispar	$29 = 28 + X0$	40
Notonectidae		
Anisopinae		
Anisops fieberi	$26 = 22 + 2m + X_1X_2O$	131
A. nivea	$26 = 22 + 2m + X_1X_2O$	131
A. sardea	$26 = 22 + 2m + X_1X_2O$	131
Bellostomatidae		
Lethocerus indicus	$26 = 22 + 2m + XY$	48

It would appear from the list (Table 1) that 5 out of 46 families have some species with prereductional meiosis, including both land and aquatic bugs, which might indicate the widespread occurrence of the primitive feature.

B. *m* Chromosomes

A pair of minute chromosomes denoted originally as *m* chromosomes by Wilson[5] characterized some taxa. Except in the Heteroptera (and in some vertebrates), the occurrence of this type of chromosome has been extremely rare. The *m* pair showed a great specialization. The general course of meiosis has been fairly constant except for the *m* pair in some groups of Heteroptera where they occurred. The heteropycnotic *m* chromosomes remained unpaired during prophase I, but at metaphase I the two invariably associated in the form of touch-and-go-pairing. They occupy characteristically the center of the hollow spindle in Coreidae, Alydidae, and Rhopalidae, generally with an XO, X_nO sex complex, while in others they might share the position depending on the sex chromosome constitution of the species having mostly XY. However, in *Physopelta* with X_1X_2Y, the *m* pair would occupy the central position as in the typical coreid bug. Anyhow, the achiasmatic *m* chromosome would undergo reductional separation at anaphase I and equational separation in anaphase II. The evolution of the *m* chromosomes may be traced from some regular form of autosomes. The *m* chromosomes were found to vary in size and behavior. In coreid bugs, like *Archimerus calcrator, Pachylis gigas*,[20] and *Petillia notatipes*[40] they were exceedingly small and could be taken as typical. On the other hand, in *Acanthocoris scabrator*, Manna[40] failed to identify them positively. The spermatogonial metaphase complement of this species contained 28 gradually seriated chromosomes of which the smallest pair in some plates appeared as an *m* pair, but during meiosis these two chromosomes neither showed negative heteropycnosis nor were far apart. Moreover, rarely in some later diplotene nuclei they seemed to be held together by an interstitial chiasma.[40] At metaphase I, this pair did not occupy the central position of the hollow spindle. Thus, they could not definitely be claimed to be the *m* pair although they did not behave like regular autosomal bivalents. Similarly, in *Hygia touchei*, Manna and Deb Mallick[50] failed to observe any small pair of chromosomes behaving as *m* chromosomes although Muramoto[126] (in another species) found them. The above example could be taken as the first step toward the evolution of the *m* chromosome. In the next step, the behavior of the smallest pair of chromosomes in *E. granulipes* was not readily distinguished. They showed no differential staining behavior during prophase I, but sometimes lay separately. The chromosomal arrangement at metaphase I was atypical for the coreid pattern. Therefore, in this species, behavior of the *m* pair deviated from that of the typical one in having no conspicuous size difference, absence of differential staining behavior, and atypical arrangement at metaphase I. In the third step, all the typical characteristics were shown by the *m* pair except that their size difference from the smallest pair of regular autosomes was not appreciable, as found in *Protenor belfragei, Anasa tristis*, etc. In the final step the *m* chromosome had all the characteristic features. Therefore, the *m* pair having minute size, negatively heteropycnotic behavior, unassociated disposition at prophase I, and achiasmate association occupying the central position of the hollow spindle at metaphase I typically seen in Coreidae, originated from a pair of smaller regular autosomes very likely through heterochromatinization. Transformation of a regular pair of autosomes as supernumerary in grasshopper[132] or incorporation into a multiple sex chromosome complex in gryllids[133] by heterochromatinization was suggested before. The origin of the *m* pair was also thought to have evolved along the same line. Though the *m* pair has been found to have great cytotaxonomic significance, it has been lost in many families, with occasional persistence in one or two species. That the *m* chromosome underwent heterochromatinization and disappeared in various families would be supported from the fact that, besides its regular occurrence in the families like the Coreidae, Alydidae, Rhopalidae, all subfamilies except Lygaeinae and Oxycareninae of Lygaeidae, Largidae, Belostomatidae, Corixidae, Naucoridae, Notonectidae, Plaeidae, Saldidae, Colobathristidae and Stenocephalidae, it was found only in *Scotinophara* sp.[134] among Podopidae, *Microtomus*[135] among Reduviidae, *Iphita limbata*[49] among Pyrrhocoridae, and *Eurystylus coelestialium* among Miridae, of different families cytologically investigated. Even among families characterized by presence of the *m* chromosomes there were some species like *Tropistethus holosericus*[136] in Rhyparochrominae of Lygaeidae, *Largus cinctus* and *L. succinctus*[20] in Largidae, *Corynocoris* (=*Merocoris) distinctus*,[22] *Ceraleptus obtusus*,[137] *Clavigralla gibbosa*,[41] *?Hygia touchi*,[50]) and *Cletus* sp.[138] in Coreidae. Its presence was uncertain in *Myrmus miriformis*,[40] *Leptocoris trivittatus*,[22] and *Harmostes reflexulus*[15] under Rhopalidae, *Lethocerus* sp.[52,140,142] and doubtful in some species of *Sigara*[128] and *Chiloxanthus pilosus*[143] and also *Saldula saltatoria*.[144] The *m* pair has been fairly stable in species where it is present. Occasionally individuals with an extra one or 2*m* besides the regular pair were reported in *Acanthocephala* and *Archimerus*.[23,123] Sometimes in some spermatogonial metaphases an extra *m* chromosome was found in *Dieuches uniguttatus*[40] which showed the touch-and-go pairing with other *m* chromosomes at metaphase I.

V. CHROMOSOME POLYMORPHISMS

In Heteroptera, chromosomal polymorphism posed a peculiar situation mainly because of the chromosomal organization. Most cytologists held an opinion that a fission and fusion mechanism played the most vital role in chromosomal evolution in Heteroptera. Structural heterozygotes for some chromosomal rearrangements like inversion and reciprocal translocation have not been reported in natural populations. In *Physopelta* the X-ray-induced exchanges took place preferentially between the X_1 and Y[102,103] while it is a normal feature in autosomes, indicating differential susceptibility of chromosomes for changes. In a natural population, structural anomalies in chromosomes of the harlequin lobe (the fifth of seven lobes) in the tropical members of Pentatomini, Halyini, and Discocephalini were reported by Bowen[145,146] and Schrader.[107-110] The cytological studies of the harlequin lobes in 21 species revealed the regular occurrence of abnormal meiosis leading to only aneuploid sperm in variable number which would very likely have little genetic function, although Schrader[110] opined that they might contribute four times as much nucleoproteins as a normal sperm.

Though structural heterozygotes for inversion and translocation have not been recorded in any one of some 1,200 species so far examined, fusion between sex chromosomes and autosomes has been suspected in belastomatids, *Lethocerus* sp.[8] and *L. indicus*. Other forms of chromosomal polymorphisms in natural populations have been discovered. The same species cytologically examined by different workers from different countries was reported to have different chromosome numbers (Table 2). These chromosomally polytypic species possibly would support their origin through chromosomal polymorphism.

The difference in chromosome numbers was due to the number of pairs of autosomes in most cases, sometimes the sex chromosome constitution, e.g., *P. apterus, C. marginatus*, etc.; presence and absence of an *m* pair, e.g., *C. schillingi*; and supernumerary sex chromosomes as in *Cimex* and Y and autosomal origin as in *Acanthocephala (Metapodius)*.

In addition, polymorphisms have been recorded in a number of species, but in most cases larger samples from the population concerned have not been studied to measure their evolutionary adaptiveness. Considerable information is, however, available on accumulation of supernumerary chromosomes in homopteran insects.[178-184] The chromosomally polymorphic species of Heteroptera involve polymorphism either of autosomes or of sex chromosomes. A precise definition of supernumerary chromosomes in animals[8] and B chromosomes in plants,[185,186] which covers all variables, has been lacking. Manna and Mazumder[132] discovered the origin of the supernumerary chromosomes by the degradation of a regular member of the genome in the grasshopper. Their nature and behavior along with cytogenetical implication in animals have been reviewed.[8,99] The behavior of the supernumerary chromosome in Heteroptera has been found to be somewhat different. They did not always show a sharp size distinction from a regular chromosome. They occur very regularly in some species of *Cimex, Acanthocephala*, etc. In Heteroptera supernumerary chromosomes other than of sex chromosomal origin have been seen only in very few species. Wilson[17] reported them in a pentatomid species. In the Lygaeidae two species, *Dieuches* sp. and *Metochus* (as *Dieuches*) *uniguttatus*[40] were found to bear a supernumerary chromosome. The extra

Table 2
**LIST OF CHROMOSOMALLY POLYTYPIC SPECIES IN
HETEROPTERA**

Taxon	Spermatogonial constitution	Ref.
Pentatomidae		
Asopinae		
Picromerus bidens	$12 = 10 + XY$	147
	$14 = 12 + XY$	119, 137
Solubea (as *Oebalus*) *pugnax*	$10 = 8 + XY$	20
	$14 = 12 + XY$	68
Thyanta pallidovirens	$14 = 12 + XY$	43
(as *custator*)	$16 = 14 + XY$	148
Lygaeidae		
Orsillinae		
Neseis hiloensis	$16 = 12 + 2m + XY$	149
approximatus	$18 = 14 + 2m + XY$	149
Pyrrhocoridae		
Pyrrhocorinae		
Iphita limbata	$20 = 14 + 6X0$	150
	$23 = 22 + X0$	151
	$23 = 12 + 2m + 9X0$	49
Pyrrhocoris apterus	$23 = 22 + X0$	20, 21
	$24 = 22 + X_1X_20$	28
Coreidae		
Coreinae		
Coreus (as *Syromastus*)	$22 = 18 + 2m + X_1X_20$	26, 147
marginatus (as *Mesocerus*	$23 = 20 + 2m + X0$	137, 152
marginatus)		
Centrocoris variegatus	$22 = 18 + 2m + X_1X_20$	40
	$21 = 18 + 2m + X0$	137
Syromastus (as *Velusaria*)	$19 = 18 + X0$	137
rhombeus guadrata	$22 = 18 + 2m + X_1X_20$	152
Cletus rusticus	$18 = 14 + 2m + X_1X_20$	153
	$17 = 14 + 2m + X0$	46
Acanthocephala (as	$21 = 20 + X0$ (1—4 sup)	2
Metapodius) *terminalis*	$22 = 20 + XY$	22
Holymania clavigera	$27 = 24 + 2m + X0$	154
	$29 = 26 + 2m + X0$	155
Rhopalidae		
Lioryssus hyalinus	$13 = 10 + 2m + X0$	137
	$15 = 12 + 2m + X0$	156
Chrosoma schillingi	11	157
	$13 = 10 + 2m + X0$	158
Berytidae		
Berytumus minor	$28 = 26 + XY$	128
	23—25 at Met I	124
Miridae		
Phylinae		
Plagiognatus arbustrom	$34 = 32 + XY$	159
	$32 = 30 + XY$	160
P. chrysanthemi	$30 = 28 + XY$	159
	$32 = 30 + XY$	160
Dicyphinae		
Dicyphus atachydis	$48 = 46 + XY$	159
	$49 = 46 + X_1X_2Y$	160
Mirinae		
Calocoris norvegius	$32 = 30 + XY$	159
	$34 = 32 + XY$	160

Table 2 (continued)
LIST OF CHROMOSOMALLY POLYTYPIC SPECIES IN HETEROPTERA

Taxon	Spermatogonial constitution	Ref.
Nabidae		
Nabinae		
Himacerus (as *Nabis*,	$18 = 16 + XY$	157, 160
Reduvius) *apterus*	$40 = 38 + XY$	152, 161
Nabis ericeterus	$20 = 18 + XY$	160
	$18 = 16 + XY$	162, 160
N. (as *Reduviolus*) *rugosus*	$18 = 16 + XY$	162
	$20 = 18 + XY$	152, 160
Reduviidae		
Harpactorinae		
Polididus armatissimus	$12 = 10 + XY$	150
	$14 = 12 + XY$	163
Stenopodinae		
Pysolampis foeda	$24 = 22 + XY$	150
	$25 = 22 + X_1X_2Y$	164
Zellinae		
Prionidus (as *Arilus*)	$26 = 22 + X_1X_2X_3Y$	34
	$28 = 24 + X_1X_2X_3Y$	43
Cimicidae		
Cimicinae		
Cimex lecturarius	$30—34 = 26 + X_1X_2Y + 2—5X$	165,166,168
	$29—36 = 26 + X_1X_2Y + 2—7X$	73, 169
Paracimex borneensis	$44—47 = 36 + (7—9X)Y$	74
P. capitatus	$41—42 = 36 + (4—6X)Y$	74
Nepidae	$= 36 + X_1X_2Y$	
Ranatrinae		
Ranatra chinensis	$46 = 44 + XY$	175
	$43 = 38 + X_1X_2X_3X_4Y$	43
R. elongata	$43 = 38 + X_1X_2X_3X_4Y$	170
	$42 = 38 + X_1X_2X_3Y$	176
Nepinae		
Laccotrephes maculatus	$43 = 38 + X_1X_2X_3X_4Y$	170
	$42 = 40 + XY$	171
Nepa rubra (as *cinerea*)	$35 = 34 + X0$	173
	$33 = 28 + X_1X_2X_3X_4Y$	71, 174
Saldidae		
Saldula saltatoria	$36 = 34 + XY$	144
	$35 = 34 + X0$	143
Gerridae		
Gerrinae		
Gerris (as *Hydrometra*)	$23 = 22 + X0$	177
paludum	$24 = 22 + XY$	46

chromosome was of medium size with regular behavior during meiosis. Its origin could not be predicted as it showed no pairing either with the autosomes or the sex chromosome. In *Dieuches* sp. two out of six males had a supernumerary chromosome each, with size and behavior similar to *M. uniguttatus*.[40] The supernumerary had 6.96% relative volume. Possibly it originated by nondisjunction of some medium-sized autosome followed by changes in size and heterochromatinization. The frequency of supernumerary individuals in the populations of the two species seemed to be very high.

In Heteroptera, chromosomal polymorphisms relating to the sex chromosome have been documented. In the Pentatomidae sex chromosomal polymorphisms from XY to X_1X_2Y and

$X_1X_2X_3Y$ were found in *Oechalia pacifica*,[70] leading to the suggestion that the fragmentation of the original X of the XY complex and in some cases also an autosome gave rise to the multiple Xs and a Y. In the Lygaeidae several instances of sex chromosomal polymorphism have been reported, including a population study in *Lygaeus hospes*.[187] The sex chromosomal variations in some individuals from XY to XY_1Y_2 were found in *L. equestris*,[136,152] *L. hospes,* and *L. pandurus*, and *Tropidothorax leucopterus*.[136] Among the Rhyparochrominae some male individuals had X_1X_2Y in *Eremocoris abietis* and *Rhyparochromus chiraga* and $X_1X_2X_3Y$ in *Trapezonotus nebulosus*.[136] In *L. hospes* 190 males yielded 7 with an XY_1Y_2 complex and the population possibly had two types of X chromosomes and 3 types of Y chromosomes.[187] The statistical testing of the observed data of 219 males with X_1Y_1, X_1Y_2, X_1Y_3, X_2Y_1, X_2Y_2 and X_2Y_3 gave a χ^2 value of 14.54 ($p<0.001$). Therefore, the mating was not random. Further, the X_1, X_2, and Y_1 were not present in equal frequency. Another congeneric species, *L. pandurus* from the same area had the same type of sex chromosomal polymorphisms and similar behavior of the sex chromosomes during meiosis. The origin of the two types of Xs and three types of Ys in these two species seemed to have taken place by structural exchanges between the X and the Y elements of normal individuals as seen from the metrical data of XY chromosomes. The metrical data of autosomes in two sex chromosomal forms were correspondingly very close to each other.

In the Coreidae, chromosomal polymorphisms in three species of *Acanthocephala* (as *Metapodius*), viz., *A. femoratus*, *A. granulosa*, and *A. terminalis* were recorded for supernumerary chromosomes.[17,22,23] All 3 species had the basic coreid type chromosome constitution of $2n = 21$, $18 + 2m + XO$, but most of the individuals had in addition 1 to 5 supernumerary Y chromosomes. The Y elements formed a chain with the single X at metaphase II and segregated regularly at the opposite pole of the X. The extra Y chromosomes were suspected to be supernumeraries because no genetical effect was reported.

In *Iphita limbata* of the Pyrrhocoridae, males showed $2n = 20$, $14 + 6XO$[150;] $2n = 23$, $22 + XO$,[151] and 23 as $12 + 2m + 9XO$ chromosomes,[49] while the female contained 16 chromosomes.[49] The species was therefore polymorphic for both sex chromosomes and autosomes. No basic pattern of evolution of the different chromosome constitutions could be suggested. In the male, 8 of 9 sex chromosomes underwent degeneration after anaphase II and very likely originated by gene amplification from the single X during embryogenesis.[49] In the family Largidae, in *Lohita grandis*, Banerjee[150] reported that males contained an unstable number of X chromosomes, although the basic number was $2n = 15$, $14 + XO$; while we[50] found $2n = 15$, $12 + 2m + XO$ with variable behavior. In the female the oogonial count was 16 to 21. At any rate, the sex mechanism was not upset in this species. The regular origin of the extra X elements by fragmentation and their elimination at metaphase II, leaving only one X, in *Lohita* could not be accounted for. Therefore, the increase in number of X elements was not only due to fragmentation but also to nondisjunction and sometimes gene amplification.

In the Cimicidae, sex chromosomal polymorphism has been discovered in some species of *Cimex* by Slack,[165-167] Darlington,[168] and Ueshima[43,74,169] and in *Paracimex* by Ueshima[43,74]. In *C. lecturarius* the diploid number in males varied between 29 and 36 and in *P. borneensis* between 44 and 47 and in *P. capitatus* between 41 and 42 chromosomes. In all these species the number of Xs was variable while the number of Ys was constant, though meiosis was regular. When the number of Xs was very high, some amount of nondisjunction possibly occurred which led to the variation in number of Xs in second division. At metaphase II the single Y was situated opposite to the Xs in the central region of the hollow spindle. The reason for such orderly orientation without any physical connection was explained in *Cimex* by the suggestion that the Y contained a strong centromere.[8,168]

Such chromosomal polymorphisms involving autosomes and sex chromosomes in different species of Heteroptera may suggest that the mechanisms of chromosomal evolution are principally the same as in species with monocentric chromosomes except for the occurrence of structural heterozygotes.

VI. INTERSPECIFIC HYBRIDIZATION

Foot and Strobell[33] were able to cross the pentatomids *Euschistus variolarius* and *E. servus*, both having $2n = 14$ with an XY:XX sex mechanism and observed besides stickiness, some nondisjunction of chromosomes at anaphase I. The hybrids were fertile in the second generation. On the other hand, the crosses of the other two pentatomid species showed a somewhat different result. Crossing of male *Thyanta pallidovirens* $(14 + XY)$ with female *T. custator* $(14 + XX)$ showed fewer meiotic disturbances due to some irregular orientation of chromosomes, while male hybrids obtained from the crossing of male *T. pallidovirens* with female *T. calceata* $(24 + X_1X_1X_2X_2)$ had $2n = 22$ with a large amount of failure in the meiotic process. When the hybrid male was crossed with normal *T. pallidvirens* females some progeny were obtained. It seemed that some of the sperm, in spite of gross meiotic failure in the F_1 hybrid, were fertile. F_1 males of reciprocal crosses of *C. lecturarius* and *C. columbarius* gave no gross irregularity during meiosis.[74,168] However, an extensive study of the hybrid males of the *C. pilosellus* complex of five species showed that the fertility of the hybrids was reduced in the F_1 generation while in the F_2 a majority were sterile.[74,169] The male hybrids, through reciprocal crosses between *Hesperocimex* $(40 + XX)$ and *H. cochmiensis* $(38 + XY)$, gave 41 as spermatogonial number, while the first metaphase contained 19 bivalents with one heteromorphic and one univalent autosome and univalent X and Y. Separation was regular, leading to two types of metaphase II, viz., $19 + XY$ and $20 + XY$. Species hybrids of *Triatoma* and a hybridization experiment in *Blissus* (Lygaeidae) indicated that because of the chromosomal homology fewer meiotic disturbances occurred when the parents had fewer differences in chromosomal number. These experiments would lead us to think that natural hybridization could be more easily possible in the Heteroptera, but such studies have rarely been made.

VII. EVOLUTION OF SEX CHROMOSOME MECHANISM

Out of 1145 species in 42 families listed by Ueshima,[43] 846 had XY, 173 had XO, 95 had X_nY, and 31 had the X_nO type of sex chromosome mechanism. Further, 13 families had solely or overwhelmingly X_nY, including the Gelastocoridae and Mesoveliidae, each known from single species. Four families had X_nO males, but the number never exceeded more than 2 Xs except in some rare instances. Seventeen families had partly or wholly the XO type, and the remaining ones exclusively or overwhelmingly XY:XX sex mechanism. XY and XO forms of sex chromosomes could be taken as the simple types from which multiple types evolved. If the size of the single X chromosome in the XO:XX type was compared between congeneric species and allied taxa, a good deal of change in size was noticed. In some families it was outstandingly large in comparison to other autosomes, while in other families it was relatively small and even smaller than some autosomes. Studies in males of different species with XY sex chromosome showed from the metrical data[40] that interchromosomal changes, deletion and duplication of parts occurred in bringing changes to X and Y chromosomes, as much have occurred in XO types as well.

The multiple sex chromosome mechanism in some species within a family must have taken place from the simple XY or XO types to X_nY and X_nO. Excluding some sex chromosomal polymorphism in some families, all the four types, XY, X_nY, XO, and X_nO were not found in any particular family. A multiple sex chromosome mechanism (X_nY) was found in the Nepidae, Reduviidae, etc., and X_1X_2O in the Coreidae, Pyrrocoreidae, and some Lygaeidae. In the origin of multiple sex chromosome mechanism, X-autosome translocation has played no significant role. Rather, fragmentation, fusion, and sometimes nondisjunction were more important factors. In the X_nY types, the Y was mostly larger than the Xs. Congeneric species with the XY type had generally a smaller Y and a larger X. Therefore, fragmentation of the single X could have given rise to more Xs. This would explain why the Y was larger than the Xs in the X_nY system. The evolution of multiple sex chromosomes through fragmentation of the original sex chromosomes has been advocated by different workers.[34,70,40,43] The X chromosome appeared to be relatively more fragile than the Y. The survival of the fragments seemed to have been facilitated by the postreductional meiotic mechanism which prevented chiasma formation between them.

In many species with an X_nO mechanism, the X components remained fused throughout or for the greater part of meiosis. In the X_nO system, the fragmentation of the original X has not taken place more than once because so far all these species showed X_1X_2O sex chromosomes.[137] Further, they generally remained fused throughout meiosis, as found in the Coreidae, Alydidae, etc., but in *Dysdercus* of the Pyrrochoridae the two X elements fused at anaphase I.[41,122] However, in *Iphita limbata* and *Lohita grandis* a highly peculiar case of the origin of extra X elements from the original single X was reported. Unlike *Lohita*, the origin of the extra X elements was not due to fragmentation but due to gene amplification.[49]

The maximum number of Xs in the X_nY type was reached in *Acholla multisinosa* and *Sinea rileyi* under Reduviidae having 5XY. Among X_nY types species with X_1X_2Y in males were greatest, which would suggest that the original X broke more often once than twice, thrice, and four times to give rise to X_1X_2Y, $X_1X_2X_3Y$, $X_1X_2X_3X_4Y$ and $X_1X_2X_3X_4X_5Y$. Species with X_1X_2Y males have been reported in *Macropygium reticulare*[191] and *Thyanta calceata*[124] among 185 species of Pentatomidae, in *Dinidor rufocintus*[96] among 6 species of Dinidoridae, in *Physopelta schlanbuschi*[106] among 16 of Pyrrhocoridae, *Arocatus suboeneus*, *Cavelerius illustris*, *Oxycarenus luctuosus*,[149] *O. hyalinipennis*,[192] *O. laetus*,[150] *Thylochromus nitidulus*,[149] and *Megalonostus* (as *Rhyparochromus*) *chirara*[136] among 330 species in the Lygaeidae, *D. lunatas* in the Dysodidae,[96] *Mezina pacifica* and *Isodermus gayi*[193] of the Aradidae, in *Coranatus fuscipennis*,[164] *Acanthaspis* sp.,[48] *Ectomocoris cordiger*, *E. ochropterus*,[164] *Rasahus thoracicus*,[43] *Oncopeltus* sp., *O. impudicus*, *Pirates* sp., *Pygolampis foeda*,[164] *Mestor megistus*,[194] *Panstrogylus herreri*, *Triatoma barberi*, *T. gestaeckeri*, *T. penninsularis*, *T. phyllosoma*, *T. protracta*, *T. rubida*, *T. sinaloensis*,[73] *T. nitida*,[195] *T. rubrofasciatus*,[196] *T. sanguisuga*,[34] *Fithchia spinulosa*, and *Rocconota annulicornis*[34] among 74 species of Reduviidae; in *Cimex lecturarius*, *C. columbarius*, *C. hemipterus* (as *rotundatus*),[168] *C. pipestrelli*, *C. insuetus*,[43] *C. stadleri*, *C. japonicus*, *C. pilosellus*, *Oeciacus hirundinis*, *O. vicarius*, *Paracimex-caledoninae*, *P. gerdheinrichi*, *P. setosus*, *Crassicimex pilosus*, *Afrocimex leleupi*, *Haematosiphon inodorus*, *Synxenoderus comosus*,[73] and *Stricticimex parvus*[43] among 43 species of Cimicidae; in *Dicyphus stachydis*, *Dryophilocoris flavoquadrimaculatus*,[160] and *Cyllecoris histricnicus*[128] among 166 species of Miridae; in *Abedus indentatus* among 12 species of Belostomatidae; and in *Laccotrephes griseus*[40] and *L. maculatus*[172] among 10 species of Nepidae.

Males with $X_1X_2X_3Y$ were found in *Sphragisticus nebulosus*[136] in the Lygaeidae, in *Harpactor fuscipes*,[40] *H. marginatus*,[150] *Sycanus* sp.[40] *S. collaris*[142], *Velinus nodipes*[197], *Oncocephalus* sp.[164], *O. nubilis*[43], *Triatoma eratyrusiforme*, *T. vitticeps*[73], *Prionidus* (as *Arilus*) *cristatus*[34], *Pselliopus cinctus*[36], *Sinea complexa*, *S. confusa*, *S. diadema* and *S. spinepes*[34] among the Reduviidae; in *Paracimex capitatus*, *P. inflatus*, *P. philippinensis*,[74] and *Hesperocimex colorarensis*[198] among the Cimicidae and in *Ranatra elongata*[176] and *R. filiformis* (as *sordidula*)[199] among the Nepidae.

Males with $X_1X_2X_3X_4Y$ were found in a few species as in *Graptopeltus japonicus*[197] among the Lygaeidae, in *Pasiropsis* sp.[140] and *Pnirontis modesta*[36] among the Reduviidae, in *Cimex adjunctus* and *C. brevis* among the Cimididae, the sole species *Gelastocoris oculatus*[34] of the Gelastocoridae, *Mesovelia furcata* under the Mesoveliidae and in *Laccotrephes maculatus*,[170] *L. kohili* (as *L. rubra*),[40] *Nepa rubra* (as *N. cinerea*),[174] *Ranatra chinesis*,[43] *R. elongata*,[170] and *R. linearis*[200] under the Nepidae, while two species with the maximum 5 Xs and a Y were found in the Reduviidae.

The evolution of a multiple sex chromosome mechanism in the Heteroptera has been mainly due to the fragmentation of the original X. The involvement of the X and the autosomes in the evolution of a multiple sex chromosome mechanism, although very rare, has occurred in the Pentatomid *Rhytiodolomia senilis* and in the Largidae, *Physopelta schlanbuschi*[43] and in simple types in the belostomatid *Lethocerus indicus* with XY males. In *Lethocerus* sp. with $2n = 4$, the sex chromosome very likely is incorporated into a pair of autosomes. Nondisjunction of the sex chromosomes must also have played some role in the evolution of multiple sex chromosome mechanisms as evidenced by the supernumerary Xs in three species of the Cimicidae and the supernumerary Ys in *Acanthocephala*. Further, the size of the X elements was same in many species with X_nY. Since, due to the postreductional behavior, pairing and chiasma formation have been suppressed in sex chromosomes in first division, the pairing of the extra X elements arising by nondisjunction is prevented.

VIII. EVOLUTION OF CHROMOSOME NUMBERS WITHIN AND BETWEEN VARIOUS GROUPS

Since different workers have investigated about 1200 species of Heteroptera cytologically under 46 families it would not be possible to discuss adequately the evolution of chromosome numbers in different taxa.[40-44] Heteropteran taxonomists are not in good agreement as to the supergeneric classification and relationships of Heteroptera. The classification of Stys and Kerzhner[121] has been followed here with some modifications.

The diploid numbers in spermatogonial metaphases of the 1200 species of Heteroptera studied range between 4 (*Lethocerus* sp.) and 80 chromosomes (*Lopidea* of Miridae) with a distinct peak at 14 chromosomes shown by 325 species belonging to 11 families and some lower ones at 16, 34, 24, etc. The diploid number of 14 (12 + XY) has been considered as the modal number for various reasons.[41]

(1) Family **Pentatomidae** — In 192 species belonging to 82 genera under Asopinae and six tribes of Pentatominae, spermatogonial numbers ranged between 6 and 27 with a peak at 14 and an XY:XX sex mechanism (Histogram 1*). The XY was changed to X_1X_2Y only in *M. reticulare* and *T. calceata* mentioned before. The arrangement and behavior of chromosomes during meiosis at first and second division showing the absence of an *m* pair were very typical. In the subfamily Asopinae, 16 species under 8 genera showed typical behavior, while numerical deviations to $2n = 12$, due to fusion in some species of *Mineus*, *Oechalia*, and *Picromerus* and to $2n = 16$ due to fission in *Podisus*, were seen.

In 176 species of 71 genera under 6 tribes in the subfamily Pentatominae, the maximum heterogeneity was shown in the Pentatomini. The spermatogonial number ranged between 6 and 27 chromosomes, having the peak at 14 in 86 species of 123 species in Pentatominae under 50 genera. Greater variation was shown by the species belonging to *Banasa*, *Rhytidolomia*, and *Thyanta*. Though the number was doubled in *T. calceata* and *B. rufirons*, the DNA values remained almost the same as that of congeneric species with $2n = 14$, suggesting chromatid autonomy.[201,202] These findings strengthened the view that fragmentation rather than polyploidy played the vital role in chromosomal evolution.

The metrical data of a few species of Pentatominae studied showed that the same autosome in different species underwent variable changes; so, also, did the X and the Y. Since, except *Spermatodes* sp. all other species had $2n = 14$, the data would definitely suggest that inter-

---

* Histograms follow the text beginning on page 214.

chromosomal changes between congeneric as well as species of different genera played a vital role besides fragmentation and fusion. No cytological characterization of different tribes was possible as there was not much variation between them. Chromosomes showed conservatism in number and meiotic behavior. Therefore, interchromosomal changes played a major role, occasionally supplemented by fragmentation and fusion.

(2) Family **Podopidae** — This elevated family from the subfamily Podopinae of the Pentatomidae, known by 8 species within 4 genera, showed typically 14, $12 + XY$ chromosomes and meiotic behavior with the sole deviation of $2n = 12$, $8 + 2m + XY$ in *Scotinomorpha* sp.[134] This species needs be checked as no *m* pair was found in the congeneric species, *S. horvathi*.[203] Speciation must have taken place by interchromosomal changes in all but *Scotinomorpha* where the fusion mechanism was suggested.

(3) Family **Dinidoridae** (formerly subfamily Dinidorinae of Pentatomidae) — This family has been known cytologically by 6 species under 3 genera. They showed the typical pentatomoidean pattern of meiosis. The modal number of 14 was present in 3 species of *Aspongopus*, while 2 species of *Megymenum* deviated to $2n = 20$, and in *Dinidor rufocinctus* $2n = 21$, $18 + X_1X_2Y$. Therefore, fragmentation in autosomes and X chromosomes played the major role and the interchromosomal changes were not so perceptible in species belonging to same genus.

(4) Family **Eumenotidae** — This newly elevated family has been known by only *Eumenotes obscura* which has $2n = 14$, $12 + XY$ with pentatomoidean chromosome number and behavior. The chromosomes appeared more voluminous in relation to the nuclear size as compared to the pentatomids.[40]

(5) Family **Urostylidae** — Two species of *Urochella* with $2n = 14$ and one each of *Urolabida* and *Urostylus* with $2n = 16$ having XY males showed the pentatomoidean pattern of meiosis, but the modal number was uncertain. The relative percentage values of chromosomes suggested both fragmentation and interchromosomal changes in speciation.

(6) Family **Cydnidae** — Ten species under three subfamilies have been investigated. Two species each of Sehirinae and Cydninae had typically $2n = 14$, $12 + XY$, while four other species of Cydninae had $2n = 12$, $10 + XY$ but also pentatomid meiosis. In the Scaptocorinae *Scaptocoris castaneus* ($24 + XY$) and *Stibaropus molginus* ($28 + X_1X_2Y$) had a higher diploid number and the latter a multiple sex chromosome mechanism. Therefore, in this family, fusion, fragmentation, and interchromosomal changes occurred more frequently in autosomes and less frequently in sex chromosomes.

(7) Family **Tingidae** — Although included in Cimicomorpha major group,[121] we have put this family under pentatomoidean stock from cytological consideration. It, unlike any other family of Heteroptera, showed exclusively the prereductional XY sex chromosomes. One species of *Agramma* under Agramminae and 14 under Tinginae, belonging to 9 genera, had characteristically $2n = 14$ with one major deviation to 12 *(Acalypta)*. The post reductional meiosis was the major difference from pentatomoid, in general, but no *m* chromosome was observed. Interchromosomal changes, fusion in one species, and prereductional meiosis could account for the evolution of the family.

(8) Family **Acanthosomatidae** — Showed the characteristic pentatomoidean meiotic pattern but the modal number was $2n = 12$, $10 + XY$. *Elasmucha grisea* had $14 + XY$, while *Acanthosoma expansum* gave $2n = 11$, $10 + XO$ males,[219] figures which need be checked. In speciation, interchromosomal rearrangement at the intra- and intergeneric level played a major role, supplemented by chromosomal elimination (Y)[219] and autosomal fragmentation in some species.

(9) Family **Scuteleridae** — 21 Species under 4 subfamilies (Eurygasterinae, Odontotarsinae, Scutellerinae, and Tetyrinae) had $10 + XY$ chromosomes with the typical pentatomoidean pattern of meiosis. In speciation, interchromosomal changes played the major role as supported by metrical data.

(10) Family **Plataspidae (Coptosomatidae)** — All the 9 species under *Brachyplatys* and *Coptosoma* had the same characteristic diploid number of $10 + XY$ chromosomes and meiotic pattern. Some interchromosomal changes at the intra- and intergeneric level played a major role in speciation as revealed by the metrical data in three species.

(11) Family **Tessaratomidae** — Only *Eusthenes saevus* having $2n = 12$, $10 + XY$ chromosomes with the pentatomoidean pattern, has been studied.

(12) Family **Lygaeidae** — Among the Heteroptera, the largest number (amounting to some 350 species under 13 subfamilies, excluding an unknown one cytologically investigated) revealed that the family has been heterogeneously constituted (Histogram 2).* The spermatogonial number ranged between 10 and 30 with the distinct modal peak at $2n = 14$ in some 131 species, and a lower peak at 16 in 112 species belonging to 9 subfamilies each. The meiotic metaphase arrangement was basically similar to the pentatomoids in Lygaeinae. In others the occurrence of the *m* pair brought little variation by sharing its position with sex chromosomes within the autosomal ring. Four out of 7 species of *Aphanus* under the Aphaninae had a modal constitution of $10 + 2m + XY$, while two had a pair of autosomes reduced for fusion, and *A. japonicus* had $17 = 10 + 2m + X_1X_2X_3X_4Y$ following fragmentation of the X element. There were 37 species in 10 genera, except *Calvelerius illustris* ($10 + 2m + X_1X_2Y$), which had in common XY and $2m$, but the autosomal number was increased by a pair in some species to the modal number of $10 + 2m + XY$. Two species of *Chauliops* in the Chauliopinae, having $16 + 2m + XY$, could be generic characteristics. In the Cyminae, nine species under Cymini had higher numbers of 28 and 30, while one species in Ontiscini and three out of four in Ninini had 22 chromosomes. All had common XY and $2m$ chromosomes, indicating autosomal number changes by fragmentation. In the Geocorinae, also, 13 species had in common XY and $2m$ chromosomes, while the autosomal number ranged between 12 and 16. Only *Engistus viduus*, representing the Henestrinae, had a modal constitution of $10 + 2m + XY$. Six species within 4 genera in the Heteropterogastrinae represented equally the two peak numbers of 14 and 16 including $2m$ and XY chromosome. All the seven species under three genera in the Ischnorhynchinae represented the modal constitution of $10 + 2m + XY$. In the Lygaeinae, all 27 species of 12 genera had XY except *Arocatus suboeneus* ($12 + X_1X_2Y$), pentatomid pattern of chromosome behavior, and a modal number generally 14, but occasionally 16 and 22, but had no *m* chromosome. Thus fragmentation, mainly in autosomes and rarely in sex chromosomes, was the main source of chromosome evolution. All 68 species of Orsillinae had XY and $2m$ in common. Species of the Metrargini had 12 autosomes, while those of Nysiini and Orsillini more commonly had 10 chromosomes with some deviation to a higher number. The numbers $2n = 17$, $14 + X_1X_2Y$ in the Oxycareninae possibly arose by fragmentation of the X and the absence of *m* pair except in a polytypic species. In the Pachygronthinae, the tribe Pachygronthini had six out of nine species with an alydid-like chromosome constitution of $2n = 13$, $10 + 2m + XO$. Except *Uttaris* ($10 + 2m + XY$), all had an XO:XX mechanism, possibly by the elimination of the Y from the lygaeid modal number of 14. Fragmentation of autosomes gave $2n = 17$ and 23 in two other species. Four species of Teracriini had the typical lygaeidean modal number of 14, $10 + 2m + XY$ chromosomes. The Rhyparochrominae is the most heterogeneous among all other subfamilies. In the Antillocorini, four species had typically 14, $10 + 2m + XY$, while one species had a pair of autosomes more. *Tropistethus* has 12 due to the absence of the *m* pair, an exception. Only *Clerada* sp. in the Cleradinii had a higher number of 24 chromosomes. In Drymini, $2n = 20$, $16 + 2m + XY$ was found in 20 species, three had 16, one had 18 and *Thylochromus nitidulus* had multiple sex chromosomes ($16 + 2m + X_1X_2Y$). Thus, fragmentation in autosomes and in the X could account for these differences in diploid constitution. Thirteen species of 6 genera under the Lethaeini had all in common XO and $2m$. The common autosomal number was reduced to 8, probably by fusion in two species of *Cryphula* and one of *Lethaeus* and increased by a pair in *Diniella*.

Thus, the common constitution of $2n = 13$, $10 + 2m + XO$ makes this tribe chromosomally similar to the Pachygronthini and even to the Rhopalidae and Alydidae of the Coreoidea through the loss of Y from the lygaeidean basic constitution of $2n = 14$, $10 + 2m + XY$. Among three species in the Megalonotini, one had the modal number of 14, while *M. chiraga* (15, $10 + 2m + X_1X_2Y$) and *S. nebulosus* ($16 = 10 + 2m + X_1X_2X_3Y$) arose possibly by fragmentation of the X. In the Myodochini, all the 42 species within 15 genera possessed $2m + XY$ with $2n = 12$, which was reduced to 10 in *Graphoraglius*, indicating that autosomal changes were significant in chromosomal evolution. The same mode of evolution could be envisaged in two species with $2n = 16$ and $2n = 14$ in the Ozophorini and five species of *Plinthisus* in the Plinthisini. In the Rhyparochromini, 23 species all showed $2m$, three forms of sex chromosomes, viz., XY in all but 4 species of *Poeantius* having XO and one, *R. angustatus* had X_1X_2Y. The basic autosomal number of 10 was reduced to 8 in all eight species and to 6 in one species. Thus, in this tribe fragmentation of X, elimination of Y and fusion of autosomes took place in chromosomal evolution. In Stygnocorini four species had $2n = 16$, $12 + 2m + XY$, and two had 18, showing a change in autosome number. In one species each of the Targaremini and Udeocorini, the basic chromosome constitution was $2n = 14$ and 16 including $2m$, and the XY was of lygaeid pattern.

In the Lygaeidae fragmentation and fusion in autosomes and fragmentation of the X and sometimes elimination of the *m* pair and the Y account for chromosomal evolution accompanied by interchromosomal changes, as indicated by metrical data.

(13) Family **Berytidae** — Among 5 species, the absence of the *m* pair, the modal number of $16 = 14 + XY$ in four and $2n = 28$, $26 + XY$ in *Berytimus minor* link them with pentatomoidean stock.

(14) Family **Colobathristridae** — The only species, *Phaenacantho* sp. $2n = 14$, $10 + 2m + XY$ studied, shows links with Lygaeidae.

(15) Family **Stenocephalidae** — *Dicranocephalus agilis* $2n = 14$, $10 + 2m + XY$ links the family with the Lygaeidae, while *D. lateralis* with $2n = 14$, $10 + 2m + X_1X_2O^{42}$ requires reconfirmation because we[50] found $2n = 10 + 2m + XY$ like *D. agilis* in an unidentified species.

(16) Family **Pyrrhocoridae** — 17 Species within 8 genera, all belonging to the Pyrrhocorinae, showed an XO and X_1X_2O sex mechanism, and the diploid numbers ranged between 13 and 27, having the peak at 16 (Histogram 3). The Y and the *m* pair were absent except in the polymorphic species *Iphita limbata* with a pair of *m*.[49] Changes are involved in both sex chromosomes and autosomes. The metrical data obtained on two congeneric species of *Dysdercus* suggest some amount of interchromosomal changes, particularly in sex chromosomes, besides the fusion-fragmentation mechanism generally applied.

(17) Family **Largidae** — The family was separated from the Pyrrhocoridae. Largids have characteristically the *m* pair, a low diploid number between 11 and 17 (Histogram 4), and an XO:XX mechanism except in *P. schlanbuschi*. Among five species, *L. grandis* showed intraindividual X chromosome fragmentation,[205] while of two species of *Physopelta* (Physopeltinae), one had $2n = 17$, $12 + 2m + X_1X_2Y$ in males. The largest X_1 and Y and the smaller X_2 in *P. schlanbuschi* possibly arose by the duplication of large X followed by X-autosome translocation.

(18) Family **Coreidae** — The family is characterized by the presence of the *m* pair, XO and X_1X_2O type sex chromosomes (Y present in *Acanthocephala* as an exception), and the modal number of $21 = 18 + 2m + XO$ within the range of 13 to 29 (Histogram 5). The subfamily Merocorinae, known by *Corynocoris distinctus* ($2n = 25$, $24 + XO$), had no *m* pair. In the Pseudophloenae, two out of four species showed no *m* pair and the diploid numbers were 13, 22, and 23. Two congeneric species of *Coriomeris* had $2n = 13$, $10 + 2m + XO$, and $2n = 22$, $18 + 2m + X_1X_2O$, which might have arisen by chromosome doubling. The Coreinae included some 75 species in 12 tribes with diverse cytological forms.

The diploid number ranged between 15 and 29, but most of the genera showed some characteristic pattern and typical arrangement of chromosomes at metaphases I and II. In the Acanthocephalini, three species of *Acanthocephala* had supernumerary Ys for which the diploid number was unstable, while *Diactor* had a modal constitution of $2n = 18 + 2m + XO$. Seven species under four genera in the Anisoscelini had typically the *m* pair and XO, with an autosomal number varying between 18 and 26. The two species of *Chariesterus* under the Chariesterini had $2n = 25$, $22 + 2m + XO$, while one species of each of the Chelinideini and Leptoscelidini had typically 20 chromosomes. In the Colpurini, among three species of *Hygia*, one had typical constitution, the second had 22, $18 + 2m + X_1X_2O$, while *?H. touchi* had uncommonly $2n = 17$ without *m* pair, indicating chromosomal heterogeneity in this tribe. In 13 species of the Corieini, the modal constitution was common, but some species had X_1X_2O and, exceptionally, *S. marginatus* var. *fundator*[137] had $2n = 23$, $18 + 2m + X_1X_2X_3O$. Both autosomes and X chromosomes varied, but the *m* pair was stable. Gonocerini had characteristically $2n = 18$, $14 + 2m + X_1X_2O$, but some species had XO. Variation of the autosomal number and absence of an *m* pair were also reported. Species of the Mitchini showed stability of the XO and *m* pairs, while autosomal numbers varied between 12, 18, and 22. Two species of *Petillia* (Petascelini) had $2n = 24 + 2m + X_1X_2O$ while two of *Acanthocoris* (Physomerini) had $2n = 24$, $22 + X_1X_2O$ with no *m* pair. Metrical study in two species, *Elasmomia granulipes* and *Homocerus* sp., with the same diploid number indicated interchromosomal changes during speciation. In the case of multiple X_nO mechanisms, the X fragmented once and attained the stable condition but it could also fragment more in some exceptional cases, as in *Lohita* and *Iphita* mentioned before.

(19) The Family **Alydidae,** elevated from the subfamily Coreidae showed characteristically the diploid number of 13, $10 + 2m + XO$ in most species under two subfamilies (Histogram 6) with the coreid pattern of chromosome arrangement during meiosis. Among 16 species in the Alydini, only *A. fasciatus* had X_1X_2O, originated possibly by the fragmentation of the X, while *M. pallescens* ($2n = 15$) and *U. variegatus* ($2n = 17$) deviated from the modal constitution of $2n = 10 + 2m + XO$ by a pair of autosomes. In the Micrelytrinae under the Leptocorisini, two species of *Leptocorisa* had $2n = 17$, $14 + 2m + XO$. The higher autosomal number arose possibly by fragmentation, while two species of Micrelytrini had the modal constitution of 13. Metrical study indicated some interchromosomal changes in *L. acuta* and *Riptortus pedestris*.

(20) Family **Rhopalidae (Corizidae)** was elevated from the Coreidae, and like the Alydidae had the same modal constitution of $13 = 10 + 2m + XO$ (Histogram 7) and meiotic chromosome arrangement, except the polytypic species *Liorhyssus hyalinus* ($2n = 15$, $12 + 2m + XO$), among 24 species studied under 5 tribes of Rhopalinae and in the Serinnethinae. Apparently interchromosomal changes between and within the genera played the vital role in speciation.

(21) Family **Aradidae** — *Mezira pacifica* ($2n = 27$, $24 + X_1X_2Y$) in the Nezirinae and *Isodermus gayi* ($2n = 23$, $20 + X_1X_2Y$) within the Isoderminae had a higher diploid number and multiple sex chromosomes, but no *m* pair. Ueshima[43] put *Dysodius* under this family.

(22) Family **Dysodiidae** — Only *Dysodius lunatus* had $2n = 31$, $28 + X_1X_2Y$ with no *m*, showing its cytological affinity to Aradidae. Ueshima[43] included it under the same family.

(23) Family **Mesovelidae** — Previously, Distant[206] put the Mesovelinae as a subfamily of Dysodiidae. *Mesovelia furcata* with $2n = 35$, $30 + X_1X_2X_3X_4X_5Y$,[208] showed cytological affinity with the Dysodiidae in high diploid number and multiple sex chromosomes with the absence of an *m* pair.

(24) Family **Joppeicidae** — Only *Joppeicus paradoxus* studied had $2n = 24$, $22 + XY$ with no *m* pair and with orthodox meiotic behavior and pentatomoidean affinity through autosomal duplication.

(25) Family **Reduviidae** — It is heterogeneously constituted.[42,43,48-51,73] A diploid number

ranging between 12 and 34 with the same frequency peaks at 22 and 23 chromosomes was found in 98 species belonging to 44 genera within 12 subfamilies (Histogram 8). An almost equal number of species had XY and $X_n Y$ sex chromosomes and three had XO.[40,50,51] The arrangement of chromosomes was mostly typical at metaphase II. The autosomal number was less variable than the X chromosome. In the Acanthaspidinae (Reduviinae), four out of five species had XY and one had 4XY. Species with XY exclusively were found in the Apiomerinae, Bactrodinae, and Emesinae, while subfamilies Harpactorinae, Piratinae, Stenopodinae, Triatominae, and Zeelinae had XY and $X_n Y$ males. *Polytoxus* sp. under Saicinae[51] and two species of *Extrychotes* (Ectrichodiinae)[40] had the uncommon XO:XX mechanism, and the latter genus had prereductional meiosis. In the Reduviidae the sex chromosomal evolution has apparently proceeded from the XY type in two directions; in one the Y was lost and in the other X was broken to give rise to a maximum of 5XY, although the 2XY type was more common. The autosomal number varied between 10 to 32 of which 20 was most common, followed by 22. Structural alterations occurred at the intra- and intergeneric levels.

(26) Family **Phyamatidae**, formerly a subfamily of Reduviidae, has been poorly studied. Of two species, one had $2n = 28$, $26 + XY$ and the other $2n = 29$ with no mention of a sex chromosome. Therefore, cytological characterization of the family was not possible.

(27) Family **Nabidae**[160]: Some five species under four genera in the Nabinae with a diploid number ranging between 18 and 40 (Histogram 9) had the modal constitution of $18 = 16 + XY$ and no *m*. The autosomal number increased by fragmentation and sometimes by doubling, but the sex chromosome number (XY) remained the same.

(28) Family **Miridae (Capsidae)** — Out of eight subfamilies, seven have been cytologically studied. The spermatogonial numbers in 167 species were found to range between 14 and 80 with the peak at $34 = 32 + XY$ (Histogram 10) absence of an *m* pair, and typical meiotic behavior. The deviation from XY to XO was recorded rarely, e.g., in the genera *Phylus* and *Plagionathus* (Phylinae), $X_1 X_2 O$ in *Calocoris rapidus* (Mirinae) and $X_1 X_2 Y$ in *Dicyphus strachydis* (Dicyphinae) and in *Cyllecoris* and *Dryophilocoris* (Orthotylini). Therefore, autosomal fragmentation, fusion, and doubling took place more frequently, while the sex chromosomal changes were limited. Four species of *Lopidea* under the Lopidini of the Orthotylinae showed $2n = 80$, the highest number reported among the Heteroptera, but no mention was made of the sex chromosome constitution.[207]

(29) Family **Anthocoridae** — Five species belonging to three genera under two subfamilies had XY and no *m* pair. The autosomal number was 28 for *Anthocoris*, 22 for *Orius*, and 30 for *Amphiareus*, indicating greater changes in autosomes, with typical meiotic arrangement.

(30) Family **Polyctenidae**, is closely related to Cimicidae[43] but cytologically less so because of the low diploid number of $2n = 6$, $4 + XY$ in *Hesperoctenes fumarius*; $2n = 12$, $10 + XY$ in *H. setosus* and $2n = 8$, $6 + XY$ in *Eotenes intermedius*. The sex mechanism remained unchanged though autosomal number changed by doubling in *H. setosus* and possibly by fragmentation in *Eotenes*, from the basic $2n = 6$.

(31) Family **Cimicidae** — The spermatogonial number ranged between 10 and 42 chromosomes with a peak at 31 in 43 species under six subfamilies. Although the sex chromosome number was variable, the autosomal numbers gave a peak 28 within a range of 8 and 40 (Histogram 11). The number of species with a multiple sex chromosome mechanism, mostly $X_1 X_2 Y$, was 25 against 18 with XY males. Some species possessed supernumerary Xs and a Y. Two species of the Primicimicinae had an XY pair, but the autosomal number differed by a pair. Species of *Cimex* differed both in autosomal and sex chromosomal number. *Paracimex* had a stable autosomal number of 38, while the number of Xs differed. The two species of *Oeciacus* had $2n = 31$, $28 + X_1 X_2 Y$. Species of the Cacodminae had variable diploid numbers and sex chromosome constitution. The lower diploid number and XY in three species might link the Polyctenidae with this family. Species of the Haematosiphoninae

showed a wide variation of both autosomal and sex chromosome numbers. Therefore, in speciation both autosomes and X chromosomes underwent considerable changes.

(32) Family **Hydrometridae** — Three species of *Hydrometra* had 18 autosomes, while the sex chromosome mechanism was XO and XY. There was no *m* pair. Some cytological affinity could be suggested with the Nabinae.

(33) Family **Veliidae** — Five species under three genera in three subfamilies showed $2n = 21$, $20 + XO$ in *Microvelia reticulata* of Microveliinae and in *Hebrovelia* sp. of Hebroveliinae and $2n = 25$, $24 + XO$ in *Velia* of Vellinae. However, *M. douglasi* had $2n = 22$, $20 + XY$[46]; *m* pair was absent in all the species. The sex mechanism could be both XO and XY as in the Hydrometridae. In both families, evolution towards XO type was suggested but autosomal numbers were different.

(34) Family **Hebridae** — Two species belonging to *Hebrus* ($2n = 19$, $18 + XO$) and *Merragata* ($2n = 27$, $26 + XO$) differed in autosomal number. The absence of an *m* pair and XO males with a diploid number around $\pm$ 20 indicated a general cytological affinity with *Gerromorpha*.

(35) Family **Gerridae** — Thirteen species under *Gerris* and one each under *Limnogonus* and *Neogerris* indicated the modal constitution as $2n = 20 + XO$ (Histogram 12). The *m* pair was absent and the meiotic arrangement of chromosomes was somewhat disorderly in some species. The occurrence of an XY pair in *G. paludum*[46] was the only deviation from the XO type which needed reinvestigation. The autosomal number deviated by a pair on either side of the modal number of 20, possibly by fragmentation and fusion mechanism.

(36) Family **Dipsocoridae** — Since only *Pachycoleus refescens*, having $2n = 21$, $20 + XO$ chromosomes, is known so far, no comment is possible except for noting the similarity to the Gerridae.

(37) Family **Saldidae (Acanthididae)** — One species each under *Chiloxanthus* and *Salda* and five species under *Saldula* revealed that the $2n$ in each genus was different as 19, 33, and 35, respectively, but with a common XO:XX sex mechanism and *m* pair. *Saldula scotica* had a high number of 47 and *S. saltatoria* had XY,[144] unlike XO found in others. This family was considered to be very primitive[208] and differed basically from Gerromorpha in the presence of *m* pair and higher diploid number.

(38) Family **Leptopodidae** — Only *Leptopus marmoratus*, having $2n = 28$, $26 + XY$,[143] has been studied.

(39) Family **Corixidae** shows $2n = 24$, $20 + 2m + XY$ (Histogram 13) with characteristic meiosis; the *m* pair was absent in one species each of *Cymatia* (Cymatinae), *Micronecta* (Micronectinae) and *Glaenocorisa* (Glaenocorisini) *C. borsdorffi* was the only exception in having $2n = 26$ instead of the modal number of 24. With the stereotype diploid constitution, evolutionary changes could only be shown better by metrical study which was lacking.

(40) Family **Notonectidae** — Two subfamilies showed different modal chromosome constitutions (Histogram 14). In the genus *Anisops* under Anisopinae, three species had uniformly $2n = 26$, $22 + 2m + XXO$ with prereductional sex chromosomes,[131] while under the Notonectinae three species of *Notonecta* had $2n = 24$, $20 + 2m + XY$ and another three had $2n = 26$, $22 + 2m + XY$. The meiotic pattern was not very different. The autosomal changes were marked by one pair which possibly changed from 22 to 20 by fusion but the *m* pair remained unchanged. The diverse cytological characters may indicate that this family is ancestral to other related ones.

(41) Family **Belostomatidae** — The two subfamilies showed diverse cytological features (Histogram 15). All but *Abedus indentatus* ($2n = 29$, $24 + 2m + X_1X_2Y$) had XY while an *m* pair was found in all but the species with low diploid number (*Lethocerus* sp. $2n = 4$, *L. americanus* $2n = 8$, $6 + XY$). A prereductional XY pair has been seen in *L. indicus* in which X and Y were not identifiable by morphological differences. In *Lethocerus* sp. $2n = 4$, no sex chromosome was found. It was suggested[8] that the sex chromosomes were incorporated

with the autosome. Thus, in the Belostominae all species except *Lethocerus* had in common the XY and a pair of *m* chromosomes, but the autosomal number was different in different genera. In *Diplonychus* it was $2n = 28$ more than the modal number of $2n = 24 + 2m + XY$. The diploid number varied more in the Lethocerinae (4 and 30), while it was less variable in the Belostominae. In this family, fragmentation and interchromosomal changes played a major role in speciation, as revealed from the metrical data.

(42) Family **Pleidae** — Three species of *Plea* showed uniformly $2n = 23$, $20 + 2m + XO$ with regular meiosis. Therefore, this family could have evolved from notonectid stock by the elimination of the Y chromosome. As only one genus has been investigated, it is not known if the diploid constitution would apply for the family as well.

(43) Family **Naucoridae** — All seven species, two under *Ambrysus* (23, $20 + 2m + XO$) of the Ambrysinae and the remaining five under Naucorinae had in common $2m + XO$, but the autosomal number varied widely in species of *Illycoris, Naucoris, Pelocoris,* and *Limnocoris*. Therefore, autosomal changes were primarily responsible in speciation.

(44) Family **Ochteridae** — Since the diploid number in *Ochterus marginatus* was approximately 50 and an unidentified species had $2n = 35$, $30 + X_1X_2Y$,[43] no comment can be made. The latter species had a chromosome constitution like Gelastocoridae, but the sex mechanism was basically different. It could also be derived from Bellostomatid stock.

(45) Family **Gelastocoridae (Nethridae, Galgulidae)** — *Gelastocoris oculatus* contained $2n = 35$, $30 + X_1X_2X_3X_4Y$, indicating its closeness to Nepidae in diploid number and sex chromosome mechanism. The meiotic arrangement of chomosomes was more like nepid and reduviid types.

(46) Family **Nepidae** seems to have characteristically multiple sex chromosome mechanism. In two subfamilies, the Nepinae and Ranatrinae, $2n = 38$ (Histogram 16). Variation of the diploid number was due to the X chromosomes. *Laccotrephes griseus, L. kohlii* (as *rubra*), *L. filiformis, R. linearis,* and *R.* sp. could be chromosomally monomorphic, while *L. maculatus, N. rubra, R. chinensis,* and *R. elongata* were sex chromosomally polymorphic. Therefore, in both the subfamilies the changes in X chromosomes played a significant role which ranged from XY to 4XY. *N. rubra* was claimed to have XO[173] through the elimination of the Y. The arrangement of meiotic metaphase chromosome was quite regular.

The evolution of chromosome number in different species under different families has been attributed to fragmentation, fusion, interchromosomal exchanges, sometimes nondisjunction, polyploidy etc. All these mechanisms operated in somewhat different ways as the chromosomes of Heteroptera were not of the conventional monocentric type. The role of interchromosomal exchanges has been amply evidenced from the metrical studies of chromosomes.[40] Fragmentation and fusion of chromosome occurred at the intra- and intergeneric levels. The role of polyploidy has been suspected[43] as the DNA value was not different in congeneric species having doubling of number. This was found in Heteroptera,[201,202] in moths,[209] and in the plant *Luzula*[201,211] having the so-called holocentric chromosomes. But counter arguments could be put forward from other forms of evidence. On the whole the same mechanism of chromosomal evolution appears to operate in Heteroptera as the species with localized centromere.

IX. CONCLUDING REMARKS

The evolutionary interrelationships between various groups of Heteroptera, suggested here from cytological findings, may have some taxonomic limitations as there has been no uniformly acceptable supergeneric classification of Heteroptera. There have been some attempts before to trace the interrelationships supported by cytological findings.[41,43,160,208,212,213] A major limitation in the present evaluation is that cytological knowledge of many families is very poor, while in others it is uneven, and information was variable and sometimes incomplete. In spite of these limitations, the present attempt may help in taxonomic disputes.

In the absence of a well-preserved fossil record, it has not been possible to suggest which of the present day families is most primitive. Manna[41] treated *Paragnightia magnifica* Evans, the well-preserved heteropteran fossil discovered from the upper Permian deposits of Lake Macquarie, Australia, as the base of the family-tree. Evans[214] included it under the family Paraknithtiidae. The fossils of Jurassic rocks were so badly preserved that their inclusion in Heteroptera was questionable.[215] The structure of the embolium of *P. magnifica* resembled that of a number of families including Halyini under Pentatominae of Pentatomidae. The modal number in Protoheteroptera was considered to be $(2n = 12 + XY)$ because this tribe showed characteristically the above the chromosome constitution.[41] In partial modification of the previous family tree,[41] in the present study, diphyletic origin has been envisaged because many heteropterists hold the view that gerromorphan[43] or cimicomorphan[160] or leptopodomorphan stock[208] generally possessed the higher diploid numbers. The embolium of the fossil form also resembled that of the Miridae, Nepidae, Belostomatidae, Nethridae, Naucoridae, Corixidae, and Notonectidae.[41] The reasons for favoring protoheteropterans with the modal number of 14, $12 + XY$ are:

1. The same chromosomal constitution is present in 325 species belonging to 11 families among 1200 species studied so far.
2. Since sex chromosomes differentiated from a pair of autosomes, the XY:XX mechanism would be primitive. Out of 1145 species,[43] 846 had an XY:XX mechanism.
3. As fragmentation has been favored in the evolution of chromosome number in Heteroptera, it would involve increase in chromosome number.
4. Some primitive families with a high diploid number had some species with a low diploid number and XY:XX sex mechanism.
5. The *m* pair which seemed to have originated later by the degradation of a pair of autosomes is absent in Pentatomoidea and in many other families, while it is present in some primitive taxa.

Thus in the diphylectic branch of the tree (Figure 1), family Pentatomorpha and other groups have been put in two groups with $2n = 14$ and 24 chromosomes. Since Protoheteroptera has been supposed to have $2n = 14$, $12 + XY$ constitution in males, one branch continued as represented by the pentatomorphan stock while the other group possibly arose by doubling of the autosomal number followed by fusion or elimination, while the sex chromosomal constitution remained as XY in males. This would make the basic chromosome constitution $2n = 24$, $22 + XY$ as found in Joppeicidae, Corixidae, and Notonectidae. The protoheteropteran chromosome constitution has been retained in a number of pentamorphan families like the Pentomidae, Eumenotidae, Podopidae, Dinidoridae, etc. Even in the pentatomorphan group some families such as the Tessaratomidae, Plataspidae, Acanthosomatidae, Scutelleridae, etc. are primitive and originated during the Jurassic period. Four major groups, viz., the Geocorisae, Hydrocorisae, Saldirrhyncha, and Amphibocorisae evolved separately, diverging from a common Protosaldidae stock. Species with high diploid number, absence of bouquet arrangement, and a single large X chromosome as found in the Saldidae were suggested as characterizing the primitive cytological constitution of Heteroptera.[208] However, the Saldoidea might form an offshoot of the Pentatomorpha.[212] From the hypothetical modal number of $2n = 14$, $12 + XY:XX$, the pattern $2n = 12$, $10 + XY$ found in the Scutelleridae, Plataspidae, Acanthosomatidae, and Tessaratomidae could be achieved by fusion of two pairs of autosomes into one or else by the loss of a pair of autosomes. This would make these families interlinked, with the Pentatomidae having a pair of autosomes less. The meiotic

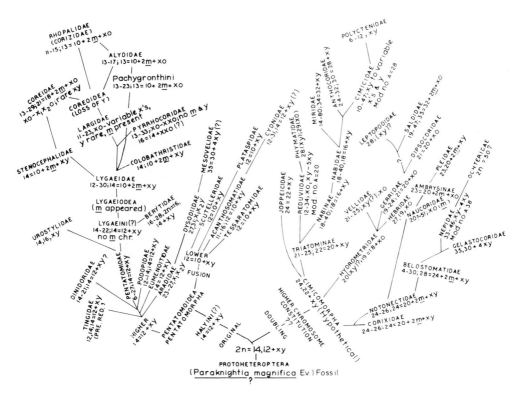

FIGURE 1. Schematic representation of the interrelationship between various groups of Heteroptera based on cytological data.

pattern and sex mechanism remained very much alike. Some primitive pentatomoids like the Podopidae, Eumenotidae, Urostylidae, and Dinidoridae retained the primitive diploid constitution as well as the common meiotic behavior of chromosomes. The inclusion of the Tingidae from the Cimicomorpha[121] in pentatomoidean stock has been done purely from cytological considerations (see Figure 1). This family differed only in the prereductional sex chromosome mechanism. If prereductional meiosis was taken to be primitive, the Tingidae would then be retaining faithfully the ancestral behavior. Its inclusion in the Cimicomorpha[121] has not been cytologically favored because the latter contained characteristically a higher diploid number. The diploid number of 14 in the Tingidae was regarded by others as a case of reduction[212] during phylogenesis within the Cimicomorpha. Among the Pentatomorpha, the primitive family Aradidae is considered to be a specialized group cytologically through which the nemomorphan Mesoveliidae could be linked. Kumar[213] concluded from the morphological and anatomical studies on the Aradidae that the Aradidoidea represented a very early offshoot in the evolution of terrestrial Heteroptera that became specialized for a sedentary and mycetophagous mode of life. From the limited cytological knowledge of this family it appears that polyploidy followed by other changes could have taken place in their origin from the pentatomorphan stock. Linking of the Dysodiidae with the Aradidae is possible, as the single species studied so far had a $28 + X_1X_2Y$ chromosome constitution somewhat like the Aradidae, and was put under subfamily Nezirinae of Aradidae.[43] The linking of the Mesoveliidae of Nepomorpha with the pentatomorphan Aradidae has been done mainly on the basis of high diploid number, multiple sex chromosome mechanism, and absence of the *m* pair obtained from only one species and therefore, could be subject to revision, although Distant[206] put the Mesoveliinae as a subfamily of Dysodiidae.

The pentatomorphan stock could broadly be divided into three major groups: (a) the

Aradidoidea with high diploid number, multiple sex chromosomes, and no m chromosome; (b) the Lygaeoidea-Coreoidea-Pyrrhocoroidea complex with presence of m pair, modal number around 14 in most families, evolution of sex mechanism toward XO and X_1X_2O as in Coreoidea and Pyrrhocoroidea; and (c) the original Pentatomoidea, retaining mostly a protoheteropteran chromosome complex of $12 + XY$ and the absence of m pair.

The Lygaeidae could reasonably be linked with the Pentatomoidea through the Lygaeinae due to the presence of the same modal constitution of $12 + XY$, the absence of the m pair, and the pentatomid meiotic chromosome behavior. The Lygaeidae constituted a heterogeneous group with diploid numbers between 10 and 30, m chromosomes mostly present, and the XY sex mechanism deviating more towards XO than X_nY mechanism. Cytologically the Berytidae could be taken as an offshoot of the Lygaeoidea linked through Lygaeinae in primitive modal constitution of $2n = 14$, $12 + XY$, and the absence of the m pair. The Colobathristidae, poorly known from a single species, showed the lygaeid modal number of $14 = 10 + 2m + XY$, which would support its placement within the Lygaeiodea. The Stenocephalidae with $2n = 14$, $10 + 2m + XY$ could link the Lygaeidae with the Coreoidia by the loss of a Y chromosome to have 13, $10 + 2m + XO$, as is generally found in the Alydidae, Rhopalidae, and in some Coreoidae. Further, the presence of the m pair in the Coreoidea and Lygaeoidea was parallel with the occurrence of tricobothria, a type of spermatheca, and other morphological features observed.[212]

The role of the m pair in heteropteran cytotaxonomy has been shown to be quite significant. The superfamily Pyrrhocoroidea[216] consisted of the Largidae and Pyrrhocoridae, while Scudder[217] considered the Largidae to be closely allied to the Lygaeidae under Lygaeoidea. The cytological data would suggest that the Largidae and Pyrrhocoridae originated independently from lygaeid stock, though both in largids and pyrrhocorids some cytological similarity could be shown in the absence of Y (except *Physopelta*). The modal number was close, but they differ in the presence of the m pair in the Largidae. In the superfamily Coreoidea, the modal number resembled the Alydidae and Rhopalidae ($2n = 13$, $10 + 2m + XO$). In the Coreidae, the autosomal number was almost double while the m pair and XO mechanism remained fairly constant. Thus the Coreidae could have originated through a lower number represented by Alydid-Rhopalid stock. The family Stenocephalidae may bridge the Coreoidea and Lygaeoidea in possessing $2n = 10 + 2m + XY$.[72] The evolution of the Coreoidea could be envisaged by the loss of Y from this complex.

Most of the aquatic and hydrophilous species of the group Cryptocerata, which included the Belostomatidae, Corixidae, Notonectidae, Naucoridae, Nepidae, and Nethridae, could be cytologically linked in spite of diversities (see Figure 1). Some of them show cytological similarities with different families of Gymnocerata. Thus, in the families from the Protoheteropteran stock, one branch through doubling of numbers may be the ancestor of other groups. Even some species of older groups have a low diploid number with the XY:XX mechanism from which other modal numbers could have originated. Therefore, doubling of the higher basic constitution to $2n = 24$, $22 + XY$ of the nonpentatomorphan major groups[121] like Gerromorpha, Dipsocomorpha, Leptopodomorpha, Nepomorpha, and Cimicomorpha has been cytologically linked. Species belonging to the Corixidae have diploid numbers between 24 and 26 with the modal number of $24 = 20 + 2m + XY$. The Notonectidae have the same modal constitution of $24 = 20 + 2m + XY$. The presence of the m pair was a deviation from the prototype of $22 + XY$. It may indicate the later origin of the family. Its general absence in the Nepomorpha is, therefore, as significant as that of Pentatomoidea. The Nepomorpha have been subdivided into Nepoidea (Nepidae and Belostomatidae), Corixoidea (Corixidae), Gelastocoreoidea (Ochteridae and Gelastocoridae), Naucoroidea (Naucoridae and Aphelocheiridae), and Notonectoidea (Notonectidae, Pleidae, and Helotrephidae).[143] Though some systematists object in putting the Corixidae under the Nepomorpha, the cytological behavior does not show much difference from the Notonectidae with which they were included in a common stock. The Notonectidae and Belostomatidae both have the *m* pair and the prevalent XY:XX mechanism. The autosomal number in the Belostomatidae differs by a pair (modal number 28, $24 + 2m + XY$). However, some species of Lethocerinae are different, having $2n = 8$, $6 + XY$ with the absence of *m* pair in *L. americanus*, $2n = 4$ in *Lethocerus* sp., prereductional sex chromosomes in *L. indicus*, etc., which might indicate the evolutionary diversification of the members under Belostomatidae. The remaining families of the Nepomorpha could be linked with the Belostomatidae. The linking of the the Nepidae with the Belostomidae could be done by presuming several cytological changes like increase in number of autosomes, introduction of multiple sex chromosome mechanism, the elimination of the *m* pair, etc. The linking of the Nepidae and Gelastocoridae was not so difficult since both families had a higher diploid number, a multiple sex chromosome mechanism, and no *m* pair. The data on the Ochteridae were not clear. If $2n = 30 + X_1X_2Y$ determined from an unidentified species[43] was taken into account, it could be linked with the Nepidae in higher diploid number, multiple sex chromosome mechanism, and the absence of *m* pair. Another species was seen with a very high number but no other information.[143] The Naucoridae, having XO and variable diploid number, could not suitably be linked with the Notonectidae or the Belostomatidae except for the presence of the *m* pair in all of them. The Pleidae and Naucoridae were related through the Ambrysinae by assuming several cytological changes. Since in the Dipsocoromorpha only one species under Dipsocoridae has been investigated, showing $2n = 21$, $20 + XY$ with no *m* pair, it has been linked with nonpentatomorphan stock by the loss of Y and a pair of autosomes. The cytological information regarding the Gerromorpha has been limited except for the Gerridae. All the families, Hydrometridae, Hebridae, Veliidae, Gerridae, and Mesoveliidae have uniformly no *m* pair, XO:XX sex mechanism (some doubtful report of XY), and the autosomal number mostly around 20 except in the Mesoveliidae.

The Cimicomorpha is heterogeneous. It has been assumed to arise from the base of the nonpentatomorphan group with $22 + XY$ chromosomes through the Nabinae and Triatominae. The Nabidae with $2n = 18$, $16 + XY$ and no *m* pair and the Miridae with $2n = 34$, $32 + XY$ could be linked by assuming the doubling of the autosomal number with further modification. The Anthocoridae with $2n = 28$, $26 + XY$, by fragmentation in autosomes could also be derived from the Nabidae. The Cimicidae show a very heterogeneous chromosome constitution, but the modal autosomal number of 28 is similar to the Anthocoridae. The relationship between the Polyctenidae having very low diploid number with XY mechanism could only be shown with the Cimicidae through some species of *Laxasbis* having $8 + XY$ chromosome. The Reduviidae have also a very heterogeneous constitution which could originate from the Triatominae with $2n = 22$, $20 + XY$.

X. ACKNOWLEDGMENT

The author expresses his sincere thanks to his collaborator Dr. S. Deb Mallick, Teacher Fellow of the University Grants Commission at the Cytogenetics Laboratory, Department of Zoology, University of Kalyani for his help in the preparation of the manuscript and providing some of his data.

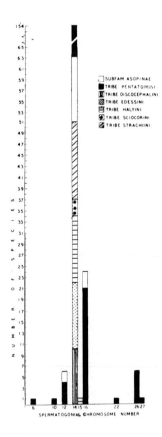

Histogram 1. Spermatogonial numbers in subfamily Asopinae and six tribes of Pentatominae of Pentatonidae.

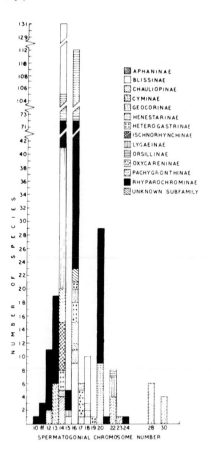

Histogram 2. Spermatogonial numbers in 13 known and 1 unknown subfamilies of Lygaeidae.

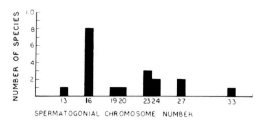

Histogram 3. Spermatogonial numbers in Pyrrhocoridae.

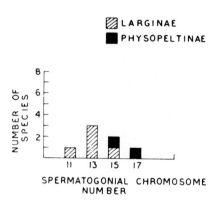

Histogram 4. Spermatogonial numbers in 2 subfamilies of Largidae.

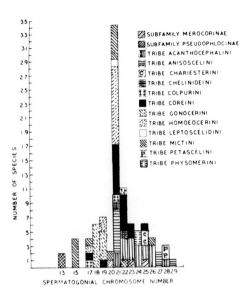

Histogram 5. Spermatogonial numbers in Merocorinae, Pseudophlonae, and 12 tribes of Coreinae of Coreidae.

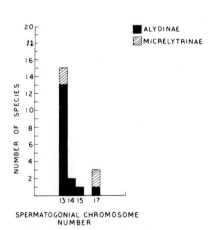

Histogram 6. Spermatogonial numbers in 2 subfamilies of Alydidae.

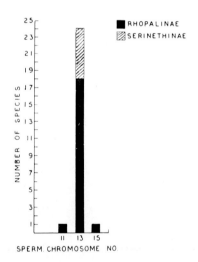

Histogram 7. Spermatogonial numbers in 2 subfamilies of Rhopalidae.

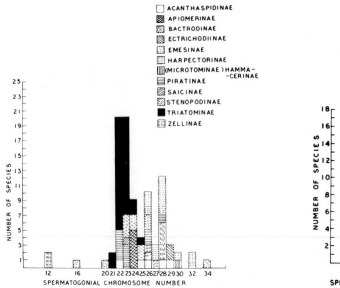

Histogram 8.   Spermatogonial numbers in 12 subfamilies of Reduviidae.

Histogram 9.   Spermatogonial numbers in Nabidae.

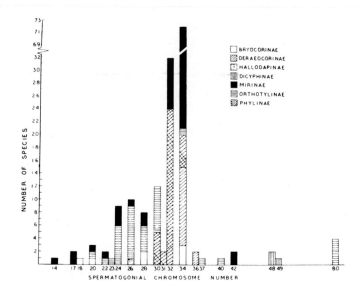

Histogram 10.   Spermatogonial numbers in 7 subfamilies of Miridae.

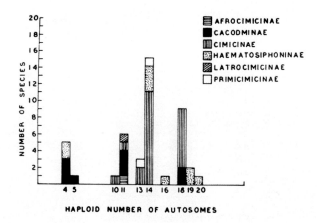

Histogram 11.   Haploid number of autosomes only in 6 subfamilies of Cimicidae.

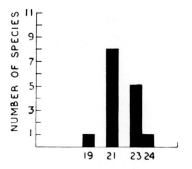

Histogram 12.   Spermatogonial numbers in Gerridae.

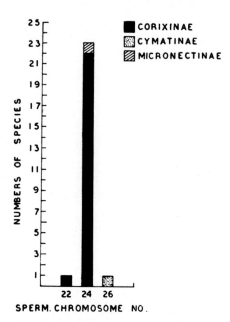

Histogram 13.   Spermatogonial numbers in 3 subfamilies of Corixidae.

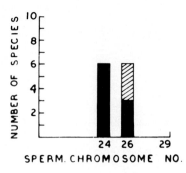

Histogram 14.   Spermatogonial numbers in 2 subfamilies of Notonectidae.

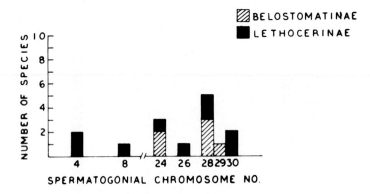

Histogram 15.   Spermatogonial numbers in 2 subfamilies of Belostomatidae.

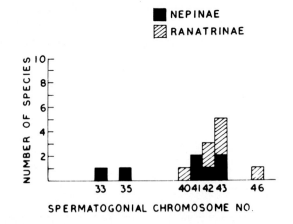

Histogram 16.   Spermatogonial numbers in 2 subfamilies of Nepidae.

REFERENCES

1. **Henking, H.,** Untersuchungen uber die ersten Entwicklungsvorgange in den Eiern der Insekten. II. Uber Spermatogenese und deren Beziehung zur Eientwicklung bei *Pyrrhocoris apterus* L., *Z. Wiss. Zool.,* 51, 685, 1891.

2. **Montgomery, T. H.,** A study of the chromosomes of the germ cells of Metazoa, *Trans. Am. Phil. Soc.,* 20, 154, 1901.

3. **Montgomery, T. H.,** Further studies on the chromosomes of the Hemiptera-Heteroptera, *Proc. Acad. Nat. Sci. Phil.,* 53, 261, 1901.

4. **Wilson, E. B.,** Studies on chromosomes. I. The behavior of idiochromosomes in Hemiptera, *J. Exp. Zool.,* 2, 371, 1905.

5. **Wilson, E. B.,** Studies on chromosomes. II. The paired microchromosomes, idiochromosomes and heterotropic chromosomes in Hemiptera, *J. Exp. Zool.,* 2, 507, 1905.

6. **McClung, C. E.,** The spermatocyte divisions of the Locustidae, *Kansas Univ. Sci. Bull.,* 1, 185, 1902.

7. **Barr, M. L. and Bertram, L. F.,** A morphological distinction between neurons of the male and female and the behaviour of the nucleolar satellite during accelerated nucleoprotein synthesis, *Nature,* 163, 676, 1949.

8. **White, M. J. D.,** *Animal Cytology and Evolution,* Cambridge University Press, London, 1973, 961.

9. **Morrill, C. V.,** The chromosomes in the oogenesis, fertilization and cleavage of coreid Hemiptera, *Biol. Bull.,* 19, 79, 1910.

10. **Hoy, W. E.,** A preliminary account of the chromosomes in the embryos of *Anasa tristis* and *Diabrotica vittata, Biol. Bull.,* 27, 45, 1914.

11. **Montgomery, T. H.,** Some observations and considerations upon the maturation phenomena of the germ cells, *Biol. Bull.,* 6, 137, 1904.

12. **Montgomery, T. H.,** Chromosomes in the spermatogenesis of the Hemiptera-Heteroptera, *Trans. Am. Phil. Soc.,* 21, 97, 1906.

13. **Montgomery, T. H.,** On the dimegalous sperm and chromosomal variation of *Euschistus,* with reference to chromosomal continuity, *Arch. Zellforsch.,* 5, 120, 1910.

14. **Montgomery, T. H.,** The spermatogenesis of an Hemipteran, *Euschistus, J. Morphol.,* 22, 731, 1911.

15. **Wilson, E. B.,** Studies on chromosomes. III. The sexual difference of the chromosome groups in Hemiptera, with some consideration on the determination and inheritance of sex, *J. Exp. Zool.,* 3, 1, 1906.

16. **Wilson, E. B.,** Differences in the chromosome groups of closely related species and varieties, and their possible bearing on the 'physiological species', *Proc. 7th Int. Congr. Zool. (Boston),* 19, 347, 1907.

17. **Wilson, E. B.,** Notes on the chromosome group of *Metapodius* and *Banasa, Biol. Bull.,* 12, 303, 1907.

18. **Wilson, E. B.,** The case of *Anasa tristis, Science,* 25, 191, 1907.

19. **Wilson, E. B.,** The supernumerary chromosomes of Hemiptera, *Science,* 26, 870, 1907.

20. **Wilson, E. B.,** Studies on chromosomes. IV. The accessory chromosome in *Syromastes* and *Pyrrhocoris,* with a comparative review of the types of sexual difference of the chromosome groups, *J. Exp. Zool.,* 6, 69, 1909.

21. **Wilson, E. B.,** The female chromosome groups in *Syromastes* and *Pyrrhocoris, Biol. Bull.,* 16, 199, 1909.

22. **Wilson, E. B.,** Studies on chromosomes. V. The chromosomes of *Metapodius,* a contribution to the hypothesis of the genetic continuity of chromosomes, *J. Exp. Zool.,* 6, 147, 1909.

23. **Wilson, E. B.,** Studies on chromosomes. VI. A new type of chromosome combination in *Metapodius, J. Exp. Zool.,* 9, 53, 1910.

24. **Wilson, E. B.,** Studies on chromosomes. VIII. Observations on the maturation phenomenon in certain Hemiptera and other forms, with considerations on synapsis and reduction, *J. Exp. Zool.,* 13, 345, 1912.

25. **Wilson, E. B.,** A chromatoid body simulating an accessory chromosome in *Pentatoma, Biol. Bull.,* 24, 392, 1913.

26. **Gross, J.,** Die Spermatogenese von *Syromastes marginatus, Zool. Jahrb.,* 20, 439, 1904.

27. **Gross, J.,** Ein Beitrag zur Spermatogenese der Hemipteren, *Verh. Dtsch. Zool. Ges.,* 9, 180, 1904.

28. **Gross, J.,** Die Spermatogenese von *Pyrrhocoris apterus* L., *Zool. Jahrb.,* 23, 269, 1907.

29. **Gross, J.,** Heterochromosomen und Geschlechtsbestimmung bei Insekten, *Zool. Jahrb.,* 32, 99, 1912.

30. **Foot, K. and Strobell, E. C.,** A study of chromosomes in the spermatogenesis of *Asana tristis, Am. J. Anat.,* 7, 279, 1907.

31. **Foot, K. and Strobell, E. C.,** The accessory chromosome of *Anasa tristis, Biol. Bull.,* 12, 119, 1907.

32. **Foot, K. and Strobell, E. C.,** A study of chromosomes, chromatin, and nucleoli in *Euschistus crassus, Arch. Zellforsch.,* 9, 47, 1912.

33. **Foot, K. and Strobell, E. C.,** The chromosomes of *Euschistus variolarius, Euschistus servus* and the hybrids of the F_1 and F_2 generations, *Arch. Zellforsch.,* 12, 485, 1914.

34. **Payne, F.,** Some new types of chromosome distribution and their relation to sex, *Biol. Bull.,* 16, 119, 1909.

35. **Payne, F.,** The chromosomes of *Acholla multispinosa, Biol. Bull.,* 18, 174, 1910.
36. **Payne, F.,** I. A further study of the chromosomes of the Reduviidae. II. The nucleolus in the young oocytes and the origin of the ova in Gelastocoris, *J. Morphol.,* 23, 331, 1912.
37. **Browne, E. N.,** The relation between chromosome numbers and species in *Notonecta, Biol. Bull.,* 20, 19, 1910.
38. **Browne, E. N.,** A study of the male germ cells in *Notonecta, J. Exp. Zool.,* 14, 61, 1913.
39. **Browne, E. N.,** A comparative study of the chromosomes of six species of *Notonecta, J. Exp. Zool.,* 14, 61, 1916.
40. **Manna, G. K.,** A study of the chromosomes during meiosis in forty three species of Indian Heteroptera, *Proc. Zool. Soc. (Bengal),* 4, 1, 1951.
41. **Manna, G. K.,** Cytology and inter-relationships between various groups of Heteroptera, *Proc. 10th Int. Cong. Entomol.,* 2, 919, 1956.
42. **Manna, G. K.,** A further evaluation of the cytology and inter-relationsips between various groups of Heteroptera, *Nucleus,* 5, 7, 1962.
43. **Ueshima, N.,** Hemiptera II: Heteroptera, in *Animal Cytogenetics,* Vol. 3: *Insects,* Bernard John, Ed., Stuttgart, 1979, 1.
44. **Makino, S.,** *An Atlas of the Chromosome Numbers in Animals,* Iowa State College Press, Ames, 1951.
45. **Makino, S.,** *A review of the chromosome numbers in animals* (rev. ed.), Hokuryukan, Tokyo, 1956.
46. **Takenouchi, Y. and Muramoto, N.,** Chromosome numbers of Heteroptera, *J. Hokkaido, Univ., Educ.,* II B, 20, 1, 1969.
47. **Deb Mallik, S. and Manna, G. K.,** Morphometrical analysis of meiotic chromosomes of three species of scutellerid Heteroptera, in *Proc. 4th All India Cong. Cytol. and Genet.,* Bhagalpur, October 17—21, 1981 (in press).
48. **Manna, G. K. and Deb Mallik, S.,** Prereductional meiosis in an XO male reduviid bug *Ectrychotes abbreviatus* Rent. (Heteroptera), *Entomon,* 5(1), 19, 1980.
49. **Manna, G. K. and Deb Mallik, S.,** Sex chromatin elimination in the polymorphic male pyrrhocorid bug *Iphita limbata, Entomon,* 6(4), 287, 1981.
50. **Manna, G. K. and Deb Mallik, S.,** Meiotic chromosome in 41 species of Heteroptera, *CIS* 31, 9, 1981.
51. **Manna, G. K. and Deb Mallik, S.,** *Polytoxus* sp., the second reduviid genus with XO males (Heteroptera), *Entomon.,* 7, 151, 1982.
52. **Chickering, A. M.,** An unusual chromosome complement in *Lethocerus, Anat. Rec.,* 37, 156, 1927.
53. **Hughes-Schrader, S.,** Cytology of Coccids (Coccoidea-Homoptera), *Adv. Genet.,* 2, 127, 1948.
54. **Brown, S. W.,** Lecanoid chromosome behaviour in three more families of the Coccoidea (Homoptera), *Chromosoma,* 10, 278, 1959.
55. **Brown, S. W.,** The Comstockiella system of chromosome behaviour in the armored scale insects (Coccoidea: Diaspididae), *Chromosoma,* 14, 360, 1963.
56. **Brown, S. W.,** Chromosomal survey of the armored and palm scale insects (Coccoidea Diaspididae and Phoenicococcidae, *Hilgardia,* 36, 189, 1965.
57. **Halkka, O.,** Chromosome studies on the Hemiptera-Homoptera Auchenorrhyncha, *Ann. Acad. Sci. Fenn.,* A IV. 43, 1, 1959.
58. **Manna, G. K.,** Some aspects of chromosome cytology, *Proc. 56th Ind. Sci. Congr.,* 2, 1, 1969.
59. **Nordenskiold, H.,** Cytotaxonomical studies in the genus *Luzula.* I. Somatic chromosomes and chromosome numbers, *Hereditas,* 37, 325, 1951.
60. **La Cour, L. F.,** The *Luzula* system analysed by X-rays, *Heredity,* 6 (Suppl. vol.), 77, 1953.
61. **Scholl, H.,** Ein Beitrag zur Kenntnis der Spermatogenese der Mallophagen, *Chromosoma,* 7, 271, 1955.
62. **Bayreuther, K.,** Holokinetische Chromosomen bei *Haematopinus suis* (Anoplura, Haematopinidae), *Chromosoma* 7, 260, 1955.
63. **Ortiz, E.,** Chromosomes and meiosis in Dermaptera, *Chromosomes Today,* 2, 33, 1969.
64. **Schrader, F.,** Heteropycnosis and nonhomologous association of chromosomes in *Edessa irrorata* (Hemiptera, Heteroptera), *J. Morphol.,* 69, 587, 1941.
65. **Nordenskiold, H.,** A study of meiosis in the progeny of X-irradiated *Luzula purpurea, Hereditas,* 49, 33, 1963.
66. **Schrader, F.,** Notes on mitotic behaviour of long chromosomes, *Cytologia,* 6, 422, 1935.
67. **Schrader, F.,** Touch-and-go pairing in chromosomes, *Proc. Nat. Acad. Sci. Wash.,* 26, 634, 1940.
68. **Hughes-Schrader, S. and Schrader, F.,** The kinetochore of the Hemiptera, *Chromosoma,* 12, 327, 1961.
69. **Troedsson, P. H.,** The behaviour of the compound sex chromosomes in the females of certain Hemiptera-Heteroptera. *J. Morphol.,* 75, 103, 1944.
70. **Heizer, P.,** The chromosome cytology of two species of the Pacific genus *Oechalia* (Pentatomidae, Hemiptera-Heteroptera), *Oechalia patruelis* Stal and *Oechalia pacifica* Stal, *J. Morphol.,* 87, 179, 1950.
71. **Halkka, O.,** Studies on the mitotic and meiotic cell division in certain Hemiptera under normal and experimental conditions, *Ann. Acad. Sci. Fenn.,* 32, 5, 1956.

72. **Lewis, K. R. and Scudder, G. G. E.,** The chromosomes of *Dicranocephalus agilis* (Hemiptera-Heteroptera), *Cytologia,* 23, 92, 1958.
73. **Ueshima, N.,** Cytotaxonomy of the Triatominae (Reduviidae: Hemiptera), *Chromosoma,* 18, 97, 1966.
74. **Ueshima, N.,** Cytology and cytogenetics, in monograph of *Cimicidae,* Usinger, R. L., Ed., Entomological Society of America, 1966, 183.
75. **Wolfe, S. L. and John, B.,** The organisation and ultrastructure of meiotic chromosome in *Oncopeltus fasciatus, Chromosoma,* 17, 85, 1965.
76. **Comings, D. E. and Okada, T. A.,** Holocentric chromosomes in *Oncopeltus:* kinetochore plates are present in mitosis but absent in meiosis, *Chromosoma,* 37, 177, 1972.
77. **Buck, R. C.,** Mitosis and meiosis in *Rhodnius prolixus.* The fine structure of the spindle and diffuse kinetochore, *J. Ultrastruc. Res.,* 18, 489, 1968.
78. **Ladowski, J. M., Mei-Ying, Wong Yu, Forrest, H. S. and Laird, C. D.,** Dispersity of repeat DNA sequences in *Oncopeltus fasciatus,* an organism with diffused centromeres, *Chromosoma,* 43, 349, 1973.
79. **Hughes-Schrader, S. and Ris, H.,** The diffused spindle attachment of Coccids, verified by the mitotic behaviour of induced chromosome fragments, *J. Exp. Zool.,* 87, 429, 1941.
80. **Ris, H.,** A cytological and experimental analysis of the meiotic behaviour of the univalent X chromosomes in the bearberry aphid *Tamalia (= Phyllaphis) coweni* (Ckll.), *J. Exp. Zool.,* 90, 267, 1942.
81. **Halkka, O.,** X-ray induced changes in the chromosomes of *Limnotettix* (Homoptera), *Chromosoma,* 16, 185, 1965.
82. **Seshachar, B. R., Rao, S. R. V., and Dass, C. M. S.,** The kinetochore in *Eurybrachis apicalis* (Homoptera: Auchenorrhyncha), *Cytologia,* 24, 335, 1959.
83. **LaChance, L. E. and Degrugillier, M.,** Chromosomal fragments transmitted through three generations in *Oncopeltus* (Hemiptera). *Science,* 166, 235, 1969.
84. **La Chance, L. E., Degrugillier, M. and Leverich, A. P.,** Cytogenetics of inherited partial stability in three generations of the large milkweed bug as related to holokinetic chromosomes, *Chromosoma,* 29, 20, 1970.
85. **La Chance, L. E., and Richard, R. D.,** Irradiation of sperm and oocytes in *Oncopeltus fasciatus* (Hemiptera:Lygaeidae): sex ratio, fertility, and chromosome aberration in the F_1 progeny, *Can. J. Genet. Cytol.,* 15, 713, 1973.
86. **La Chance, L. E. and Riemann, J. G.,** Dominant lethal mutations in insects with holokinetic chromosomes: I. Irradiation of *Oncopeltus* (Hemiptera:Lygaeidae) sperm and oocytes, *Ann. Entomol. Soc. Am.,* 66, 813, 1973.
87. **Manna, G. K.,** On the possible involvement of genes in the induction of chromosome aberrations. *Proc. 4th All India Congr. Cytol. Genet.,* 1981, 143.
88. **Mendes, L. O. T.,** Observacoes citologicas em *Dysdercus.* Cadeias de cromosomios em tecido somatico de *Dysdercus mendesi* Bloete (Hemiptera-Pyrrhocoridae), *Bragantia* 9, 53, 1949.
89. **Rao, S. R. V.,** The kinetochore problem in Hemiptera, *Curr. Sci.,* 27, 303, 1958.
90. **Parshad, R.,** Cytological studies in Heteroptera. II. Chromosome complement and meiosis in the males of three Pyrrhocorid bugs, *Cytologia,* 22, 127, 1957.
91. **Parshad, R.,** Structure of the Heteropteran kinetochore. The behaviour of the long chromosomes in some lygaeid and coreid bugs during mitosis and meiosis, *Cytologia,* 23, 25, 1958.
92. **Lima-de-Faria, A.,** Genetics, origin and evolution of kinetochore, *Hereditas,* 35, 422, 1949.
93. **Desai, R. N.,** Monocentric nature of the chromosomes of *Ranatra* (Heteroptera) verified by the induced fragmentation experiments, *Experientia,* 25, 1170, 1969.
94. **Ruthmann, A. and Permantier, Y.,** Spindel und Kinetochoren in der Mitose und Meiose der Baumwollwanze *Dysdercus intermedius* (Heteroptera), *Chromosoma,* 41, 271, 1973.
95. **Ruthmann, A. and Dahlberg, R.,** Pairing and segregation of the sex chromosomes in X_1X_2-males of *Dysdercus intermedius* with a note on the kinetic organisation of Heteroptera chromosomes, *Chromosoma,* 54, 89, 1976.
96. **Schrader, F.,** The role of the kinetochore in the chromosomal evolution of the Heteroptera and Homoptera, *Evolution,* 1, 134, 1947.
97. **Schrader, F.,** Data contributing to an analysis of metaphase mechanics, *Chromosoma,* 3, 22, 1947.
98. **Gassner, G. and Klemetson, D. J.,** A transmission electron microscope examination of hemipteran and leipdopteran gonial centromeres, *Can. J. Genet. Cytol.,* 16, 457, 1974.
99. **John, B. and Lewis, K. R.,** The meiotic system, *Protoplasmatologia,* 6, 1, 1965.
100. **Rhoades, M. M. and Vilkomerson, H.,** On the anaphase movement of chromosomes, *Proc. Natl. Acad. Sci. U.S.A.,* 28, 433, 1942.
101. **Piza, S. de T.,** Normally dicentric insect chromosomes, *Proc. 10th Int. Congr. Entomol.,* 2, 945, 1958.
102. **Manna, G. K. and Dey, S. K.,** X-ray induced chiasma like configurations between the sex chromosomes of the Pyrrhocorid bug, *Physopelta schlanbuschi, Perspectives in Cytology and Genetics,* Manna, G. K., and Sinha, U., Eds., Hindasia, Delhi, India, 3, 1981, 367.

103. **Manna, G. K. and Dey, S. K.,** Differential sex chromosome aberrations induced by X-rays and their alterations by penicillin in male Heteropteran bug, *Physopelta schlanbuschi,* in *Cytogenetics in India,* Roy, R. P., and Sinha, U., Eds., Hindasia, Delhi, 1982, in press.

104. **Dey, S. K. and Manna, G. K.,** X-ray induced chromosome aberrations in testes of nymph to new adult of *Physopelta schlanbuschi* (Heteroptera), *Proc. 4th All India Congr. Cytol. Genet.,* Bhagalpur, 1981, 114.

105. **Barik, S. K.,** X-Ray Induced Chromosome Aberrations in the Testes of Nymphs and Adult Plant Bug *Lygaeus hospes* and Their Alterations by Penicillin, Ph.D. Thesis, University of Kalyani, West Bengal, 1979.

106. **Ray-Chaudhuri, S. P. and Manna, G. K.,** Evidence of a multiple sex chromosome mechanism in a pyrrhocorid bug *Physopelta schlanbuschi* Fabr., *Proc. Zool. Soc. (Calcutta),* 8, 65, 1955.

107. **Schrader, F.,** The cytology of regular heteroploidy in *Loxa* (Pentatomidae-Hemiptera), *J. Morphol.,* 76, 157, 1945.

108. **Schrader, F.,** The elimination of chromosomes in the meiotic divisions of *Brachystethus rubromaculatus* Dallas, *Biol. Bull.,* 90, 19, 1946.

109. **Schrader, F.,** Autosomal elimination and preferential segregation in the harlequin lobe of certain *Disocephalini* (Hemiptera), *Biol. Bull.,* 90, 265, 1946.

110. **Schrader, F.,** Cytological and evolutionary implications of aberrant chromosome behaviour in the harlequin lobe of some Pentatomidae (Heteroptera), *Chromosoma,* 11, 103, 1960.

111. **Ogawa, K.,** Chromosome studies in the Myriapoda. VI. A study on the sex chromosomes in allied species of chilopods, *Annot. Zool. Jpn.,* 25, 434, 1952.

112. **Ogawa, K.,** Chromosome studies in the Myriapoda. VIII. Behaviour of X-ray induced chromosome fragments of *Thereuonema hilgendorfi, Zool. Mag. Tokyo,* 64, 167, 1955.

113. **Ogawa, K.,** Chromosome studies in the Myriapoda. IX. The diffuse kinetochores verified by X-ray induced chromosome fragments in *Esastigmatobius longitarsis* Verhoeff, *Zool. Mag. Tokyo,* 64, 291, 1955.

114. **Ogawa, K.,** Chromosome studies in the Myriapoda. XV. On individually different three karyotypes found in *Otocryptops sexspinosus* (Say) (Chilopoda) (Preliminary report), *Zool. Mag. Tokyo,* 70, 176, 1961.

115. **DuPraw, E. J.,** *DNA and Chromosomes,* Holt, Rinehart and Winston, New York, 1970.

116. **Geitler, L.,** Die Analyse des Kernbaus und der Kernteilung der Wasserlaufer *Gerris lateralis* und *Gerris lavustris* (Hemiptera, Heteroptera) und die Somadifferenzierung, *Z. Zellforsch.,* 26, 641, 1937.

117. **Geitler, L.,** Uber den Bau des Ruhekerns mit besonderer Beruksichtigung der Heteroperen und Dipteren, *Biol. Zentralbl.,* 58, 152, 1938.

118. **Geitler, L.,** Die Entstehung der polyploiden Somakerne der Heteropteren durch Chromosomenteilung ohne Kernterlung, *Chromosoma,* 1, 1, 1939.

119. **Geitler, L.,** Das Heterochromatin der Geschlechtschromosomen bei Heteropteren, *Chromosoma,* 1, 197, 1939.

120. **Geitler, L.,** Das Washstum des Zellkerns in tierischen und pflanzlichen Geweben, *Ergebn. biol.,* 18, 1, 1941.

121. **Stys, P. and Kerzhner, I.,** The rank and nomenclature of higher taxa in recent Heteroptera, *Acta Entomol. Bohem.,* 72, 65, 1975.

122. **Manna, G. K.,** Sex mechanism in *Dysdercus, Curr. Sci.,* 26, 187, 1957.

123. **Wilson, E. B.,** Polyploidy and metaphase patterns, *J. Morphol.,* 53, 443, 1932.

124. **Wilson, E. B.,** Studies of chromosomes. VII. A review of the chromosomes of *Nezara* with some more general considerations, *J. Morphol.,* 22, 71, 1911.

125. **Piza, S. de T.,** Comportamento dos cromosomios na meiosis de *Euryphthalmus rufipennis* Laporte (Hemiptera-Pyrrhocoridae), *Luiz de Querioz,* 3, 27, 1946.

126. **Muramoto, N.,** A list of chromosome numbers of heteropteran insects of Japan, *CIS,* 14, 29, 1973.

127. **Jande, S. S.,** Pre-reductional sex chromosome in the family Tingidae, (Gymnocerata-Heteroptera), *Nucleus,* 3, 209, 1960.

128. **Southwood, T. R. E. and Leston, D.,** *Land and Water Bugs of the British Isles,* Frederick Warne, London, 1959.

129. **Harley, K. L. S. and Kassulke, R. C.,** Tingidae for biological control of *Lantana camara* (Verbenaceae), *Entomophaga,* 16, 384, 1971.

130. **Toshioka, S.,** On the chromosomes of certain Heteroptera, *Oyo-dobutsugaku-Zasshi,* 6, 109, 1934.

131. **Jande, S. S.,** Chromosome studies in the family Notonectidae (Cryptocerata-Heteroptera), *Chromosoma,* 12, 318, 1961.

132. **Manna, G. K. and Mazumder, S. K.,** Evolution of karyotype in an interesting species of grasshopper, *Tristia pulvinata, Cytologia,* 32, 236, 1967.

133. **Manna, G. K.,** Heterochromatinization, a common cause for the origin of diversified chromosomal constitution in some Orthopteroid species, *Proc. Natl. Acad. Sci., India,* 42 B, 55, 1972.

134. **Jande, S. S.,** M-chromosome in a Pentatomid bug, *Scotinophara* sp., *Curr. Sci.,* 29, 436, 1960.

135. **Piza, S. de T.,** Comportamento dos cromosomios na espermatogenese de *Microtomus conspicillaris* (Drury), *Rev. Agric.,* 32, 53, 1957.

136. **Pfaler-Collander, E. V.,** Vergleichend-Karyologische Untersuchungen an Lydaeiden, *Acta Zool. Fenn.,* 30, 1, 1941.

137. **Xavier, A, da C. M.,** Cariologia comparada de laguns Hemipteros Heteropteros (Pentatomideos e Coreideos), *Mem. Estud. Mus. Zool. Univ. Coimbra,* 163, 1, 1945.

138. **Parshad, R.,** Cytological studies in Heteroptera. IV. Chromosome complement and meiosis in 26 species of the Pentatomidae, Lygaeidae and Coreidea with a consideration of the cytological bearing on the status of these three superfamilies, *Res. Bull. Panjab Univ.,* 133, 521, 1957.

139. **Das, C. C.,** Studies on the structure and behaviour of the chromosomes of *Ranatra elongata* (Fabr.) (Nepidae: Hemiptera-Heteroptera), *La Cellule,* 59, 205, 1958.

140. **Ekblom, T.,** Chromosomenstudien bei der Spermatogenese des *Myrmus miriformis, Z. Zellforsch. Mikros. Anat.,* 21, 510, 1934.

141. **Chickering, A. M.,** Spermatogenesis in the Belostomatidae. III. The chromosomes in the male germ cells of a *Lethocerus* from New Orleans, Louisiana, *Pap. Mich. Acad. Sci.,* 15, 357, 1932.

142. **Jande, S. S.,** Chromosome number and sex mechanism in nineteen species of Indian Heteroptera, *Res. Bull. (NS) Panjab Univ.,* 10, 415, 1959.

143. **Cobben, R. H.,** Evolutionary trends in Heteroptera. Part I. Eggs, architecture of the shell, gross embryology and eclosion, *Center Agric. Publ. and Document,* Wageningen, 1968, 475.

144. **Takenouchi, Y. and Muramoto, N.,** A survey of the chromosomes in 20 species of Heteropteran insects (in Japanese), *J. Hokkaido Univ. Educ.* II B, 18, 1, 1967.

145. **Bowen, R. H.,** Notes on the occurrence of abnormal mitosis in spermatogenesis, *Biol. Bull.,* 43, 184, 1922.

146. **Bowen, R. H.,** Studies on insect spermatogenesis. II, *J. Morphol.,* 37, 81, 1922.

147. **Yosida, T. H.,** A chromosome survey in twenty species of Heteropteran insects with special reference to the morphology of sex chromosomes. I, *La Kromosomo,* 2, 57, 1946 (in Japanese).

148. **Wilson, E. B.,** *The Cell in Development and Heredity,* Macmillan, New York, 1925.

149. **Ueshima, N., and Ashlock, P. D.,** Cytotaxonomy of the Lygaeidae (Heteroptera), *Univ. Kansas Sci., Bull.,* 1980.

150. **Banerjee, M. K.,** A study of the chromosomes during meiosis in 28 species of Hemiptera (Heteroptera, Homoptera), *Proc. Zool. Soc. (Calcutta),* 11, 9, 1958.

151. **Rajasekarsetty, M. R.,** Cytology of *Iphita limbata* (Pyrrhocoridae, Heteroptera), *Caryologia,* 16, 1, 143, 1963.

152. **Schachow, S. D.,** Abhandlungen uber haploide Chromosomengarnituren in den Samendrusen der Hemiptera, *Anat. Anz.,* 75, 1, 1932.

153. **Toshioka, S.,** On the chromosomes in Hemiptera-Heteroptera. V. On the chromosomes of three species of *Cletus, Dobutsugaku Zasshi,* 47, 637, 1935 (in Japanese).

154. **Piza, S. de T.,** A note on chromosomes in three corrid bugs, *Rev. Agric.,* 31, 32, 1956.

155. **Parshad, R.,** Chromosome number and sex mechanism in twenty species of the Indian Heteroptera, *Curr. Sci.,* 16, 125, 1957.

156. **Piza, S. de T.,** Nota previa sobre a meiosis de *Corizus (Lyorhyssus) hayalinus* (Hemiptera-Corizidae), *Luiz de Queiroz,* 3, 141, 1946.

157. **De Meijere, J. H. C.,** Uber einige europaische Insekten, besonders gunstig zum Studium der Reifungsteilungen nebst einigen Zusatzen zur Azetokarmin-Methode, *Zool. Anz.,* 88, 209, 1930.

158. **Mikolajski, M.,** Chromosome complement and meiosis in the males of *Chorosoma schillingi* (Schum) (Heteroptera, Rhopalidae), *Zool. Pol.,* 20, 155, 1970.

159. **Slack, H. D.,** Cytological studies on five families of Hemiptera-Heteroptera, Thesis, University of Edinburgh, Edinburgh, 1938.

160. **Leston, D.,** Cytotaxonomy of Miridae and Nabidae (Hemiptera), *Chromosoma,* 8, 609, 1957.

161. **Yosida, T. H.,** A chromosomal survey in 12 species of Hemiptera, *Iden-no-sogokenkyu,* 1, 85, 1950 (in Japanese).

162. **Mikolajski, M.,** Multiple sex chromosome mechanism in *Nabis* Linn. (Heteroptera, Nabidae), *Zool. Pol.,* 14, 15, 1964.

163. **Jande, S. S.,** Chromosome mechanism in *Polididus armatissimus* (Reduviidae, Heteroptera), *Experientia,* 16, 440, 1960.

164. **Jande, S. S.,** Chromosome number and sex mechanism in twenty seven species of Indian Heteroptera, *Res. Bull. (NS) Panjab Univ.,* 10, 215, 1959.

165. **Slack, H. D.,** Chromosome number in *Cimex, Nature,* 142, 358, 1938.

166. **Slack, H. D.,** The chromosomes of *Cimex, Nature,* 143, 78, 1939.

167. **Slack, H. D.,** Structural hybridity in *Cimex, Chromosoma,* 1, 104, 1939.

168. **Darlington, C. D.,** The genetical and mechanical properties of the sex chromosomes. V. *Cimex* and Heteroptera, *J. Genet.,* 39, 101, 1939.

169. **Ueshima, N.,** Supernumerary chromosomes in the human bed bug, *Cimex lectularius* Linn. (Cimicidae: Hemiptera), *Chromosoma,* 20, 311, 1967.

170. **Dass, C. M. S.,** Meiosis in two members of the family Nepidae (Hemiptera-Heteroptera), *Caryologia,* 4, 77, 1952.

171. **Bawa, S. R.,** Studies on insect spermatogenesis. I. Hemiptera-Heteroptera. The sex chromosome and cytoplasmic inclusions in the male germ cells of *Laccotrephes maculatus* Fabr. and *Sphoerodema rusticum* Fabr., *Res. Bull. Panjab Univ.,* 39, 181, 1953.

172. **Sharma, G. P. and Parshad, R.,** The morphology of chromosomes in *Laccotrephes maculatus* Fabr. (Hemiptera-Heteroptera), *Res. Bull. Panjab Univ.,* 72, 67, 1955.

173. **Spaul, E. A.,** The gametogenesis of *Nepa cinerea, C. R. Soc. Biol.,* 92, 1476, 1922.

174. **Steopoe, I.,** La spermatogenese chez la *Nepa cinerea, C. R. Paris Soc. Biol.,* 92, 1436, 1925.

175. **Shikata, T.,** Studies on the chromosomes of *Ranatra chinensis* Mays (in Japanese), *Jpn. J. Genet.,* 22, 94, 1949.

176. **Srivastava, M. D. L., and Das, C. C.,** Sex chromosomes of *Ranatra elongata, Nature,* 172, 505, 1953.

177. **Wilke, G.,** Die Spermatogenese von *Hydrometra lacustris* L., *Jena Z. Naturw.,* 42, 669, 1907.

178. **Hughes-Schrader, S.,** The chromosomes of *Nautococcus schraderae* Vays. and the meiotic division figures of male Llaveine coccids, *J. Morphol.,* 70, 261, 1942.

179. **Nur, U.,** A supernumerary chromosome with an accumulation mechanism in the lecanoid genetic system, *Chromosoma,* 13, 249, 1962.

180. **Nur, U.,** Sperm, sperm bundles and fertilization in a mealy bug *Pseudococcus obscurus* Essig (Homoptera: Coccoidea), *J. Morphol.,* III, 173, 1962.

181. **Nur, U.,** Population studies of supernumerary chromosomes in a mealy bug, *Genetics,* 47, 1679, 1962.

182. **Nur, U.,** Harmful supernumerary chromosomes in a mealy bug population, *Genetics,* 54, 1225, 1966.

183. **Nur, U.,** The effect of supernumerary chromosomes on the development of mealy bugs, *Genetics,* 54, 1239, 1966.

184. **Nur, U.,** Mitotic instability leading to an accumulation of B-chromosomes in grasshoppers, *Chromosoma,* 27, 1, 1969.

185. **Battaglia, E.,** Cytogenetics of B-chromosomes, *Caryologia,* 17, 245, 1964.

186. **Jones, R. N.,** B-chromosome systems in flowering plants and animal species, *Internat. Rev. Cytol.,* 40, 1, 1975.

187. **Barik, S. K., Patra, S., Deb Mallick, S., and Manna, G. K.,** The occurrence of a supernumerary Y-chromosome in a lygaeid bug, *Lygaeus hospes,* in *Perspectives in Cytology and Genetics,* Manna, G. K., and Sinha, U., Eds., 3, 131, 1981.

188. **Ueshima, N.,** Distribution, host relationships and speciation in the genus *Paracimex* (Cimicidae: Hemiptera), *Mushi,* 42, 15, 1968.

189. **Ueshima, N.,** Chromosome behaviour of the *Cimex pilosellus* complex (Cimicidae:Hemiptera), *Chromosoma,* 14, 511, 1963.

190. **Leinard, D. E.,** Biosystematics of the "*Leucopterus* complex" of the genus *Blissus, Bull. Conn. Agric. Exp. Sta.,* 677, 5, 1966.

191. **Srivastava, M. D. L.,** Compound sex chromosome mechanism and regularly occurring meiotic aberrations in the spermatogenesis of *Macropygium reticulare* (Pentatomidae-Hemiptera), *La Cellule,* 58, 259, 1957.

192. **Menon, P. S.,** On the multiple sex chromosome mechanism in a lygaeid, *Oxycarenus hyalinipennis* (Costa), *Experientia,* 11, 483, 1955.

193. **Ueshima, N.,** Chromosome study of *Thyanta pallidovirens* (Stal) in relation to taxonomy, *Pan. Pac. Entomol.,* 39, 149, 1963.

194. **Barth, R.,** Estudos anatomicos e histologicos sobre a subfamilia Triatominae (Hemiptera:Reduviidae). VI. Estudo comparativo sobre a espermiocitogenese das especies mais importantes, *Mem. Inst. Oswardo Cruz,* 54, 599, 1956.

195. **Schreiber, G. and Pellegrino, J.,** Ete ropicnosi di autosomi come possible mecanismo di speciazione, *Sci. Genet.,* 3, 215, 1950.

196. **Manna, G. K.,** Multiple sex chromosome mechanism in a reduviid bug, *Conorhinus rubrofasciatus (De Geer), Proc. Zool. Soc. (Bengal),* 3, 355, 1950.

197. **Yosida, T. H.,** A chromosome survey in twenty species of Heteropteran insects, with special reference to the morphology of sex chromosomes. II, *La Kromosomo,* 3—4, 139, 1947 (in Japanese).

198. **Ryckman, R. E. and Ueshima, N.,** Biosystematics of the *Hesperocimex* complex (Hemiptera:Cimicidae) and avian hosts (Piciformes:Picidae:Passerformes:Hirundinidae), *Ann. Entomol. Soc. Amer.,* 57, 624, 1964.

199. **Parshad, R.,** Cytological studies in Heteroptera. I. The behaviour of chromosomes during the male meiosis of *Ranatra sordidula* with general consideration of the evolution of compound sex chromosomes in Nepidae, *Caryologia,* 8, 349, 1956.

200. **Steopoe, I.,** La spermatogenese chez *Ranatra linearis, C. R. Paris Soc. Biol.,* 96, 1030, 1927.

201. **Schrader, F. and Hughes-Schrader, S.,** Polyploidy and fragmentation in the chromosomal evolution of various species of *Thyanta* (Hemiptera), *Chromosoma,* 7, 469, 1956.

202. **Schrader, F. and Hughes-Schrader, S.,** Chromatid autonomy in *Banasa* (Hemiptera-Pentatomidae), *Chromosoma,* 9, 193, 1958.

203. **Toshioka, S.,** On the chromosomes in Hemiptera-Heteroptera. II, *Dobutsugaku Zasshi,* 46, 34, 1934 (in Japanese).
204. **Piza, S. de T.,** Prova cruciacis da discentricidade dos cromosomios dos Hemipteros, *Luiz de Queiroz,* 10, 156, 1953.
205. **Banerjee, M. K.,** Chromosome elimination during meiosis in the males of *Macroceroea (Lohita) grandis* (Gray), *Proc. Zool. Soc., (Calcutta),* 12, 1, 1959.
206. **Distant, W. L.,** Rhynchota, in *Fauna of British India,* London, 5, 1902—1910.
207. **Akingohungbe, A. E.,** Chromosome numbers of some North American mirids (Heteroptera:Miridae), *Can. J. Genet. Cytol.,* 16, 251, 1974.
208. **Ekblom, T.,** Chromosomenuntersuchungen bei *Salda littoralis* L., *Callocoris chenopodii* Fall., und *Mesovelja furcata* Muls. & Rey, sowie Studien uber die Chromosomen bei verschiedenen Hemiptera-Heteroptera in Hinblick auf phylogenetische Betrachtungen, *Chromosoma,* 2, 12, 1941.
209. **Suomalainen, E.,** On the chromosomes of the geometrid moth genus *Cadaria, Chromosoma,* 16, 166, 1965.
210. **Mello-Sampayo, T.,** Differential polyteny and karyotype evolution in *Luzula,* a critical interpretation of morphological and cytophometric data, *Genet. Iber.,* 13, 1, 1961.
211. **Halkka, O.,** A photometric study of the *Luzula* problem, *Hereditas,* 52, 81, 1964.
212. **Leston, D.,** Chromosome number and the systematics of Pentatomorpha (Hemiptera), *Proc. 10th Int. Congr. Entomol.,* 2, 911, 1956.
213. **Kumar, R.,** Aspects of the morphology and relationships of the superfamilies Lygaeoidea, Piesmatoidea and Pyrrhocoridea (Hemiptera:Heteroptera), *Entomol. Month. Mag.,* 103, 251, 1967.
214. **Evans, J. W.,** A re-examination of an Upper Permian insect, *Paraknightia magnifica* Ev., *Rec. Australian Museum,* 22, 246, 1950.
215. **Brues, C. T., Melander, A. L. and Carpenter, F. M.,** Classification of insects, *Bull. Mus. Comp. Zool.,* Harv. Univ., 108, 1, 1954.
216. **Southwood, T. R. E.,** The structure of eggs in the terrestrial Heteroptera and its relationship to the classification of the group, *Trans. Roy. Entomol. Soc., London,* 108, 163, 1956.
217. **Scudder, G. G. E.,** The female genitalia of the Heteroptera: morphology and bearing on classification, *Trans. Roy. Entomol. Soc. London,* 111, 405, 1959.
218. **Leston, D.,** Testis, follicle numbers and higher systematics of Miridae (Hemiptera-Heteroptera), *Proc. Zool. Soc. London,* 137, 89, 1961.
219. **Muramoto, N.,** A chromosome study in eighteen Japanese heteropterans (in Japanese), *La Kromosomo,* 91, 2896, 1973.
220. **Muramoto, N.,** A study of the C-banded chromosomes in some species of heteropteran insects, Proc. Jpn. Acad., 56(3) (Ser. B), 125, 1980.
221. **Muramoto, N.,** Notes on the Giemsa-banded chromosomes in some heteropteran insects, Proc. Jpn. Acad., 56(3) (Ser. B), 101, 1978.

Chapter 9

TRENDS OF CHROMOSOME EVOLUTION IN THE PLANT KINGDOM

Arun Kumar Sharma and Archana Sharma

TABLE OF CONTENTS

I. INTRODUCTION

The evolution of chromosomes presents a fascinating, albeit complex problem. A major aspect is the enigma of the origin of the chromosome structure of the eukaryota. The range of eukaryotic chromosome structure is too vast to have originated from the genophore of prokaryotes, which is hardly more complex than a mere DNA molecule. The issue is further complicated by the absence of any fossil record between the prokaryotes and eukaryotes, the former having arisen about three billion years ago. There is no direct evidence of the origin of the eukaryotic system from the simple prokaryota except on the basis of the universality of the genetic code.

Recently, the concept of the prokaryotic ancestry of higher organisms has run into difficulties, following the discovery of intervening sequences in the chromosomes of plants and animals. The widespread occurrence of such large sequences and the splicing enzyme and other factors related to mRNA processing clearly distinguish the eukaryotic chromosome from the prokaryotic genophore. The concept of the evolution of such complexity through progressive specialization of the gene-bearing structure of viruses and bacteria is invalidated by the lack of selective advantage in such a process. The intervening sequences inserted in a well-organized DNA molecule of prokaryota are bound to be eliminated through selection pressure. On the other hand, the origin of interruptions or "intron" sequences through mutation is difficult to visualize in view of the required time scale against the background of spontaneous mutation rate of the genes. All these factors point to the possibility that the two groups had evolved simultaneously in two different directions. The universality of the genetic code may further suggest that these two groups, in the initial phase, might have been subjected to a similar chemical environment which led to the development of the coding system. The primitive eukaryote might have resembled the prokaryote except in the presence of intervening sequences with associated enzymes in the former. Such an evolutionary process appears to be more tenable until evidences to the contrary are received from other studies, such as of the stromatolites and from molecular hybridization. With this premise, the present discussion will be restricted principally to the evolution of chromosome structure in plants. The Euglenophyta, in view of its chlorophyll content and photosynthetic efficiency, will be included within this discussion. Evolution within the animal kingdom, as is well known, has followed a different pattern and most of it has been covered in the different groups discussed in the previous articles of this two-volume work.

II. EVOLUTION IN LOWER PLANT GROUPS

The algal system represents a very good example of evolution of complexity from a primitive structure. In the euglenoid forms, nuclear division involves duplication of chromatids and their separation into daughter nuclei. However, the absence of centromere and spindle is characteristic of the group. High chromosome numbers have been reported in several genera, up to even 177 in *Peranema trichophora*.[1,2] The chromosomes, though organized into discrete units, with the capacity for duplication and separation, yet could not be resolved into specialized functional segments. The high chromosome number recorded in such an otherwise primitive system is a special feature. A similar, slightly advanced mitotic system is found in the dinoflagellates.[3] The chromosome number is also quite high but is characterized by a clearcut nuclear membrane. Notwithstanding such features, dinoflagellates do not possess the basic protein in the nucleus, the presence of which is characteristic of the eukaryotic systems. The chromosomal framework is composed of continuous DNA fibrils, appearing beaded at the ultrastructural level. This group has been dealt with extensively in an earlier article by Y.S.R.K. Sarma. The term *mesokaryota*, as applied to Dinophyceae, on the basis of its chromosome structure, is quite appropriate. The typical flagellated condition of these groups may be a limiting factor in their progressive evolutionary complexity.[4]

The presence of high chromosome numbers in such an undoubtedly primitive group with very little differentiation may be an index of the redundancy of the genes. An understanding of the extent of the redundancy is essential. Progressive chromosome evolution over the dinophycean system is manifested in the green algae. In the Conjugales, the chromosomes are well defined, spindle structure is well organized, and the separation of chromosomes to the two poles is well regulated. But the level of functional differentiation of chromosome segments is rather poor. Nucleolar organizers are present, but centromeric function appears to be dispersed throughout the chromosome length. However, the use of the term *diffuse* or *polycentric* is, to some extent, vague since it is difficult to resolve two adjacent centromeres in a polycentric system. The fact remains that there is parallel movement of chromosomes along the spindle. In Chlorophyceae in general the chromosome number is rather low ($n = 4$ to 10) as compared to other groups of algae. In Conjugales, despite the record of $n = 84$ in certain species of *Spirogyra*,[2] the number is not very high. But in view of the diffuse centromere the extent to which such high numbers reflect polyploidy or fragmentation is yet ambiguous. Chromosomes with a diffuse centromere have the unique advantage of survival of fragments. It is remarkable that such a group of algae, with highly organized photosynthetic system reflecting a highly complex chloroplast structure, has maintained its evolutionary manifestations within comparatively low chromosome numbers. Evidently, imperceptible gene mutations and structural changes have played very important roles in their evolution. Surprisingly, Siphonales, with its multiple nucleate system, maintains the individuality of its nuclei with discrete chromosome groups. Such behavior is in strong contrast to higher plants, which show well-defined uninucleate cells in a complex tissue system. In angiosperms the tapetal cells are mostly multinucleate but they often show nuclear fusion resulting in very high chromosome numbers. An analogous situation exists in the endosperm cells, which are free nuclear in the early phase. Fusion of nuclei is also not uncommon. Absence of any fusion in a multinucleate system, such as Siphonales, is thus in strong contrast to the frequent fusion observed in multinucleate cells of angiosperms. Explanation of such behavior may be sought for in the genetic system of Siphonales. Since the group as a whole is multinucleate, it may have an inbuilt genetic resistance to fusion of nuclei which would result in high polyploid chromosome numbers within the multinucleate cells. Since the organism is undifferentiated, such resistance may give it an adaptive advantage against the rigors of selection. On the other hand, in the higher plants the body system is highly differentiated and each cell contains a nucleus with a constant number of chromosomes. The multinucleate tissues, like the tapetum and endosperm, are exceptions which have evolved to fulfill certain nutritive requirements of the organ. Such organs form a negligible fraction of the body structure of the entire organism where uninucleate condition is the rule. Evidently the nuclei are not fully adapted to a multinucleate system in such exceptional cases. As a corollary to it, the built-in genetic system needed for the maintenance of nuclear integrity in the siphonaceous types is not so well organized in the higher plants, where such cases are exceptions. The difference between the two systems is a clear index of certain specialized features in genetic architecture of algae which are absent in otherwise highly specialized advanced plant groups.

The fungal chromosome, on the whole, does not yield any definite clue on the evolution of mitosis or chromosome structure. On the other hand it shows a high degree of complexity often comparable to higher angiosperms. Of the three fungal species studied extensively, in *Saccharomyces cerevisiae* both haploid and diploid numbers are stable, in *Aspergillus nidulans* the haploid is stable and the diploid unstable, while in *Neurospora crassa* the haploid number only is found. Diploids are extremely transient. Autopolyploids are more frequent and tri- and tetraploids have been recorded. Aneuploids and aneusomies show the meiotic behavior expected in eukaryotes. Structural alterations identified include mainly qualitative ones like inversion and translocations — reciprocal, nonreciprocal, and insertional. A specific mechanism for the relatively frequent loss of duplicated segments postulates the presence of special tandemly repeated DNA sequences common to all or several chromosome ends.[5,6]

Extensive work on chromosomes of *Neurospora*, the Ascomycete, shows a well-defined nucleolar organizer at the tip of the chromosome arm, often comparable to higher plants and animals, together with significant complexity in the chromosome structure. The chromosome cannot only be studied as a discrete unit, but its chemical nature has also been unravelled, comparable to that of higher eukaryota. The occurrence of nucleosomes in the form of octamers in the chromatin has been resolved, surrounded by a DNA supercoil of 200 base pairs. Five different types of octamers have been recorded, viz., H2a, H2b, H3, and H4 in the nucleosome proper, and H1 in the linker region. In the DNA itself, 140 base pairs are highly conserved, whereas the rest of the variable number serve as linker. In species of *Saccharomyces* the split gene nature has been fully established together with intervening sequences. In several other genera like *Aspergillus* and *Neurospora*,[6-9] a similar situation exists. A feature of special significance in the fungal chromosomes is the presence of lesser repeats as compared to higher organisms. In *Neurospora crassa* nearly 90% of the sequences are unique[10,11] whereas in the phycomycete *Achlya* sp., the repeats are slightly greater in number.[12] In this respect they are relatively more advanced than the prokaryote where the entire sequence is almost unique. Together with the complexity of nucleosomal pattern the typical uninemic and multirepliconic structure of eukaryote chromosome is also seen in fungi.[13,14] On the basis of the DNA value, nearly 20 to 40 thousand average genes have been estimated to be present in the fungi.[6] However, the very characteristic difference between the nuclear systems of higher eukaryotes and fungi is the presence of a nuclear envelope in the latter even when the spindle has been organized through spindle pole bodies, and loci associated with termination and initiation of microtubules are observed in both.[15-19]

The evolution of the meiotic behavior of chromosomes finds parallel to that of higher eukaryotes including even the presence of a synaptonemal complex between paired homologous chromosomes at pachytene.[20-22] Variations in meiotic behavior have been recorded in polyploids in *Xylaria, Allomyces*, and *Cyathus*,[23-26] and aneuploids in *Aspergillus* and *Saccharomyces*.[27,28] An inbuilt system for loss of additional chromosomes is seen in *Aspergillus* and *Neurospora*.[5,29-31]

All these data on the ultrastructural, molecular, and structural level of chromosomes indicate clearly that a high degree of specialization has been reached in this otherwise simple eukaryote in chromosome structure and behavior. In the presence of a comparatively lower number of repeated sequences, the fungi are slightly more advanced than the prokaryota, but in the complexities of chromosome structure, they resemble the most complex eukaryotic system. The presence of a continuous nuclear membrane, on the other hand, is a feature unique to this group. Such complexity in lower eukaryotes like fungi requires special attention. It is true that even with such complex genetic architecture, fungi have not contributed to higher forms in evolution. They may represent a degenerated or at least a blind line in the evolutionary system. The extent to which such a high level of specialization, in an otherwise simple thallophytic system with hardly any scope for differentiation, is responsible for a blind pathway in evolution, is a matter of conjecture. Possibly the saprophytic and parasitic adaptation of the group is related to this complexity in genetic architecture. Though the fungal chromosome does not provide any significant clue to chromosome evolution in higher eukaryotes, yet the value of understanding its complexity in relation to its specialized adaptive feature is undeniable.

The bryophytes occupy a unique position in the phylogenetic system in view of their potential in the origin of or derivation from the vascular cryptogams. The plant body, though well organized and differentiated into photosynthetic and conducting systems, is but a gametophyte containing a haploid chromosome number. It is the only category of plants where the chromosomes, though present in a single set, are in a well-differentiated morphological system exposed to the rigors of selection characteristic of terrestrial environment. It is also the only group where the sporophyte, containing the diploid chromosome number, has attained a certain degree of complexity but yet is dependent on the haploid system. The extent to which such an unusual interaction between genotype and environment has affected chromosome evolution in this group is yet to be assessed from an in-depth analysis of the chromosomes of bryophytes as a whole. Unfortunately, of the 14 to 15 thousand species, only about three thousand have been cytologically studied, involving principally chromosome counts. The later techniques for chromosome study, including pretreatment and banding patterns, have yet to be adopted in large scale for identification of chromosome segments in bryophytes. As such, in spite of its uniqueness, the *modus operandi* of chromosomal evolution in this group still remains ambiguous.

Bryophytes as a whole show comparatively low chromosome numbers, the lowest ($n = 4$) being recorded in *Takakia*. The most prevalent chromosome number, $n = 7$, is mainly restricted to Bryopsida, whereas in Hepaticopsida or Anthoterotopsida numbers are multiples of 9 and 5, respectively. In spite of the low number in *Takakia*, its relationship with the rest of the hepatics is still a matter of debate.[32] In hepatics the role of polyploidy is relatively slight except in species of *Riccia*, where it has been suggested as associated with xerophytic adaptation and weediness.[32]

In bryophytes as a whole, intra- and interspecific polyploidy play a significant role in evolution. The relative effect of structural changes in evolution is not yet clear due to the absence of detailed analysis of chromosome architecture. However, aneuploidy and reduction in number have been claimed to have been brought about by structural changes involving principally unequal reciprocal translocations rather than loss of whole chromosomes which may prove to be lethal.[32]

The gametophyte plant body is supposed to provide interesting information regarding evolution of sex chromosomes, which were first reported in *Sphaerocarpus donnelii*[33] and *Ceratodon perperus*.[34] In the former, the *x* chromosome was found to be the largest and *y* the smallest. However, later studies on dioecious species as a whole in several bryophytes have shown that the autosomal complement too plays a very important role in the expression of sex, since irradiation damage in *x* chromosome may result in maleness even in absence of *y*. Evidently, autosomes to a great extent carry the maleness, though a male mobility gene has been claimed in *y*.[35] In *Sphaerocarpus donnelii* variable numbers of sex chromosomes have been recorded in diploid, triploid, and tetraploid sporophytes without affecting their normalcy,[36] indicating the importance of autosomes in the balance of sex determination. Taking all these factors into account, the term "sex associated chromatin" has been attributed to the so-called sex chromosomes of bryophytes.[37] Thus, the nature of the heterochromatin found in bryophytes is not fully established.[38]

A chromosomal feature of special significance is the presence of microchromosomes. There has been considerable debate as to their function and homology, but their regular separation in meiosis and functional attributes differentiate them from accessory chromosomes. In general, they are almost half the size of the smallest chromosome of the complement of mosses.[39] Their association with sex determination has been claimed.[40] The origin of these chromosomes, which contain a significant amount of heterochromatin, is still a matter of debate. If they are regarded as very small normal chromosomes, their origin from multiple sex chromosomes is ruled out though such an origin may hold good for accessory chromosomes. However, the origin of the latter may involve other mechanisms as well.[41] An analogous condition is found in some insects as well.

Fundamental aspects of chromosome evolution were manifested in *Pleurozium schreberi* where centromeric activity was recorded in more than one point.[42] *Pleurozium* was regarded as representing an intermediate step between holocentric with diffuse centromeres and the monocentric states. The primitive centromere was visualized as associated with a simple chromosome block undifferentiated by any constriction. The fusion of such blocks may lead to evolution of larger chromosomes, some with the centromeric activities localized at the two ends, as in *Pleurozium*.[42] Such a fusion may lead to large chromosomes with centromeres at the telomeric ends. A concept of far-reaching implications is that fusion of two such telocentrics may lead to the origin of metacentrics as demonstrated in Commelinaceae.[43]

The evolution of such metacentrics through this process may also explain the molecular architecture of centromeres with duplicate chromomeres.[44] Thus the origin of symmetry from asymmetry in the chromosome architecture is also another feature in evolution.

III. EVOLUTION IN HIGHER PLANT GROUPS

The pteridophytes present a striking contrast to bryophytes, where every major group has a very high chromosome number. The high degree of polyploidy, noted in several genera belonging to primitive groups, indicates their relative antiquity. This phenomenon, coupled with the fact that very low chromosome numbers have been recorded in several genera, suggests a distant homology with the cytological state in the flowering plants.[45] Extensive hybridization and polyploidy have evidently given rise to the modern fern genera.[46-54] Even in primitive orders like Psilotales, Equisetales, and Ophioglossales the chromosome numbers are unusually high. The climax occurs in the last-mentioned group, where the chromosome number is more than 1200.[55] As the antiquity of these groups is undoubted, ferns are characterized by a high chromosome number accumulating in ancient forms. The case of *Ophioglossum* is an example where a high degree of polyploidy has been reached without lethality.[53] Such a level of tolerance undoubtedly indicates the tremendous resilience of certain primitive species against environmental upheavals.

From an evolutionary point of view, the ferns exhibit a remarkable adaptation in their chromosome behavior. Autopolyploidy is maintained in many cases through bivalent formation. Evidently the capacity of multivalent formation has been eliminated through gene mutation. Diminution in chromosome size associated with polyploidy has been recorded extensively in higher plants.[56-59] In ferns, such diminution in size has been suggested to be related to physiological readjustment enabling the organism to sustain successive polyploidization.[45] Decrease in chromosome size with polyploidy has, however, been attributed to either compaction of spirals or elimination of heterochromatic segments[58] or reduction in the lamellar number.[60]

Despite the important role of aneuploidy in the evolution of fern genera, reproductive isolation through a genetic barrier is the crucial factor in maintaining species stability in successful groups. In pteridophytes, in general, a remarkable contradiction is often recorded, especially in species of Psilotales, Lycopodiales, Equisetales, and Ophioglossales. On one hand, radical reduction in chromosome number has been induced through different cytological mechanisms including fragmentation and translocation.[53] On the other hand, there are species with very high chromosome numbers in otherwise primitive groups. But a contradiction is noted in genera like *Selaginella* ($n = 9$), *Isoetes* ($n = 10$), *Hymenophyllum* ($n = 13$), and *Osmunda* ($n = 12$), where chromosome number is rather low. It is quite likely that in species where very high chromosome number has been recorded, the counterparts with low numbers are eliminated.[45]

The relative roles of low and high chromosome number in evolution are remarkably manifested in pteridophyte evolution. The genus *Selaginella*, with comparatively low $n = 9$, has almost 800 species, whereas *Equisetum* ($n = 108$) has only two dozen. The low evolutionary capacity of *Equisetum*, so manifested, may be due to its high load of polyploidy.[45]

As such high chromosome number is the general rule in ferns, it is true that a pteridophytes can hardly be regarded as an innovative source of the evolution of major flora on the earth.[45] The high chromosome number in ferns might have forstalled any adventure in evolution, but the basic factor leading to such a high chromosome number is still a problem unsolved. The relative simplicity of most fern genera, with body structure much less differentiated than that of the phanerogams, is not at all commensurate with the number of genes expected to be present in such high numbers of chromosomes. It is true that differentiation, as controlled by number of genes, is not necessarily reflected in the chromosome complement[61] and extra genetic material is attributed to redundancy. In ferns, however, such redundancy does not remain restricted to the segmental level but may involve entire chromosomes. The question is whether all such high chromosome numbers are necessary for existence and vital functioning of all the fern species where they are found. Unfortunately, experimental studies on ferns involving conventional methods of chromosome elimination or recent techniques for repeat DNA analysis are yet to be carried out. In absence of these data the problem of uniqueness of most of the fern chromosomes will still remain unsolved.

The chromosomes of gymnosperms present certain unusual features in their evolution. In the living gymnosperms, more than one haploid number has been recorded in only two families, such as 8, 9, 11, and 13 in Cycadaceae and 9 to 19 in Podocarpaceae. The remaining families are characterized by single base numbers such as $x = 12$ in Ginkgoaceae, Cephalotaxaceae, Taxaceae, and Pinaceae; 13 in Auraucariaceae; 10 in Sciadophyteaceae, 11 in Taxodiaceae and Cupressaceae; 7 in Ephedraceae; 21 in Welwitschiaceae; and 22 in Gnetaceae. Structural alterations as well as polyploidy have been suggested to be responsible for evolution. Structural changes involved are quantitative ones, segmental interchanges, para- and pericentric inversions, all principally leading to position effect.

The principal controversy centers around the direction followed by the chromosomal changes. The cycads, though undoubtedly a primitive order, contain mostly acrocentric chromosomes whereas the conifers or podocarps have a large number of metacentrics. An analogous situation has not been observed in the angiosperms,[56,62] where most primitive families exhibit a symmetrical karyotype. As such, evolution has been assumed to be from symmetry to asymmetry in such families. In Commelinaceae, on the other hand, chromosome evolution has been shown to involve principally Robertsonian translocation leading to the origin of metacentrics from acro- and telocentrics.[43]

In view of the presence of acrocentrics in cycads, it is claimed that chromosome and morphological evolution have proceeded along two different directions in gymnosperms.[56,63,64] However, if the suggestion that centric fusion has been the principal feature in evolutionary advance in gymnosperms is accepted, the contradiction between morphological and chromosomal evolution is resolved. As supporting evidence, telocentrics have also been reported in several angiosperms including rye.[65-67] All these facts indicate that both centric fission and fusion as well as fragmentation have played very important roles in the evolution of chromosomes, at least in certain families of gymnosperms.[68] In cycads evolution may have principally involved fission of chromosome ends. The direction of the change and its stabilization have evidently adaptive implications in space and time.

In the angiosperms chromosome studies have been carried out extensively both in dicotyledons[69] and monocotyledons.[70] The changes so far recorded involve both the structure and number of chromosomes. Karyotype orthoselection has been almost absent in the majority of the flowering plants.[71,72] The chromosomal alterations occur at all levels of taxonomic hierarchy including species, genera, and families. Moreover, B chromosomes[73] and isochromosomes have arisen in certain genera through misdivision of centromere. Diffuse centromere has also been located in some taxa mainly belonging to Juncaceae and Cyperaceae. It is thus difficult to outline the specific chromosomal changes which have occurred during the evolution of angiosperms even though they have served in solving problems of taxonomic dispute in many cases.[74] In general, however, polyploidy, aneuploidy, and structural changes of chromosomes have been operative in the evolution of different taxa. The tendency in evolution has been from symmetry to asymmetry accompanied by diminution in chromosome size. Chromosome changes either induced through fragmentation, translocation, and inversion, or through interspecific hybridization as well as polyploidy or aneuploidy, leading to numerical alterations, have been stabilized in several taxa following vegetative and sexual apomixis. The importance of structural alterations of chromosomes in the evolution of higher plants is being gradually appreciated with advances of methodology for the study of chromosome structure. Subsequent to the use of pretreatment chemicals in clarifying chromosome details, the banding techniques have opened up new possibilities in delineating chromosome phylogeny. In plants, the banding patterns have not yet been as extensively applied as in animals.[75-78] The grasses, considered to be one of the most successful of the flowering groups, are the result of chromosome evolution in all its facets.

IV. GENERAL COMMENTS ON EVOLUTION OF CHROMOSOME STRUCTURE

Chemical complexity of the chromosome structure in eukaryotes, including plants, has involved different components of chromosomes. There has been an evolution of inbuilt genetic system to control all aspects of differentiation, which is sequential and phasic. The structure as visualized at present is fibrillar in nature, made up of fibrils 25- to 30 Å units in width, laid in several folds to give an ultimate diameter of 100 Å. It is a deoxyribonucleoprotein fiber in which condensed and decondensed segments alternate, and to a great extent the condensed segments are comparable to the classical "chromomeres". A condensed segment in an uncoiled state may be comparable to the entire genophore length of a prokaryote. The chromosome structure has further been resolved into nucleosomal subunits which are octamers of five types of histones, with one forming the linker. The DNA thread coils around the histone and the conserved sequences are clearly differentiated as compared to variable ones. From the protein standpoint the nucleosomes to some extent represent repeats. Further complexity is caused by the fact that genes are not concrete uninterrupted units but are composed of intervening sequences or introns and exons. Both basic and nonbasic proteins are known to enter into the composition of the chromosomes. The interrelationship of all the components including nucleic acids and proteins determines the volume of the chromosome in all phases of development and differentiation. Such a complex structure certainly is far removed from a simple prokaryotic genophore.

The discovery of multiple copies of sequences in chromosomes, otherwise referred to as repetitive ones, has uncovered further complexity in chromosome structure. Such repeated sequences may be highly homogeneous, being otherwise termed *satellite DNA*, or moderate and minor repeats, depending on the number of sequences present. Palindromes or inverted repeats are usually located at intercalary or termination points. A chromosome may have a highly homogeneous repeat at one locus and minor or moderate repeats interspersed between the unique sequences.

The total amount of DNA in the genome far exceeds that needed in coding for structural genes, a situation which causes the *C-value paradox*. In plants, specially in the grasses, a very high percentage of DNA has been found to be repetitive, being up to 75% in rye. In the human system only about 56,000 essential genes have been estimated, though the DNA can accommodate more than six million genes. However, there are a few cases, e.g., *Aspergillus nidulans*, where 97 to 98% DNA has been found to be unique.[79] The function of these repeats has not yet been fully established. Several authors have suggested their function in regulation to be loci for accumulation of mutation and to provide flanking support to the chromatin.[80,81] During fusion, yeast repetitive sequences are involved in gene conversion with special reference to RNA genes; in *Saccharomyces purpuratus* RNA hybridization is seen to be effective in nonrepetitive sequences linked with repeats.

The repeats have been located in different regions of chromosomes and on the basis of their location and attributed function several terms have been coined. Their presence in introns has been clearly established at intercalary, terminal, or initiation points. In yeast phenylalanine, the repeats are rather short.[82,83] In the animal system they occur as an insertion element.[84] The first insertion sequences were recorded in maize,[85] which are later referred to as *transposons*. Such sequences, when mobile, have adjacent repeats as demonstrated in yeast.[86] In *S. cereviseae* the position of the *Ty-I* gene was altered following culture for a month, and it was associated with a high degree of repeats. The importance of *spacers* is well established, especially in 5s rRNA genes. The term *selfish DNA* has been applied to the repetitive DNA sequences on the presumption that their amplification and survival are principally due to their replicating capacity in a selected cellular environment.[87] This term also includes middle repeats, introns, and intergenic DNA. The properties of amplification and dispersion are shared by both selfish and ignorant DNA, but the latter is sequence-independent, as in the case of 5s rRNA genes. *Incidental DNA* visualizes the origin of such repeats through high mutation pressure.[88] Such amplified, nonspecific, and dispersed DNAs have also been referred to as *nucleotypic DNA*.[89-91] Such DNA controls cell and chromosome volumes, generation time, and other chromosomal factors. The term *supplementary DNA* has been given to such amplified DNA sequences between related plant species, containing both repeats and nonrepeats and conferring a certain range of tolerance.[92] In view of the fact that all the different types of repetitive sequences, to which various epithets have been applied, have in common the properties of amplification, dispersion, and mobility, the term *Dynamic DNA* has been proposed.[93,93a] It exerts a dynamic control over all nucleotypic functions, including cell division and chromosomal integrity. The multiple copies provide scope for the generation of genetic variability, whereas mobility and dispersion are strategies to withstand total obliteration during evolution. Such sequences are essential for influencing genes required for physiological adaptation. Thus, the evolution of sequence complexity of DNA with special reference to repeats has been one of the principal features in chromosome evolution in eukaryota associated with the dynamic influence of genes on various aspects of metabolism.

In addition to the sequence complexity, the chromosomes of eukaryota exhibit a high degree of polymorphism with regard to their protein and DNA components. Variability in the amount of DNA during organogenesis, especially achieved through endoreplication, has been recorded in several plant systems during their ontogeny.[94,95] Examples are the metabolic DNA of *Petunia hybrida*,[96] transient DNA satellite in pith tissue,[97] and in differentiating region of *Vicia faba*.[98] Similar variability has been recorded in the protein components as well.[99-101] These studies reveal the variability in the amounts of histone and nonhistone proteins associated with DNA during the different phases of organ development. Endoreplication in chromosomes has often been resorted to meet the need for differentiation without increase in the number of cells, an essential requirement in differentiated systems. Analogous situations have been met with extensively in animal systems.[102-104] These data show that the composition of chromosomes undergoes a dynamic change during organogenesis while maintaining the basic genetic skeleton.

All these evidences, gathered from varied lines of approach indicate clearly that chromosome evolution in the eukaryotic system has led to a tremendous complexity in chromosome structure, together with a high potential for flexibility. The extreme complexity of the chromosome structure can be understood in view of the contradictory functions attributed to chromosomes in higher organisms. Such functions include sequential and phasic growth in a harmonious way, allowing continuity but maintaining scope for discontinuity, through recombinations and mutations. In its own framework specific and vital functions such as replication, translation, recombination, and mutation are attributed to certain genes within the architecture. It is the highest complex organic molecule possible, the composition of which is guided by the very basic principle of the genes and for the genes.

Chromosome evolution has been the end result of two separate mechanisms. On the one hand, there have been progressive changes leading to complexity of a vast magnitude. This complexity involves the DNA with its repetitive and nonrepetitive sequences as well as the proteins, both basic and nonbasic, of which the former enters the basic architecture of nucleosomes. In keeping with this complexity there has been an evolution of an inbuilt system of flexibility, as represented by the dynamic DNA as well as the variability of the components during organogenesis. The simultaneous progress of complexity and flexibility has endowed the eukaryotic chromosomes with the capacity to exert a supreme control over all aspects of reproduction and differentiation.

ACKNOWLEDGMENT

The authors would like to express their gratitude to Miss Sarmistha Sen for her ungrudging assistance in the preparation of this manuscript.

REFERENCES

1. **Leedale, G. F.**, Division and Cytology in the Euglenineae, Ph.D. Thesis, University of London, 1957.
2. **Godward, M. B. E.**, Ed., *The Chromosomes of Algae*, Edward Arnold, London, 1966.
3. **Soyer, M. O. and Haapala, O. K.**, Division and function of dinoflagellate chromosomes, *J. Microscopia*, 19, 137, 1974.
4. **Dodge, J. D.**, The Dinophyceae, in *The chromosomes of algae*, Godward, M. B. E., Ed., 1966, 96.
5. **Newmeyer, D. and Galeazi, D. R.**, The instability of *Neurospora* duplication Dp(1L — 1R) H4250 and its genetic control, *Genetics*, 85, 461, 1977.
6. **Fincham, J. R. S., Day, P. R., and Radford, A.**, Fungal genetics, *Botanical Monographs*, Blackwell Scientific, Oxford, 1979, 4.
7. **Felden, R. A., Sanders, M. M. and Morris, N. R.**, Presence of histones in *Aspergillus nidulans*, *J. Cell Biol.*, 68, 430, 1976.
8. **Goff, C. G.**, Histones of *Neurospora crassa*, *J. Biol. Chem.*, 251, 4131, 1976.
9. **Thomas, J. O. and Furber, V.**, Yeast chromatin structure, *FEBS Letters*, 66, 274, 1976.
10. **Brooks, P. R. and Huang, P. C.**, Redundant DNA of *Neurospora crassa*, *Biochem. Genet.*, 6, 41, 1972.
11. **Dutta, S. K.**, Transcription of non-repeated DNA in *Neurospora crassa*, *Biochim. Biophys. Acta*, 324, 428, 1973.
12. **Hudspeth, M. E. S., Timberlake, W. E., and Goldberg, R. B.**, DNA sequence organization in the water mold *Achlya*, *Proc. Nat. Acad. Sci. U.S.A.*, 74, 4432, 1977.
13. **Petes, T. D., Byers, B., and Fangman, W. L.**, Size and structure of yeast chromosomal DNA, *Proc. Nat. Acad. Sci. U.S.A.*, 70, 3072, 1973.
14. **Petes, T. D. and Williamson, D. H.**, Replicating circular DNA molecules in yeast, *Cell*, 4, 249, 1975.
15. **Kubai, D. F.**, Mitosis and fungal phylogeny, *Nuclear Division in the Fungi*, Brent Heath, I., Ed., Academic Press, New York, 1978, 177.
16. **Fuller, M. S.**, Mitosis in Fungi, *Int. Rev. Cytol.*, 45, 113, 1976.
17. **Pickett-Heaps, J. D.**, The evolution of the mitotic apparatus: an attempt at comparative ultrastructural cytology of dividing plant cells, *Cytobios*, 31, 257, 1969.
18. **Borisy, G. G., Peterson, J. B., Hyams, J. S., and Ris, H.**, Polymerization of microtubules into the spindle pole body of yeast *J. Cell. Biol.*, 67, 38a (abst.) 1975.
19. **Brent Heath, I.**, Experimental studies of mitosis in the fungi, in *Nuclear Division in the Fungi*, Brent Heath, I., Ed., Academic Press, New York, 1978, 7.
20. **Byers, B. and Goetsch, L.**, *Proc. Nat. Acad. Sci., U.S.A.*, 72, 5056, 1975.
21. **Raper, J. R. and Hoffman, R. M.**, *Schizophyllum commune*, in *Handbook of Genetics*, King, R. C., Ed., Plenum Press, New York, 1974, 597.

22. **Zickler, D.,** Development in the synaptenemal complex and the 'recombination nodules' during meiotic prophase in the seven bivalents of the fungus *Sordaria macrospora* Anersus, *Chromosoma*, 61, 289, 1977.
23. **Lu, B. C.,** Polyploidy in the Basidiomycete *Cyathus stercoreus*, *Amer. J. Botany*, 51, 343, 1964.
24. **Rodgers, J. D.,** *Xylaria curta:* the cytology of ascus, *Canad. J. Bot.*, 46, 1337, 1968.
25. **Leopold, U.,** Tetrad analysis of segregation in autopolyploids, *J. Genet.*, 54, 427, 1956.
26. **Emmerson, R. and Wilson, C. M.,** Interspecific hybrids and the cytogenetics and cytotaxonomy of Euallomycees, *Mycologia*, 46, 393, 1954.
27. **Käfer, E. and Upshall, A.,** The phenotypes of the eight disomics and trisomics of *Aspergillus nidulans*, *Heredity*, 64, 35, 1973.
28. **Parry, E. M. and Cox, B. S.,** The tolerance of aneuploidy in yeast, *Genet. Res., Camb.*, 16, 333, 1971.
29. **Barry, E. G.,** Chromosome aberrations in *Neurospora* and the correlation of chromosomes and linkage groups, *Genetics*, 55, 21, 1967.
30. **Roper, J. A. and Nga, B. H.,** Mitotic nonconformity in *Aspergillus nidulans:* the production of hypodiploid and hyperdiploid nuclei, *Genet. Res. Camb.*, 14, 127, 1969.
31. **Turner, B. C.,** Euploid derivatives of duplication from a translocation in *Neurospora*, *Genetics*, 85, 439, 1977.
32. **Smith, A. J. E.,** *The British and Irish Moss Flora*, Cambridge University Press, London, 1978.
33. **Allen, R. F.,** A cytological study of heterothallism in flax rust, *J. Agric. Res.*, 49, 765, 1934.
34. **Heitz, E.,** Geschlechtschromosomen bei einem Laubmoos, *Ber. Dt. Bot. Ges.*, 50, 204, 1932.
35. **Lorbeer, G.,** Struktur und Ihalt der Geschlechtschromosomen, *Ber. Dt. Bot. Ges.*, 59, 369, 1941.
36. **Allen, C. E.,** The genetics of bryophytes, *Bot. Rev.*, 11, 260, 1945.
37. **Berrie, G. K.,** *Bull. Soc. Bot. Fr.*, 121, 129, 1974.
38. **Newton, M. E.,** *J. Bryol.*, 9, 327, 1977.
39. **Wigh, K.,** Accessory chromosomes in some mosses, *Hereditas*, 74, 211, 1973.
40. **Tatuno, S.,** Weitere Untersuchungen über die Vergleichung der Heterochromasie bei einigen europäischen und amerikanischen Arten der Marchantiales, *Cytologia*, 25, 214, 1960.
41. **Jones, R. N. and Brown, L. M.,** Chromosome evolution and DNA variation in *Crepis*, *Heredity*, 36, 91, 1976.
42. **Vaarama, A.,** *Ann. Bot. Soc. Vanamo*, 28, 1, 1954.
43. **Jones, K.,** Aspects of chromosome evolution in higher plants, *Adv. Bot. Res.*, 6, 119, 1978.
44. **Lima-de-Faria, A.,** The relation between chromosomes, replicons, operons, transcription units, genes, viruses, palindromes, *Hereditas*, 81, 249, 1975.
45. **Manton, I.,** *Problems of Cytology and Evolution in the Pteridophyta*, Cambridge University Press, London.
46. **Manton, I. and Reichstein, T.,** *Ber. Schweiz. Bot. Ges.*, 71, 370, 1961.
47. **Roy, S. K. and Manton, I.,** *J. Linn. Soc. (Bot.)*, 59, 343, 1966.
48. **Sleep, A. and Reichstein, T.,** *Bauhinia*, 3, 299, 1967.
49. **Bir, S. S.,** Cytological observations on some Indian species of *Cystopterus* Bernh., *Nucleus*, 14, 56, 1971.
50. **Vida, G.,** Evolution in Plants, *Symp. Biol. Hungarica*, 12, 51, 1972.
51. **Brownsey, P. J.,** An evolutionary study of the *Asplenium lepidum* Complex, Ph.D. thesis, University of Leeds, 1973.
52. **Klekowski, E. J. Jr. and Hickok, L. G.,** *Am. J. Botany*, 61, 422, 1974.
53. **Lovis, J. D.,** Evolutionary patterns and processes in ferns, *Adv. Bot. Res.*, 4, 229, 1978.
54. **Walker, T. G.,** Chromosome and evolution in pteridophytes, in *Chromosomes in Evolution of Eukaryotic Groups*, Volume II, Sharma, A. K. and Sharma, A., Eds., CRC Press, Boca Raton, Fla., 1984, 103.
55. **Abraham, A., Ninan, C. A. and Mathew, P. M.,** *J. Indian Bot. Soc.*, 51, 339, 1962.
56. **Stebbins, G. L., Jr.,** *Chromosomal Evolution in Higher Plants*, Edward Arnold, London.
57. **Sharma, A. K. and Sharma, A.,** Chromosomal alterations in relation to speciation, *Bot. Rev.*, 25, 514, 1959.
58. **Sharma, A. K.,** Polyploidy and chromosome size, in *Chromosomes Today*, 3, 248, 1972.
59. **Sharma, A. K.,** A new look at chromosome and its evolution, *Proc. Nat. Acad. Sci. U.S.A.*, 42B, 12, 1976.
60. **Darlington, C. D.,** *Chromosome Botany*, George Allen and Unwin, London, 1964.
61. **Sharma, A. K.,** Evolution and taxonomy of monocotyledons, in *Chromosomes Today*, 2, Oliver and Boyd, London, 1969.
62. **Stebbins, G. L.,** *Variation and Evolution in Plants*, Columbia University Press, New York, 1950.
63. **Khoshoo, T. N.,** Cytological evolution in the gymnosperms — karyotype, *Proc. Sum. Sch. Bot.*, 1960, 119.
64. **Woolhouse, H. W.,** Ed., *Adv. Bot. Res.*, 6, 1978.
65. **Prakken, R. and Müntzing, A.,** A meiotic peculiarity in rye simulating a terminal centromere, *Hereditas*, 28, 441, 1942.

66. **Jones, G. H.,** Further correlations between chiasmata and U-type exchanges in rye meiosis, *Chromosoma,* 26, 105, 1969.
67. **Strid, A.,** *Bot. Not.,* 121, 153, 1968.
68. **Marchant, C.,** Chromosome patterns and nuclear phenomena in the Cycad Families Stageriaceae and Zamiaceae, *Chromosoma,* 24, 100, 1968.
69. **Raven, P. H.,** The basis of angiosperm phylogeny:cytology, *Ann. Miss. Bot. Gard.,* 62, 724, 1975.
70. **Sharma, A. K.,** Chromosome evolution in the monocotyledons — an overview, in *Chromosomes in the Evolution of Eukaryotic Groups,* Volume II, CRC Press, Boca Raton, Florida, 0000, 00.
71. **Riley, H. P. and Majumder, S. K.,** *The Aloineae: a Biosystematic Survey,* University Press of Kentucky, Lexington, 1979, 177.
72. **Sharma, A. K.,** Chromosomes and distribution of monocotyledons in the Eastern Himalayas, in *Tropical Botany,* Larsen, K., Ed., 327, 1979, Academic Press, New York.
73. **Jones, R. N.,** B chromosome system in flowering plant and animal species, *Int. Rev. Cytol.,* 40, 1, 1975.
74. **Sharma, A. K.,** Cytology as an aid in taxonomy, *Bull. Bot. Soc. Bengal,* 18, 1, 1964.
75. **Hecht, F., Wyandt, M. E., and Magenis, R. E. M.,** The human cell nucleus, quinacrine and other differential stains in the study of chromatin and chromosomes; in *The Cell Nucleus,* Vol. 2, Busch, H., Ed., Academic Press, New York, 1974, 32.
76. **Vosa, C. G.,** Heterochromatic recognition and analysis of chromosome variations in *Scilla sibirica, Chromosoma,* 43, 269, 1973.
77. **Sharma, A. K.,** Orcein, banding in chromosomes, *Frontiers of Plant Sciences,* P. Parija Felicitation Vol., Orissa, Botanical Society, 1977, 181.
78. **Lavania, U. C. and Sharma, A. K.,** Trypsin-orcein banding in plant chromosomes, *Stain Techn.,* 54, 261, 1979.
79. **John, B. and Miklos, C. L. G.,** Functional aspects of satellite DNA and heterochromatin, *Int. Rev. Cytol.,* 58, 114, 1979.
80. **Federoff, N. V.,** On Spacers, *Cell,* 16, 697, 1979.
81. **Sharma, A. K.,** Chromosome banding and repeated DNA, Birbal Sahni Gold Medal Lecture, *J. Ind. Bot. Soc.,* 54, 1, 1975.
82. **Ogden, R. C., Beckmann, J. S., Abelson, J., Kang, H. S., Soll, D., and Schmidt, O.,** *In vitro* transcription and processing of a yeast tRNA gene containing an intervening sequence, *Cell,* 17, 399, 1979.
83. **Knapp, G., Ogden, R. C., Peebles, C. L., and Abelson, J.,** Splicing of yeast tRNA precursors: structure of the reaction intermediates, *Cell,* 18, 37, 1979.
84. **Lewin, B.,** *Gene Expression in Eukaryotic chromosomes,* 2, 1, 1980, Wiley-Interscience, New York.
85. **McClintock, B.,** Chromosome organization and genic expression, *Cold Spring Harb. Symp. Quant. Biol.,* 16, 13, 1951.
86. **Cameron, J. R., Loh, E. Y. and Davis, R. W.,** Evidence for transposition of dispersed repetitive DNA families in yeast, *Cell,* 16, 739, 1979.
87. **Orgel, L. E. and Crick, F. H. C.,** Selfish DNA: the ultimate parasite, *Nature,* 284, 604, 1980.
88. **Jain, H. K.,** Incidental DNA, *Nature (London),* 288, 647, 1980.
89. **Evans, G. M. and Rees, H.,** Mitotic cycles in dicotyledons and monocotyledons, *Nature (London),* 233, 350, 1971.
90. **Bennett, M. D.,** Nuclear characters in plants, *Brookhaven Symp. Quant. Biol.,* 25, 344, 1973.
91. **Price, H. J.,** Evolution of DNA content in higher plants, *Bot. Rev.,* 42, 27, 1976.
92. **Hutchinson, J., Narayan, R. K. J., and Rees, H.,** Constraints upon the composition of supplementary DNA, *Chromosoma,* 78, 127, 1980.
93. **Sharma, A. K.,** Additional genetic elements in chromosomes, *Nucleus,* 21, 113, 1976.
93a. **Sharma, A. K.,** Additional genetic materials in chromosomes, in Kew Chromosome Conference 2, George Allen and Unwin, 1983.
94. **Nägl, W.,** Molecular and structural aspects of the endomitotic chromosome cycle in angiosperms, in *Chromosomes Today,* Darlington, C. D. and Lewis, K. R., Eds., 3, 17, 1972.
95. **Sharma, A. K.,** Change in chromosome concept, *Proc. Indian Acad. Sci.,* 87B, 161, 1978.
96. **Essad, S., Vallade, J. and Cornu, A.,** Variations interphasique de la teneur en DNA et du volume nucleiaires du zygote de *Petunia hybrida* consequences metabolique, *Caryologia,* 28, 207, 1975.
97. **Parenti, R., et al.,** Transient DNA satellite in dedifferentiating pith tissue, *Nature, New Biol.,* 246, 237, 1973.
98. **Pelc, S. R.,** Metabolic DNA in ciliated protozoa, salivary gland chromosomes and mammalian cells, *Int. Rev. Cytol.,* 32, 327, 1972.
99. **Innocenti, A. M.,** Cyclic change of histone DNA ratios in dedifferentiating nuclei of metaxylem cell line in *Allium cepa* root tip, *Caryologia,* 28, 225, 1975.
100. **Sau, H., Sharma, A. K. and Chaudhuri, R. K.,** DNA, RNA and protein content of isolated nuclei from different plant organs, *Indian J. Exptl. Biol.,* 18, 1519, 1980.

101. **Banerjee, M. and Sharma, A. K.,** Variations in DNA content, *Experientia,* 35, 42, 1979.
102. **Klimenko, A. I., Malyshev, H. B., and Nikitin, V. N.,** Heterogeneity and the synthesis of nonhistone protein of liver cell chromatin in the postnatal ontogenesis of white rats, *Dokl. Acad. Nauk., SSSR,* 225, 714, 1975.
103. **Holmgren, P., Rasmuson, B., Johansson, F., and Sundquist, G.,** Histone content in relation to amount of heterochromatin and developmental stage in species of *Drosophila, Chromosoma,* 54, 99, 1976.
104. **Mulherkar, R., Sharma, A., and Talukder, G.,** Chromosomal and cytochemical changes in human fibroblast cultures: a review, *J. Sci. Ind. Res.,* 38, 322, 1979.

Chapter 10

CHROMOSOME EVOLUTION IN PRIMATES WITH SPECIAL REFERENCE TO HOMINOIDEA

Archana Sharma and Geeta Talukder

TABLE OF CONTENTS

I. INTRODUCTION

Of the different groups among the eukaryotes, the maximum interest in the analysis of chromosome evolution has been evinced in the primates, principally due to the obsession of *Homo sapiens* in tracing his ancestry. A number of excellent reviews have been brought out at intervals, usually following the development of each new set of techniques for further differentiation of the chromosome segments, from differential banding patterns, to replication patterns of DNA, to high-resolution identification of localized chromosome components.[1-5] These methods have been applied to the understanding of the phylogenetic relationships and evolutionary trends within and between groups and, when possible, to postulating ancestral karyotypes. However, information has accumulated and is still accumulating on the primates so rapidly that any review is quickly outdated and lists of chromosome numbers remain incomplete.[6-8] This article will, therefore, attempt to give the general trends of chromosomal evolution within this group, with special reference to the extensively studied Hominoidea rather than to evolution within individual taxa. The system of classification related most closely to the cytological data has been followed (Table 1).

II. CHROMOSOME STUDIES WITHIN DIFFERENT GROUPS (EXCEPT HOMINOIDEA)

A. Suborder Prosimii

The members of the family Lemuridae have been studied exhaustively and the different banding pattern techniques have verified earlier conclusions. Chromosome numbers range from 44 to 60 in lemurs and 54 to 58 in hapalemurs.[2-12] The sex chromosomes are usually acrocentric in subfamily Lemurinae and microchromosomes have been reported. *Cheirogalus majus* from subfamily Cheirogalinae gives a somatic number of $2n = 66$, all chromosomes being acrocentric except for one pair of submetacentrics.[13] Among the members of family Indriidae, earlier regarded as a subfamily under Lemuridae, *Propithecus verreauxi* has $2n = 48$.[13]

Earlier studies on chromosomal evolution in Madagascar lemurs,[14,15] on being repeated using R, G, and Q bands, showed a correspondence among all clearly identifiable chromosomes of *Microcebus murinus, Lemur fulvus fulvus, L. f. collaris, L.f. albocollaris, L. macaco* and *L. mongoz*. Based on rearrangement of some of the chromosomes, the karyotype of *Microcebus murinus* seems to be closest to that of the hypothetical common ancestor. It shows variation in the number of acrocentrics, with a karyotype of $2n = 66$ or 68.[9,16,17,18] The karyotype of *L.f. fulvus*, which is nearly identical to that of *Microcebus murinus*, may have given rise to those of other forms by several rearrangements.[19] The ancestral karyotype was possibly composed only of acrocentrics.

Banding pattern analysis indicates apparent phylogenetic relationships between the respective ancestors of *Lemur catta, Hapalemur simus, H. griseus occidentalis, H. griseus* ssp. and *H. griseus griseus*. Their karyotypes can be derived from that of *L. fulvus fulvus* by rearrangements affecting mainly chromosomes 1 and X. The different subspecies of *H. griseus* show intense staining of juxtacentromeric heterochromatin, characterized by intense Q and T banding in acrocentrics, which was absent in the other species.[20] It was suggested that the ancestors of *L. catta* separated first from a line common with *Hapalemur*. This cleavage gives an intermediate status to *L. catta* between the genera *Lemur and Hapalemur*, supported by taxonomists who had separated *L. catta* from *Lemur*. Hamilton and Buettner-Janusch,[21] using only one banding technique, did not, however, agree with this interpretation.

Robertsonian translocations, which are rather frequent in these genera, are an inefficient mechanism to separate clearly two species. Conversely, a single intrachromosomal rearrangement or a single interchromosomal rearrangement of the reciprocal translocation type

Table 1
TAXONOMIC CLASSIFICATION OF THE LIVING PRIMATES

Suborder: Prosimii

Infra-order	Superfamily	Family	Subfamily	Genera	No. of species
Tupaiformes	Tupaioidea	Tupaidae	Tupaiinae	*Tupaia*	9
				Dendrogale	2
				Urogale	1
			Ptilocercinae	*Ptilocercus*	1
Lorisformes	Lorisoidea	Lorisidae		*Loris*	1
				Nycticebus	2
				Arctocebus	1
				Perodicticus	1
		Galagidae		*Galago*	6
Lemuriformes	Lemuroidea	Lemuridae	Lemurinae	*Lemur*	5
				Hapalemur	2
				Lepilemur	1
			Cheirogaleinae	*Cheirogaleus*	2
				Microcebus	2
Tarsiiformes	Tarsioidea	Indriidae		*Indri*	1
				Avahi	1
				Propithecus	2
		Daubentoniidae		*Daubentonia*	1
		Tarsiidae		*Tarsius*	3
		Callithricidae	Callithericinae	*Callithrix*	3
				Leontideus	3
				Cebuella	1
			Callimiconinae	*Callimico*	1

Suborder: Anthropoidea

Infra-order	Superfamily	Family	Subfamily	Genera	No. of species
Platyprhini		Cebidae	Aotinae	*Aotus*	1
				Brachyteles	1
				Callicebus	3
			Pithecinae	*Pithecia*	2
				Chiropotes	2
				Cacajao	3
			Alouattinae	*Alouatta*	5
			Cebinae	*Saimiri*	2
				Cebus	4
			Atelinae	*Ateles*	4
				Lagothrix	4
Catarrhini	Cercopithecoidea	Cercopithecidae	Papiinae	*Macaca*	13
				Cercocebus	5
				Papio	7
				Theropithecus	2
			Cercopithecinae	*Cercopithecus*	22
				Erythrocebus	1

Table 1 (continued)
TAXONOMIC CLASSIFICATION OF THE LIVING PRIMATES

Infra-order	Superfamily	Family	Subfamily	Genera	No. of species
		Colobidae	Colobinae	*Presbytis*	14
				Pygathrix	1
				Rhinopithecus	2
				Simias	1
				Colobus	5
				Nasalis	1
		Hylobatidae		*Hylobates*	6
				Symphalangus	1
	Hominoidea	Pongidae		*Pongo*	1
				Pan	2
				Gorilla	1
		Hominidae		*Homo*	1

probably plays an important role, having been observed in the separation of nearly all the species. The X chromosome, often considered as a most stable element, is quite variable in the lemurs.

Within the family Lorisidae, the somatic numbers range from $2n = 62$ in the genera *Perodicticus* and *Loris* to $2n = 50$ in *Nycticebus* and $2n = 52$ in *Arctocebus*.[1,2,10,21-23] In the two latter genera, the karyotypes are similar, with all submetacentric chromosomes. The two former genera, however, have different proportions of submetacentrics and acrocentrics. The genus *Galago*, in the related family Galagidae, gave $2n = 38$ and $2n = 62$ with variability in the number of submetacentrics and acrocentrics.[10,24] The chromosomal complement of *G. crassicaudalis* is similar to that of *Perodicticus potto*, indicating a possible link between Lorisinae and Galagidae. In the genus *Tarsius* under family Tarsiidae, the somatic number is $2n = 80$ (14sm + 66ac).[22a,25]

B. Suborder Platyrrhini

The family Cebidae has been studied extensively. The different species under the genus *Ateles* have almost identical karyotypes with $2n = 34$ (30sm + 2ac), metacentric X, and acrocentric Y, and *Lagothrix* has $2n = 62$.[1,7,10,11,21,26,27] *Aotus* shows $2n = 50$ or 54.[1,11,27] *Callicebus moloch* has $2n = 46$ (4m + 16sm + 24ac) with both sex chromosomes submetacentric. Its karyotype resembles the karyotypes of *Cebus, Aotus*, and *Alouatta caraya* and is almost identical with *Cacajao rubicundus*.[21] *Callicebus torquatus* has only $2n = 20$ (10sm + 10ac).[a] The genus *Pithecia* shows $2n = 46$. The species of *Alouatta* have somatic numbers ranging from 44, 52, to 53 (4m + 12 to 6sm + 26 to 30ac).[1,28,29,30]

The species of *Cebus* contain a uniform chromosome number of $2n = 54$ (4m + 14 to 20sm + 28 to 34ac). Sex chromosomes are variable; *C. capucinus* shows intraspecific variability both in $2n$ number and in types of chromosomes.[2,10,11,21,31,32] Four species of *Cebus* studied differ by some Robertsonian translocations and possess very large intercalary or terminal heterochromatic segments, representing more than 10% of the whole karyotype, as shown by banding patterns.[33,34]

The genus *Saimiri* contains uniformly $2n = 44$ chromosomes (4m + 26 to 28sm + 10 to 12ac) with submetacentric X and acrocentric Y.[2,11,21,35] The genus *Callimico* is classified by itself by taxonomists in the Callimiconieae on the basis of morphological considerations.

Its chromosomes ($2n = 48$, 4m + 24sm + 18ac, smX, and acY) are related to both Callithricidae and Cebidae, forming a link between the two groups, similar to *Saimiri*.[21,36]

The genera under Callithricidae show a general homogeneity in chromosome numbers, ranging from $2n = 44$ to 46 in *Callithrix, Saguinus* and *Leontidus* (4m + 26 to 28sm + 10 to 14ac) with submetacentric X and variable Y, indicating their close affinities and the role of centric fusion in reducing the numbers.[11,36-42]

C. Suborder Catarrhini
1. Superfamily Cercopithecoidea
Subdivision of this family has been made differently under different systems of classification, and the one given in the table tallies most closely with the considerable amount of cytological data available.[45-48]

Morphological comparison of the chromosomes in catarrhine species indicates roughly four groups:[3,49,50]

1. The group including different species of *Cercopithecus* has metacentric, submetacentric, and acrocentric chromosomes; a pair of acrocentrics marked by a broad achromatic region; and $2n = 48$ to 72.
2. The Papinae group (*Macaca, Papio, Theropithecus,* and *Circocebus*) have all metacentric and submetacentrics marked by a wide achromatic region and constant $2n = 42$.
3. The Colobinae (*Colobus, Presbytis, Nasalis, Rhinopithecus*) and Hylobitinae (*Hylobatus, Nomascus,* and *Symphalangus*) have almost all metacentric and submetacentric chromosomes; a pair marked by a wide achromatic region; and $2n = 44$ to 52.
4. Anthropoid apes (*Pan, Gorilla* and *Pongo*) and *Homo* have metacentric, submetacentric, and acrocentric chromosomes; no marker chromosomes; and $2n = 46$ to 48.

The two species *Macaca mulatta* and *Cercopithecus aethiops*, though members of the same family, are not closely related. Yet one set of chromosomes can be rearranged to match the other with a reasonable degree of precision. Apparently very few rearrangements other than gross translocations have occurred in the euchromatic chromosomes of these Old World primates.[51] Manfredi-Romanini[52] and Chiarelli[53] from spectrophotometric studies suggested that different species of cercopithecine monkeys possess different amounts of DNA per nucleus, which may be attributed to differences in heterochromatin.

The karyotypes of *Papio papio* and *P. anubis* are almost identical. A minor change in the T-staining of a short segment could be detected between the two *Papio* species and *Macaca mulatta*. A paracentric inversion was observed between these three and *M. fascicularis*.[54] Garver et al.,[55] from a comparison of chromosome banding and gene mapping, found numerous similarities within the Cercopithecoidae, some of which could be extended to *Homo* and the Pongidae for some chromosomes. De Grouchy et al.[56] obtained comparable results. Dutrillaux et al.[57] showed an almost complete, if not a complete, homology between *Papio papio*, Pongidae and man. De Vries et al.[58] reported an identical karyotype for *Macaca mulatta* and *M. fascicularis*.

Marker chromosomes, characteristic of almost all primates of the Old World[3] are identified by differential banding patterns and also alterations in DNA replication pattern. For example, in *Macaca mulatta*, a pair of chromosomes bearing large secondary constriction regions replicate asynchronously with the rest of the karyotype and may be regarded as markers.[59] Heterogeneity of silver staining in the marker chromosome was also seen in this species.[60] The retention of similar karyotypes by the two *Papio* species and *Macaca mulatta*, in spite of their clear genetic distinction, indicated by differences in their morphology, geographical distribution, and biochemical characters[61] may be partially due to the complete absence of acrocentrics in them, leading to the absence of any future Robertsonian translocations. This interpretation does not take into account the large number of pericentric inversions involved in the evolution of the Hominidae, none of which are found here. The strong stability of the cercopithecoid species includes the resemblances to the African green monkey and the apparently complete correlation with man and the Pongidae.[54]

In contrast, the karyotypes of the gibbons (Hylobatidae), usually classified between the Cercopithecidae and the Pongidae, have undergone many chromosomal rearrangements.[62] Further, two closely related species, like *Hylobatus lar* and *H. concolor*, also have very different karyotypes. Insofar as the Hylobatidae are regarded to have diverged from a common line some time after the Cercopithecidae, the karyotypes of the former appear to have evolved much more quickly than those of the latter. Consequently there is no absolute parallel between the number of chromosomal rearrangements, scale of time, and divergence of phenotypic characters. An explanation for the variability of chromosomal evolution within families may be individual polymorphisms, which differ in different groups. The relationship between variable heterochromatic segments and the possibility for a given chromosome to undergo changes may even be related.

2. Evolutionary Trends in Cercopithecoidea

Analysis of different types of banding patterns shows a remarkable amount of conservatism within the group. An almost complete analogy was found among the karyotypes of man, the Pongidae, and several species of Cercopithecidae, as well as, also, of some Platyrrhini with that of man.[54,57,63,64]

Cebus capucinus shows a karyotype probably not very different from a very ancestral primate and its chromosomes can be compared directly with those of man, and with those of *Pan* and the Cercopithecidae in between.[65,66,67] Both *Cebus capucinus* and *Microcebus murinus* are regarded to have maintained the ancestral karyotypes among primates, from which the chromosomes may have changed by end-to-end translocations and fissions.[68]

A nearly complete analogy was found in the banding patterns of *Macaca mulatta* and *M. fascicularis* and man. This analogy is almost complete if the heterochromatin is excluded, indicating a common origin of the Platyrrhini and Catarrhini. It is probable that their common ancestor was not very far from them, and the occurrence of a great number of intermediate stages in the differentiation of the two infra-orders of simians is not necessary.

A sympatric evolution of populations, accompanied by wider distributions and accumulation of several chromosomal rearrangements without leading to a direct genetic barrier, can be visualized in the Cercopithecinae and the ancestral *Homo*.[69]

Among primates, some 200 structural rearrangements have been identified and related to human chromosomes. Dutrillaux[69] has reconstituted the sequence of changes, proposing a phylogeny. The types of rearrangements have shown a high specificity with regard to the group studied. For example, pericentric inversions have been frequent during the evolution of the Pongidae and man; Robertsonian translocations among the Lemuridae, and fissions among the Cercopithecinae. The number and rate of occurrence of these chromosomal alterations were also variable ranging from only two changes observed separating 17 species of Papioninae belonging to several genera to more than 50 chromosomal alterations separating 19 species or subspecies of Cercopithecinae. In many cases, therefore, these alterations in chromosomes, to be effective in speciation, have to depend on other factors. Blood protein analysis, at 29 loci, could not identify the branching order.[69a]

Position effect could not be related to chromosomal alterations or presence or absence of heterochromatin. The sequence of replication time of DNA located in the bands was not modified through any chromosomal rearrangements.[70] Dutrillaux[69] stressed the nonrandomness of the alterations and proposed a concept of reverse and convergent chromosomal mutations.

III. CHROMOSOME STUDIES IN HOMINOIDEA

A. General

All the Hominoidea studied so far show acrocentric chromosomes in different numbers. The number of major chromosome arms in the karyotype of a species or the fundamental number (NF), regarded as more significant than chromosome number in tracing affinities,[7] is surprisingly uniform for the three major groups of primates. For example, in Prosimi it ranges from 62 to 70; in Platyrrhini from 66 to 82; and in Catarrhini from 84 to 132. The higher range for the Catarrhini is probably due to the rarity of acrocentric chromosomes and consequent inability to correctly determine "major" arms. This uniformity shows the importance of centric fusion in primate chromosome evolution.[68,71] Even in man, a case of centric fusion without abnormal phenotypic expression was recorded, indicating that the event is not unknown.[72]

The more recently evolved species usually have more acrocentrics, regardless of chromosome number, than those that evolved first. It may be due to the fact that the older forms have had more time for the inversions and translocations required to convert the acrocentrics into other types.[1,73-75] A comparison of the human karyotype with the karotypes of the other two primates groups indicates a greater similarity to the Pongidae, which bear 48 chromosomes.[71,73,76,77] The general morphology of the chromosomes is similar in the three genera of Pongidae. The principal types are meta- and acrocentrics. However, the human karyotype differs in gross morphology in the replacement of two pairs of acrocentrics of the great apes by a single metacentric pair.[11,49,56,73,76,78-83] The number and length of acro- and metacentrics vary and do not indicate complete homology of the *Homo sapiens* karyotype with any particular genus of the apes. The Y is acrocentric in *Homo* and *Gorilla* and metacentric in *Pan* and *Pongo*.

B. Comparison of Human Karyotype with Other Primates

1. Banding Pattern Analysis

Analyses of differential banding patterns have enabled investigators to suggest counterparts for each chromosome of the human complement in *Gorilla, Pan,* and *Pongo*.[73,77,84-89] Very few of the chromosomes, except the X, show exactly the same banding patterns.

The human karyotype alone has a secondary constriction on the long arm. The relative length of the two arms of chromosome 1 in man is the reverse of those of *Gorilla, Pan,* and *Pongo*.[90] Human chromosome (HSA) 2 replaces two acrocentric chromosomes of each of the other species. Almost all the chromosomes of man and the higher primates can be derived from each other through structural alterations involving principally pericentric inversions, some paracentric inversions, and a few telomeric fusions. Translocations and other complex changes are more rare. Deletions and duplications of chromosome segments, involving major genes, have not been identified, but variations have been recorded in certain cases. Differences between man and *Gorilla,* Man and *Pan,* and *Gorilla* and *Pan* have been attributed to inversions involving six to eight chromosomes.[89] Nine to ten inversions separate these species from *Pongo*.[73] Pericentric inversions are known in human populations but they are relatively rare. They are not regarded as a major mechanism involved in speciation in mammals due, presumably, to the reduced fertility of heteozygous carriers.[90] However, pericentric inversions in both homozygous and heterozygous states have been recorded in the great apes that are compatible with fertility and which have become widespread within populations.[91-93] Reciprocal translocations, involving exchange of DNA between two different chromosomes, usually are not involved in explaining the evolution of the chromosomes of the higher primates.[83,84,89,94,95]

The difference in the chromosome number of man from the members of the Pongidae could be explained by fusion of two different acrocentric chromosomes, about the length of the D chromosome, to form one large submetacentric chromosome — the human number 2. Fusion of acrocentric chromosomes is well known in the present day human population.[96] It usually involves centric fusion with loss of the short arms and is assumed to be the mechanism for these interspecific differences as well.[68,71] However, this fusion is more correctly telomeric since the short arms are not lost. The HSA 2 is apparently the result of such a telomeric fusion where the activity of one centromere has been suppressed. Such instances can also be recorded in present day populations.

Different groups of researchers, working on primate karyotypes, reached somewhat different conclusions about the evolutionary changes involved in the chromosomes. Since many of the numbering systems were based on that used for man, all of them were later combined into a standard system. The same system was used for *Pan troglodytes* (PTR) and *P. paniscus* due to the close similarity of their karyotypes. The classifications included illustrations of chromosomes stained to show *G, Q,* and *R* bands and a diagram comparing the banding patterns of chromosomes regarded as homologous. Following discussions on comparative aspects of the primates and man, only human equivalent numbers are now used.

This joint report (1975) failed to homologize the Y chromosomes of *Gorilla, Pan*, and *Pongo*, which show greater interspecific variability than any other chromosome. The banding pattern of the *Gorilla* Y chromosome is more complex than that of man, which in turn is more elaborate than the two other great apes. The *Gorilla* Y chromosome takes up moderately bright stain with quinacrine and differs from the human Y in having a very dull band on each arm. It is the only mammal, other than man, known to have a brilliant Q-stained region on the Y chromosome. *Pan* has a very small Y chromosome, not clearly banded by any of the available methods. The Y chromosome of *Pongo* is longer than that of *Pan* but has no brilliant Q-fluorescence.[85,86,97,98] The homology of genetic information and Y chromosome can also be inferred from the spermatozoa. Through scanning electron microscopy, a close similarity in the morphology of human and *Pan* spermatozoa has been unequivocally demonstrated, while those of the other primates are considerably different.[99] Chromosome banding patterns of the gibbon (*Hylobates*) also have been studied in some detail but very few chromosomes, other than X, resemble those of the higher primates.[73]

2. Distribution of F-bodies

F-bodies are used extensively in identifying chromosome abnormalities in man and in recognizing X- and Y-bearing spermatozoa. The most brilliantly fluorescent region in the human karyotype is on the Y chromosome, which is usually seen as an F-body in somewhat less than 50% of spermatozoa.[5]

Pan and *Gorilla* show brilliant fluorescent regions in some of the their chromosomes.[64,71,76,85,86,100-102] In *Pan*, they are restricted to autosomes 14, 15, 17, 22, and 23. The Y chromosome is a small submetacentric, staining palely. *Gorilla*, on the other hand, shows brilliant fluorescence, both in autosomes 2, 12, 13, 14, 15, 16, 22, and 23 and at the terminal segment of the Y chromosome. In *Pongo*, none of the chromosomes fluoresce brilliantly.[103,104]

The presence of F-bodies in the spermatozoa of *Pan* and *Gorilla* and their absence in *Pongo* can be explained on the basis of Q-banding patterns. In *Pan*, since Y has no brilliant fluorescence, the F-bodies are presumably due to autosomal material. The situation in *Gorilla* is complicated by the presence of the fluorescent regions on both Y and some autosomes. Spermatozoa have been seen with two F-bodies in *Gorilla* in 18.6% cases, compared with 1.25% in man.[105] Nondisjunction of the Y chromosome is unlikely to account for this high incidence, nor could it explain the spermatozoa with three or four F-bodies, using the same type of frequency distribution analysis as applied to *Pan*. The fact that 40% of spermatozoa show no visible F-body rules out the possibility of a single homozygous fluorescent region in *Gorilla*.

3. Chromosomal Abnormalities

Chromosomal abnormalities, their types and frequency, are often utilized in tracing the phylogenetic relationship of man and primates. In hominoid species they include structural rearrangements and segmental or complete aneuploidy.[80,92,94,100,103] Different types of sex chromosomal abnormalities, autosomal trisomies, and pericentric inversions have been reported in higher primates as well. However, records are necessarily incomplete due to biased sampling and early loss of abnormal individuals in the wild condition. A rearranged chromosome corresponding to the human 12 is present in *Pongo* with some differences in geographically isolated populations.[73] Inversions tend to provide selective advantage to the individuals through reduction in the frequency of crossing-over, thus isolating specific groups of alleles.

Position effect may be a relevant factor in the evolution of human karyotype, though Couturier and Dutrillaux have objected to it due to lack of any alteration in DNA replication patterns.[70]

4. Repetitive and Sat DNA

The recognition of different types of repeated DNA has led to the use of this parameter in studying the evolution of man and the hominoid apes.[93,106-116] The homology between repeated DNA of various species is suggested as an exponential function of the time since their divergence, indicating that repeated DNA homology reflects evolutionary relationships. Comparative $C_o t$ analyses of DNA from a range of primates show very striking differences in the complexity of their repeated DNA. The percentage of the total repeated DNA varies from 29 in *Macaca* to 47 in *Hylobates*, but the genomes are almost uniform, around 3.4×10^{12} daltons.[117] To account for variations in primates, such genomes may be regarded as maintaining a steady state relationship. The percentage of repeated sequences exhibiting high and low stability after reassociation was seen to be similar in all the genomes and attributed to selective advantage in maintaining a balance of divergent and nondivergent DNA.[109] Apparently the rate of addition of new DNA to the genome has remained fairly constant, around 0.12% of the human genome for each million years.[109,115,116] A mechanism for creating such new DNA is by saltation of a short nucleotide sequence family (sat) DNA.[118]

Secondary constrictions in human chromosomes may occur on the short arms of all five pairs of acrocentrics and may come closer by association of nucleoli. In these regions 18s and 28s rRNA genes are localized[106]. Ag-As staining shows a variable number of secondary constrictions on acrocentric HSA chromosomes.[60,107] PTR and *Hylobatus* correspond exactly to those located by in situ hybridization with rRNA. PPY has more and GGO has fewer sites stained by silver. 18 and 28s ribosomal RNA genes form a rather special class due to their presence in hundreds of copies per cell, as is true of most genes. Extensive redistribution of the rRNA genes has occurred on acrocentric chromosomes in *Pan*, in man on five pairs, and in *Gorilla* on two pairs.[74]

Detailed comparisons show the presence of heteromorphic chromosomal regions, many of which contain highly repetitive DNA sequences.[104] Quinacrine brilliant regions are found only in *Pan, Gorilla*, and man, a Q-brilliant Y being present only in the latter two species.

Human DNA contains several satellite DNA fractions, of which three have been studied with respect to higher primate evolution. They are concentrated within centromeric heterochromatin in different proportions in different chromosomes. In the human Y, sat DNA occurs in a distal heterochromatic segment in addition to the centromere. The chromosome regions containing sat DNA vary in size according to the amount of the latter and are inherited. The age of satellite III has been estimated to be about 30 million years, of sat I 18 million, and of sat II 6 million.

Human sat DNAs I, III, and IV generally anneal to corresponding sequences on fixed metaphase chromosomes of great apes, although to only a small extent in *Pongo*.[73] Human sat III is related by hybridization to DNA sequences in constitutive heterochromatin in all higher apes except *Hylobates*, showing that they share a common ancestor.[97] The chromosomal regions involved also exhibit differential staining by the Giemsa pH 11 procedure.

In *Hylobates* and *Macaca* both the staining reaction and the related sat DNA sequences have not been detected, suggesting that they arose after the separation of the ancestral form of higher primates. Human sat-I and II cannot be identified in *Pan* by molecular hybridization and are therefore assumed to have originated subsequent to divergence of *Pan*.[97] This assumption is supported by the lesser divergence of sat DNAs I and II and by the fact that the chromosomal features concerned are absent from the homologous chromosomes of *Pan*. Thus chromosome 1 in *Pan* lacks a prominent heterochromatic block near the centromere, which contains sat II in man. If human sat II-like sequences are present in *Pan*, their amount must be very small. This finding is supported by the relatively rare occurrence of 5-methylcytosine which is abundant in human and *Gorilla* sat II.[98]

The Y chromosome in *Gorilla* does show a distal fluorescent region. However, information about sat DNA in this species is relatively meager except for the presence of a sequence related to human sat III DNA, and no record of sat I is yet available.

Apparently, the known sat DNAs of man can be dated in evolution to a period not earlier than the man-gibbon divergence. Palaeontological evidence indicates it to have occurred about 30 to 40 million years ago. The four known sat DNAs of man contribute approximately 0.14, 0.6, 0.8, and 2% of the human genome, giving a total of 3.54%. This amounts to an additional 0.09 to 0.12% of DNA per million years. This value is very close to that given by Kohne[116] and Gummerson[109] for the rate of addition of new DNA to primate genomes. Probably a significantly large proportion of the new DNA, if not all, is added as simple sequence DNA. Constitutive heterochromatin obviously generates novel DNA. The fact that each species has a different and characteristic sat DNA pattern indicates a rapid spread of new DNA through a species.[118,118a]

At other times, the phenotypic changes accompanying speciation might favor alterations in constitutive heterochromatin, achieving a new homeostatic balance. Thus, those underlying saltatory processes may be encouraged which would lead to the generation of new sat DNA and perhaps the elimination of certain pre-existing ones. Further alterations in heterochromatin after stabilization of a new phenotype would be disadvantageous. According to such a mechanism, evolutionary alterations in sat DNA may denote important speciation events. The occurrence of two sat DNAs with different degrees of sequence divergence and thus of different ages exclusively in man (sat I and II), may indicate at least two important evolutionary steps that have occurred during human evolution, after diverging from *Pan*.[74,104] Further quantitative variations, inherited as polymorphism and involving quantitative differences in sat DNA on homologous chromosomes, may have led to the wide range of normal phenotypes in modern man.

Later techniques involving differential restriction enzyme patterns have made it possible to detect minute quantities of sat DNA in human chromosomes not identifiable through hybridization.[119] Sat sequences on different chromosomes have been suggested to have evolved independently from a common precursor.[120] Hae III restriction pattern, characteristic of human sat III, has been observed in all members of the Hominoidea but only in *Macaca* of lower primates. Hybrids of sat III with hominoid DNA are stable, indicating that this sequence has been conserved for 25 to 30 million years.[121,121a]

Certain sequences in sat DNA contain sequences with greater chromosome specificity than total satellite DNA, as for example, the sat DNAs specific to Y chromosomes.[120] Attempts have been made to correlate these sequences in the larger primates.[121,122]

5. Gene Loci

The gene map of chromosome 1 is particularly rich in man an many genes have also been assigned to the same chromosome of Pongidae. A comparison of the structure of this chromosome with those of apparently homologous chromosomes in *Hylobates* and *Cercopithecus* has made it possible to trace its origin further back in time, some 50 million years ago, to the common ancestor of the Catarrhini.[87,88]

C. Evolutionary Trends in Hominoidea

Every human chromosome has been observed to have a recognizable counterpart in the four present-day species of great apes. A certain amount of reorganization and alteration of the genetic material has obviously occurred during the evolution of the higher primates and man. A fusion resulted in the human $2n = 46$ chromosomes as compared to $2n = 48$ in the great apes. The most frequently observed differences have arisen through pericentric inversions. However, in view of the relatively meager information available on gene mapping on chromosomes of higher primates, this statement cannot be taken for granted. Miller[74] has suggested that the pericentric inversions which have become fixed in primate evolution have not affected the activity of the genes on these chromosomes.

An analysis of the chromosomes through differential banding patterns confirms the relatively great evolutionary distance between the gibbon *(Hylobates)* and the higher primates. *Hylobates* does not have any of the acrocentric chromosomes present in the great apes and man. In the evolution of the latter species, a series of pericentric inversions can be visualized to convert metacentric chromosomes into acrocentrics.[73] The karyotype of *Hylobates* cannot be converted through any such alterations to the other large apes, showing that it had separated at a much earlier age from the other genera.

The incidence of more than one inversion at long intervals has been postulated to account for the appearance of a polymorphic pair of chromosomes in present-day *Pongo* populations. Alternatively, the marked differences between the karyotypes of *Hylobates* and *Pongo* may indicate that the chromosomes of the ancestral *Pongo* form did not evolve from those of the *Hylobates* ancestor by inversion, but rather by a different mechanism, such as reciprocal translocation.[73]

Hylobates differs from the higher primates also in having a single site for 18 and 28s rRNA genes, compared to the multiple sites in the others. Such a redistribution of rRNA sites could be attributed to a series of reciprocal translocations. Another indication of separation of *Hylobates* from the great apes is that the former has few, if any, of the DNA sequences the others share with human sat III DNA. By comparing densitometric patterns of the banded chromosomes of *Homo, Pan,* and *Macaca,* Saxena and Seth[95] suggested that both reciprocal translocation and pericentric inversion have been effective in generating reproductive isolation in these genera. Chromosome studies also show that human, *Gorilla, Pan,* and *Pongo* chromosomes are remarkably similar with about 98 to 99% of the 500 or so bands homologous in them. However, *Pongo* has fewer chromosomes with banding patterns similar to those of the human than *Gorilla* and *Pan,* and no Q-brilliant regions and no terminal Q or C bands. *Pongo* has a somewhat different distribution of T-bands and of DNA sequences corresponding to the human satellite IV DNA, as well as eight rRNA sites compared to a maximum of five in the other species.

Sat DNA of *Pan* shares so many of the physical properties of human sat III DNA that the original sat sequence was presumably present in a common progenitor of both these primates.[122a] Following in situ hybridization, sat III DNA is found to be not only homologous with the *Pan* chromosome, but a physically similar sat DNA is also present in *Macaca.*[123] The evolutionary divergence of man and *Pan* has been suggested to have taken place after the origin of the original sat DNA indicating thereby that they had a common hypothetical ancestor and were subjected to equal selective pressures in the same direction to some extent.[4,74,95]

Pan was regarded as more closely related to man than *Gorilla* due to the presence of 12 chromosomes considered to have virtually identical general banding patterns. The more restricted regional banding methods, however, suggest that *Gorilla* is more closely related to man than *Pan*. Only these two species have brilliantly fluorescing regions on chromosome 4 and the Y, a large C-band on numbers 9 and 16, and regions with high concentrations of 5-methylcytosine. *Gorilla*, but not *Pan*, has DNA sequences in common with human sat II DNA. The proteins of *Pan* and *Homo* are very similar, on the other hand.[124,124a]

The proposed order in which the progenitors of the various primate species diverged has been suggested as *Hylobates, Pongo, Pan, Gorilla, Homo*.[4] The time interval between the divergence of *Hylobates* and *Pongo* is greater than that between the other species. Various evolutionary schemes have been proposed based on chromosome analysis.

A hypothetical R-banded karyotype of a primate progenitor, based on the presumptive inversions that separate the primate species,[83] suggests that the progenitor of the human race separated before the ancestral *Pan* and *Gorilla* diverged from one another. On comparisons based particularly on the evolution of 2p and 2q in the various primates, including *Pan paniscus*, a scheme was constructed in which *Gorilla* is though to have diverged before *Pan*.[79,79a] A phylogenetic tree, based on the extent of reassociation and thermal stability of nonrepeated sequences of DNA from the various species shows on the other hand that *Gorilla, Pan*, and human diverged at approximately the same time.[5,125]

Paleoecological, palaeogeographical, and biochemical studies in human and nonhuman primates have shown that the branching of apes and *Homo* from the hominine ancestral stock took place about 12 to 15 million years ago,[126] while that of *Macaca* arose about 40 million years ago. Rearrangements, both at the chromosomal and the DNA levels, have been important factors in the origin of the species and their polymorphism along with associated phenomena like recombination, selection, and isolation. The question of whether Darwinian selection or selective neutrality is the major force in molecular evolution could be resolved by an understanding of the basis for the reduced evolutionary rate of the primates.[127]

Somatic cell hybridization methods have been used to show that one or more genes on almost every human chromosome has a counterpart on the purported ape-homologue.[74] Comparison with rhesus and African green monkey chromosomes shows that genes carried by human chromosome 1 have been associated with a chromosome segment having the same banding pattern for perhaps 35 million years or more.

Observations on terminal Q- and T-bands, heterochromatic regions, and juxtacentromeric Q-bands show that the karyotype differences between the five closely related species involve principally rearrangements of genetic material.[96] An additional mechanism seems to be *de novo* synthesis as well as loss of chromosome material.[5]

A tentative phylogenetic tree of the primates from the prosimians to man established from banding pattern analysis of karyotypes of more than 60 species of primates shows that the entire euchromatic material, i.e., the nonvariable *R*- and *Q*-bands, appears to be identical in all species of monkeys, apes, and in humans. Quantitative and qualitative variations all involve heterochromatin. The types of chromosome rearrangements reconstructed from species differences in chromosome structure vary from one subgroup to the next.

Dutrillaux[71] had, through a comparison between Pongidae and man, constructed a probable karyotype of Proconsul, a fossil species regarded as the last common ancestor of the gorilla, chimpanzee, and man. A karyotype suggested for the last common ancestor of all the existing simians could include the same elements as man for chromosomes 13, 14, 21, 22, and X and the same elements as man or *Cebus* for 4, and those also exist unmodified in many other species of Catarrhini. It could have the same elements as *Cebus*, corresponding to HSA 1, 2q, 5, 6, 7, 8q, 9, 10, 11, 12, 15, and 17 (heterochromatin excepted) and the same elements as *Pongo* for the elements corresponding to HSA 2p, 2q, 9, and 18. The short arm of 8 and long arm of 16 can be represented by acrocentrics.[4]

In conclusion, chromosome evolution in the primates has been characterized by a remarkable degree of conservatism. Even high resolution banding pattern methods could detect only qualitative chromosomal rearrangements like pericentric inversions, reciprocal/Robertsonian translocations, etc. Most evidences indicate that *Pan, Gorilla*, and *Homo* had separated from a common ancestor at about the same time. Such chromosomal conservatism may have been further aided by the fact that the group is relatively young in the geological time-scale, and had not been subjected to any catastrophic environmental changes.

REFERENCES

1. **Bender, M. A. and Chu, E. H. Y.**, The chromosomes of primates, in *Evolutionary and Genetic Biology of Primates*, Brettner-Janusch J., Ed., Academic Press, New York, 1963.
2. **Egozcue, J.**, Primates, in *Comparative Mammalian Cytogenetics*, Benirschke, K., Ed., Springer-Verlag, Berlin, 1969.
3. **Chiarelli, A. B.**, *Comparative Genetics in Monkeys, Apes and Man*, Academic Press, London, 1971.
4. **Dutrillaux, B.**, Chromosomal evolution in Primates: tentative phylogeny from *Microcebus murinus* to man, *Hum. Genet.*, 48, 251, 1979.
5. **Datta, S., Sharma, A. and Talukder, G.**, Chromosomal alteration in human evolution, *Nucleus*, 24, 114, 1981.
6. **Borgaonkar, D. S.**, A list of chromosome numbers in primates, *J. Hered.*, 57, 60, 1966.
7. **Hsu, T. C. and Benirschke, K.**, *An Atlas of Mammalian Chromosomes*, Vol. 1 and 2, Springer-Verlag, New York, 1967—1968.
8. **Chiarelli, B., Egozcue, J., and Monchietto, M.**, *Caryological Atlas of Living Primates*, 1968.
9. **Chu, E. H. Y. and Swomley, B. A.**, Chromosomes of lemurine lemurs, *Science*, 133, 1925, 1961.
10. **Chu, E. H. Y. and Bender, M. A.**, Chromosome cytology and evolution in primates, *Science*, 133, 1399, 1961.
11. **Chiarelli, A. B.**, Some chromosome numbers in primates, *Mammal. Chrom. Newsl.*, 6, 3, 1961.
12. **Egozcue, J.**, Chromosome variability in the Lemuridae, *Am. J. Phys. Anthrop.*, 26, 341, 1967.
13. **Chu, E. H. Y. and Bender, M. A.**, Cytogenetics and evolution of primates, *Ann. N.Y. Acad. Sci.*, 102, 253, 1962.
14. **Rumpler, Y.**, Chromosomal studies in the systematics of Malagasy lemurs, in *Lemur biology*, Plenum Press, New York, 1975, 25.
15. **Sasaki, M., Oshimura, M., Takahashi, E., and Kondo, N.**, A comparative banding analysis of chromosomes in three species of lemurs (Primates, Lemuridae), *Genetica*, 45, 253, 1975.
16. **Boer, L. E. M. de**, Studies on the cytogenetics of prosimians, *J. Hum. Evol.*, 2, 271, 1973.
17. **Rumpler, Y. and Albignac, R.**, Cytogenetic study of the endemic Malagasy lemurs; Subfamily Cheirogaleinae Gregory 1915, *Am. J. Phys. Anthrop.*, 38, 261, 1973.
18. **Wurster-Hill, D. H.**, Chromosomes of the Lemuridae, *J. Hum. Evol.*, 2, 259, 1973.
19. **Rumpler, Y. and Dutrillaux, B.**, Chromosomal evolution in Malagasy lemurs. I. Chromosome banding studies in the genuses *Lemur* and *Microcebus, Cytogenet. Cell Genet.*, 17, 268, 1976.
20. **Rumpler, Y. and Dutrillaux, B.**, Chromosome evolution in Malagasy lemurs. III, *Cytogenet. Cell Genet.*, 21, 177, 1978.
21. **Hamilton, A. E. and Buettner-Janusch, J.**, Chromosomes of lemuriformes. III, The genus *Lemur*: karyotypes of species, subspecies and hybrids, *Ann. N.Y. Acad. Sci.*, 293, 125, 1977.
22. **Bender, M. A. and Mettler, L. E.**, Chromosome studies of primates, *Science*, 128, 196, 1958.
22a. **Klinger, H. P.**, The somatic chromosomes of some primates *(Tupaia glis, Nycticebus coucang, Tarsius bancanus, Cercocebus aterrimus, Symphalangus syndactylus), Cytogenetics*, 2, 140, 1963.
23. **Manna, G. K. and Talukdar, M.**, Analysis of somatic chromosome complements of both sexes of two primates, the slender loris, *Loris tardigradus* and Rhesus monkey, *Macaca mulatta, Mammalia*, 32, 118, 1968.
24. **Egozcue, J.**, The meiotic chromosomes of the lesser bushbaby *(Galago senegalensis), Mammal. Chrom. Newsl.*, 9, 92, 1968.
25. **Chiarelli, A. B. and Egozcue, J.**, The meiotic chromosomes of some primates, *Mammal. Chrom. Newsl.*, 9, 85, 1968.
26. **Egozcue, J. and Hagemenas, F.**, The chromosomes of the hooded spider monkey *(Ateles geoffroy cucullatus), Mammal. Chrom. Newsl.*, 8, 12, 1967.
27. **Chiarelli, A. B. and Barberis, L.**, Some data on the chromosomes of Prosimiae and of the New World monkeys, *Mammal. Chrom. Newsl.*, 22, 216, 1966.

28. **Egozcue, J. and Vilarasau de Egozcue, M.,** The chromosome complement of the howler monkey *(Alouatta caraya* Humboldt 1912), *Cytogenetics,* 5, 20, 1966.
29. **Egozcue, J. and Vilarasau de Egozcue, M.,** The somatic chromosomes, in *Biology of the Howler Monkey (Alouatta caraya),* Malinow, M. R., Ed., *Bibl. Primatol.,* No. 7, S. Karger, Basel, 1967.
30. **Hsu, T. C.,** Foreword, *Mammal. Chrom. Newsl.,* 15, 98, 1965.
31. **Egozcue, J. and Vilarasau de Egozcue, M.,** Chromosome evolution in the Cebidae, in *Progress in Primatology,* Starck, D., Schneider, R. and Kuhn, H. J., Eds., Gustav Fischer Verlag, Stuttgart, 1967.
32. **Egozcue, J. and Vilarasau de Egozcue, M.,** The chromosome complement of *Cebus albifrons* (Erxleben 1777), *Folia Primat.,* 5, 285, 1967.
33. **Dutrillaux, B., Couturier, J., Viegas-Pequignot, E., Chauvier, G., and Trebbau, P.,** Presence d'une heterochromatine particulaire dans le caryotype de deux *Cebus: C. capucinus* et *C. nigrivittatus, Ann. Genet.,* 21, 142, 1978.
34. **Torres de Caballero, O. M., Ramirez, C. and Yunis, E.,** Genus *Cebus,* Q and G band karyotypes and natural hybrids, *Folia primat.,* 26, 310, 1976.
35. **Egozcue, J., Vilarasau de Egozcue, M., and Hagemenas, F.,** The chromosomes of two species of *Saimiri, S. juruanus* and *S. boliviensis nigriceps, Mammal. Chrom. Newsl.,* 8, 14, 1967.
36. **Egozcue, J., Perkins, E. M., and Hagemenas, F.,** Chromosomal evolution in marmosets, tamarins and pinches, *Folia primat.,* 9, 81, 1968.
37. **Bender, M. A. and Mettler, L. E.,** Chromosome studies of primates. II, *Callithrix, Lentocebus* and *Callimico, Cytologia,* 25, 400, 1960.
38. **Benirschke, K. and Brownhill, L. F.,** Further observations on marrow chimerism in marmosets, *Cytogenetics,* 1, 245, 1962.
39. **Benirschke, K. and Brownhill, L. F.,** Heterosexual cells in testes of chimeric marmoset monkeys, *Cytogenetics,* 2, 331, 1963.
40. **Wohnus, J. F. and Benirschke, K.,** Chromosome analysis of four species of marmosets *(Callithrix jacchus, Tamarinus mystax, T. nigricollis, Cebuella pygmaea), Cytogenetics,* 5, 94, 1966.
41. **Benirschke, K., Anderson, J. M., and Brownhill, L. E.,** Marrow chimerism in marmosets, *Science,* 138, 513, 1962.
42. **Egozcue, J., Chiarelli, A. B., and Sarti-Chiarelli, M.,** The somatic and meiotic chromosomes of *Cebuella pygmaea* with special reference to the behaviour of the sex chromosomes during spermatogenesis, *Folia primat.,* 8, 50, 1968.
43. **Chiarelli, A. B.,** Comparative morphometric-analysis of primate chromosomes. II. The chromosomes of the genera *Macaca, Papio, Theropithecus* and *Cercocebus, Caryologia,* 15, 401, 1962.
44. **Chiarelli, A. B.,** Primi resultati di ricerche di genetica e cariologia comparata in primati e loro interese evolutivo, *Rivista di Antropologia,* 50, 87, 1963.
45. **Chiarelli, A. B.,** Caryology and taxonomy of the Catarrhine monkeys, *Amer. J. Phys. Anthrop.,* 24, 155, 1966.
46. **Buettner-Janusch, J.,** A problem in evolutionary cytogenetics, nomenclature and classification of baboons, genus *Papio, Folia primat.,* 4, 288, 1966.
47. **Hamerton, J. L., Klinger, H. P., Mutton, D. E., and Lang, E. M.,** The somatic chromosomes of the Hominoidea, *Cytogenetics,* 2, 240, 1963.
48. **Klinger, H. P., Hamerton, J. L., Mutton, D. and Lang, E. M.,** The chromosomes of the Hominoidea, in *Classification and Human Evolution,* Washburn, S. L., Ed., Aldine Publishing, Chicago, 1963.
49. **Chiarelli, A. B.,** Comparative chromosome analysis between man and chimpanzee, *J. Hum. Evol.,* 1, 389, 1972.
50. **Chiarelli, A. B.,** Speculations on the karyological relationship of Old World primates, *J. Hum. Evol.,* 2, 307, 1973.
51. **Stock, A. D. and Hsu, T. C.,** Evolutionary conservatism in arrangement of genetic material; a comparative analysis of chromosome banding between the Rhesus macaque (2n = 42, 84 arms) and the African green monkey (2n = 60, 120 arms), *Chromosoma,* 43, 211, 1973.
52. **Manfredi-Romanini, A. M.,** Quantitative relative determination of the nuclear DNA in the Old World primates (cited by Chiarelli, A. B., *Cytologia,* 33, 1, 1968), 1967.
53. **Chiarelli, A. B.,** Chromosome polymorphism in species of the genus *Cercopithecus, Cytologia,* 33, 1, 1968.
54. **Dutrillaux, B., Biemont, M. C., Viegas-Pequignot, E., and Laurent, C.,** Comparison of the karyotypes of four Cercopithecidae: *Papio papio, P. anubis, Macaca mulatta,* and *M. fascicularis, Cytogenet. Cell Genet.,* 23, 77, 1979.
55. **Garver, J. J., Estop, A., Pearson, P. L., Dijkman, T. M., Wijnen, L. M. M., and Meera Khan, P.,** in *Chromosomes Today,* Vol. 6, De la Chapelle, A. and Sorsa, M., Eds., Elsevier-North Holland, Amsterdam, 1977, 191.

56. **De Grouchy, J., Finaz, C., and Nguyen, V. C.,** Comparative banding and gene mapping in primate evolution: evolution of chromosome during fifty million years, in *Chromosomes today,* Vol. 6, 183, De la Chapelle, A. and Sorsa, M., Eds., Elsevier-North Holland, Amsterdam, 1977.

57. **Dutrillaux, B., Viegas-Pequignot, E., Couturier, J., and Chauvier, G.,** Identity of euchromatic bands from man to the Cercopithecidae *(Cercopithecus aethiops, C. sabaeus, Erythrocebus patas, Miopithecus talapoin), Hum. Genet.,* 45, 283, 1978.

58. **De Vries, G. F., Geleijnse, M. E. M., De France, H. F., and Hogendoorn, A. M.,** Lymphocyte cultures of *Macaca mulatta* and *M. fascicularis. Lab. Anim. Sci.,* 25, 33, 1975.

59. **Huang, C. C., Habbitt, H., and Ambrus, J. L.,** Chromosomes and DNA synthesis in the stumptail monkey *(Macaca speciosa)* with special regard to marker and sex chromosomes, *Folia Primat.,* 11, 28, 1967.

60. **Rani, R., Sharma, T. K., Nand, R., and Ghosh, P. K.,** Heterogeneity of silver staining in the marker chromosome of Rhesus monkey *(Macaca mulatta), J. Hum. Evol.,* 10, 409, 1981.

61. **Romero-Herrara, A. E., Lehmann, H., Castillo, O., Joysey, K. A., and Friday, A. E.,** Myoglobin of the orangutan as a phylogenetic enigma, *Nature,* 261, 162, 1976.

62. **Dutrillaux, B., Rethore, M. O., Aurias, A., and Goustard, M.,** Analyse du caryotype de deux especes de Gibbons *(Hylobates lar* et *H. concolor)* par differentes techniques de marquage, *Cytogenet. Cell Genet.,* 15, 81, 1975.

63. **Dutrillaux, B., Viegas-Pequignot, E., Dubos, C., and Masse, R.,** Complete or almost complete analogy of chromosome banding between the baboon *(Papio papio)* and man, *Hum. Genet.,* 43, 37, 1978.

64. **Lejeune, J., Dutrillaux, B., Rethore, M. O., and Prieur, M.,** Comparison de la structure fine des chromatides d' *Homo sapiens* et de *Pan troglodytes, Chromosoma,* 43, 423, 1973.

65. **Rumpler, Y. and Dutrillaux, B.,** Chromosomal evolution in Malagasy lemurs. I. Chromosome banding studies in the genera *Lemur* and *Microcebus, Cytogenet. Cell Genet.,* 17, 268, 1976.

66. **Rumpler, Y. and Dutrillaux, B.,** Chromosomal evolution in Malagasy lemurs. III. Chromosome banding studies in the genus *Hapalemur* and the species *Lemur catta, Cytogenet. Cell Genet.,* 21, 201, 1978.

67. **Egozcue, J., Hagemenas, F., and Vilasarau de Egozcue, M.,** Chromosome studies in Catarrhinae monkeys, *Mammal. Chrom. Newsletter,* 9, 65, 1968.

68. **Dutrillaux, B.,** Very large analogy of chromosome banding between *Cebus capucinus* (Platyrrhini) and man, *Cytogenet. Cell Genet.,* 24, 84, 1979.

69. **Dutrillaux, B. D.,** Chromosomal evolution in primates and other mammals, *Proc. 7th Int. Chr. Conf.,* Oxford, 1980, 20.

69a. **Kawamoto, Y., Shotake, T. and Nozawa, K.,** Genetic differentiation amoung three genera of family Cercopithecidae, *Primates,* 23, 272, 1982.

70. **Couturier, J. and Dutrillaux, B.,** Conservation of replication chronology of homeologous chromosome bands between four species of the genus *Cebus* and man, *Cytogenet. Cell Genet.,* 29, 233, 1981.

71. **Dutrillaux, B.,** Sur la nature et l'origine des chromosomes humains, *Monogr. des Annales de Génétique l'Expansion Scientifique,* Paris, 1975.

72. **Moorhead, P. S., Mellman, W. J., and Wenar, C.,** A familial chromosomal translocation associated with speech and mental retardation, *Am. J. Hum. Genet.,* 13, 32, 1961.

73. **Miller, D. A.,** Evolution of primate chromosomes, *Science,* 198, 1116, 1977.

74. **Miller, D. A.,** Book review: *The Phylogeny of Human Chromosomes,* Seuanez, H. N., Ed., Springer, Berlin, in *Nature,* 284, 86, 1980.

75. **Warburton, D. and Atwood, K. C.,** Evolutionary processes controlling the chromosomal distribution of DNA, in *Comm. Rep. Hum. Gene Mapp.,* 5, 1979.

76. **Miller, D. A., Firschein, J. L., Dev, V. G., Tantravahi, R., and Miller, O. J.,** The gorilla karyotype: chromosome lengths and polymorphisms, *Cytogenet. Cell Genet.,* 13, 536, 1974.

77. **Warburton, D., Firschein, I. L., Miller, D. A., and Warburton, F. E.,** Karyotype of the chimpanzee, *Pan troglodytes,* based on measurements and banding pattern: comparison to the human karyotype, *Cytogenet. Cell Genet.,* 12, 453, 1973.

78. **Chiarelli, A. B.,** Is chromosome banding a really new tool to prove genetic homology between man and apes?, *J. Hum. Evol.,* 2, 337, 1973.

79. **Dutrillaux, B., Rethore, M. O., Prieur, M., and Lejeune, J.,** Analysis of the structure of chromatids of *Gorilla gorilla* and comparison with *Homo sapiens* and *Pan troglodytes, Humangenetik,* 20, 343, 1973.

79a. **Yunis, J. J. and Prakash, O.,** The origin of man: a chromosomal pictorial legacy, *Science (Wash.),* 215, 1525, 1982.

80. **Dutrillaux, B., Rethore, M. O., and Lejeune, J.,** Analysis of the karyotype of *Pan paniscus:* comparison with other Pongidae and man, *Humangenetik,* 28, 113, 1975.

81. **De Grouchy, J., Turleau, C., and Finaz, C.,** Chromosomal phylogeny of the primates, *Ann. Rev. Genet.,* 12, 289, 1978.

82. **Hamerton, J. L., Fraccaro, M., Decarli, L., Nuzzo, F., Klinger, H. P., Hulliger, L., Taylor, A., and Lang, E. M.,** Somatic chromosomes of *Gorilla, Nature (Lond.),* 192, 225, 1961.

83. **Turleau, C., De Grouchy, J., and Klein, M.,** Chromosomal phylogeny of man and the anthropomorphic apes *(Pan troglodytes, Gorilla gorilla, Pongo pygmaeus).* Tentative reconstitution of the common ancestral karyotype, *Ann. Genet.,* 15, 225, 1972.

84. **Bobrow, M. and Madan, K.,** A comparison of chimpanzee and human chromosomes using the Giemsa II and other chromosome banding techniques, *Cytogenet. Cell Genet.,* 12, 107, 1973.

85. **Egozcue, J., Caballin, R., and Goday, C.,** Banding patterns of the chromosomes of man and the chimpanzee, *Humangenetik,* 18, 77, 1973.

86. **Ecozcue, J., Caballin, R., and Goday, C.,** Q- and G-banding patterns of the chromosomes of some primates, *J. Hum. Evol.,* 2, 289, 1973.

87. **Finaz, C., Nguyen, V. C., Freral, J., and De Grouchy, J.,** Fifty million year evolution of chromosome 1 in the primates; evidence from banding and gene mapping, *Cytogenet. Cell Genet.,* 18, 160, 1977.

88. **Finaz, C., Nguyen, V. C., Cochet, C., Frazai, J., and De Grouchy, J.,** Natural history of chromosome 1 in the primates, *Ann. Genet.,* 20, 85, 1977.

89. **De Grouchy, J., Turleau, C., Roubin, M., and Chavin Colin, F.,** Chromosomal evolution of man and the primates, in *Chromosome Identification-Technique and Application in Biology and Medicine,* Caspersson, T. and Zech, L., Eds., Academic Press, New York, 1973, 23.

90. **De Grouchy, J. and Turleau, C.,** in *Clinical Atlas of Human Chromosomes,* Wiley, New York, 1977.

91. **White, M. J. D.,** in *Animal Cytology and Evolution,* 3rd ed., Cambridge University Press, Cambridge, 1973.

91a. **Schweizer, D., Ambros, P., Andrle, M., Rett, A., and Fiedler, W.,** Demonstration of specific heterochromatic segments in the orangutan *(Pongo pygmaeus)* by a dristamycin/DAPI double staining technique, *Cytogenet. Cell Genet.,* 24, 7, 1979.

92. **Seuanez, H., Fletcher, J., Evans, H. J., and Martin, D. E.,** A polymorphic structural rearrangement in the chromosomes of two populations of orangutan, *Cytogenet. Cell Genet.,* 17, 327, 1976.

93. **Seuanez, H. N., Evans, H. J., Martin, D. E., and Fletcher, J.,** An inversion of chromosome 2 that distinguishes between Bornean and Sumatran orangutans. *Cytogenet. Cell Genet.,* 23, 137, 1979.

94. **Turleau, C., De Grouchy, J., Chavin Colin, F., Martelmans, J., and Van Den Bergh, W.,** Pericentric inversion of chromosome 3, homozygous and heterozygous and transposition of centromere of chromosome 12 in a family of orangutans. Implication for evolution, *Ann. Genet.,* 18, 227, 1975.

95. **Saxena, M. B. and Seth, P. K.,** A critical appraisal of chromosomal homology between man and *Macaca mulatta, Nucleus,* 22, 116, 1979.

96. **Vogel, F. and Motulsky, A. G.,** in *Human Genetics: Problems and Approaches,* Springer-Verlag, Berlin, 1979.

97. **Jones, K. W.,** Repetitive DNA sequences in animals, particularly primates, in *Chromosomes Today,* Pearson, P. L. and Lewis, K. R., Eds., Vol. 5, John Wiley & Sons, New York, 1976, 305, 329.

98. **Schnedl, W., Dev, V. G., Tantravahi, R., Miller, D. A., Erlanger, B. F., and Miller, O. J.,** 5-Methylcytosine in heterochromatic regions of chromosomes: chimpanzee and gorilla compared to the human, *Chromosoma,* 52, 59, 1975.

99. **Martin, D. E., Gould, K. G., and Warner, H.,** Comparative morphology of primate spermatozoa using scanning electron microscopy. I. Families Hominidae, Pongidae, Cercopithecidae and Celeidae, *J. Hum. Evol.,* 4, 287, 1975.

100. **Lin, C. C., Chiarelli, B., De Boer, L. E. M., and Cohen, M. M.,** A comparison of the fluorescent karyotypes of the chimpanzee *(Pan troglodytes)* and man, *J. Hum. Evol.,* 2, 311, 1973.

101. **Pearson, P. L., Bobrow, M., and Vosa, C. G.,** Technique for identifying Y-chromosomes in human interphase nuclei, *Nature (Lond.),* 226, 78, 1970.

102. Paris Conference (1971) Supplement (1975), Standardization in Human Cytogenetics, Birth defects: original article series, Vol. 11(9), The National Foundation, New York, 1975.

103. **Seuanez, H., Robinson, J., Martin, D. E., and Short, R. V.,** Fluorescent (F) bodies in the spermatozoa of man and the great apes, *Cytogenet. Cell Genet.,* 17, 317, 1976.

104. **Seuanez, H. N.,** *The phylogeny of human chromosomes,* Springer-Verlag, Berlin, 1979.

105. **Sumner, A. T., Robinson, J. A., and Evans, H. J.,** Distinguishing between X, Y and XY bearing human spermatozoa by fluorescence and DNA content, *Nature (New Biol.),* 229, 231, 1971.

106. **Andrle F., Fiedler, W., Retta, A. P., and Schweizer, D.,** A case of trisomy 22 in *Pongo pygmaeus, Cytogenet. Cell Genet.,* 24, 1, 1979.

107. **Gosden, J., Mitchell, A. R., Seuanez, H., and Gosden, C. M.,** The distribution of sequences complementary to human sat DNAs I, II and IV in the chromosomes of chimpanzee *(Pan troglodytes),* gorilla *(Gorilla gorilla)* and orangutan *(Pongo pygmaeus), Chromosoma,* 63, 252, 1977.

108. **Gosden, J., Laurie, S., and Seuanez, H.,** Ribosomal and human autologous repeated DNA distribution in the orangutan *(Pongo pygmaeus), Cytogenet. Cell Genet.,* 21, 1, 1978.

109. **Gummerson, K. S.,** The evolution of repeated DNA in primates, Ph.D. thesis, Johns Hopkins University, Baltimore, Md., 1972.

110. **Henderson, A., Warburton, D., and Atwood, K. C.,** Localization of rDNA in the chimpanzee *(Pan troglodytes)* chromosome complement, *Chromosoma,* 46, 435, 1974.

111. **Henderson, A. S., Atwood, K. C., and Warburton, D.,** Chromosomal distribution of rDNA in *Pan paniscus, Gorilla gorilla beringei* and *Symphalangus syndactylus*: comparison to related primates, *Chromosoma,* 59, 147, 1976.

112. **Henderson, A. S., Warburton, D., McGraw-Ripley, S., and Atwood, K. C.,** The chromosomal location of rDNA in selected lower primates, *Cytogenet. Cell Genet.,* 19, 281, 1977.

113. **Henderson, A. S., Warburton, D., McGraw-Ripley, S., and Atwood, K. C.,** The chromosomal location of rDNA in the Sumatran orangutan *(Pongo pygmaeus albei), Cytogenet. Cell Genet.,* 23, 213, 1979.

114. **Hoyer, B. H., Van De Velde, N. W., Goodman, M., and Roberts, R. B.,** Examination of hominoid evolution by DNA sequence homology, *J. Hum. Evol.,* 1, 645, 1972.

115. **Jones, K. W., Corneo, G., Ginelli, E., and Prosser, J.,** The chromosomal location of human sat DNA I, *Chromosoma,* 49, 161, 1973.

116. **Kohne, D. E., Chiscan, J. A., and Hoyer, B. H.,** Evolution of primitive DNA sequences, *J. Hum. Evol.,* 1, 627, 1972.

117. **Britten, R. J. and Davidson, E. H.,** Repetitive and non-repetitive DNA sequences and a speculation on the origins of evolutionary novelty, *Am. Rev. Biol.,* 46, 111, 1971.

118. **Britten, R. J. and Kohne, D. E.,** Saltatory events in replication, *Carnegie Inst. Wash. Yearb.,* 68, 83, 1968.

118a. **Jefferys, A. J., and Barrie, P. A.** Sequence variations and evolution of nuclear DNA in man and the primates, *Philos. Trans. R. Soc. Lond.,* 292B, 133, 1981.

119. **Beauchamp, R. S., Bostock, C. J., Buckland, R. A., and Mitchell, A. R.,** The organization of human sat-III sequences in human chromosomes, in *Committee Reports, Human Gene Mapping,* 5, 1979.

120. **Cookie, H. J. and Noel, B.,** DNA analysis shows that chromosomes 15p+ and 13p+ arise by translocation of the Y chromosome distal heterochromatin, in *Committee Reports, Human Gene Mapping,* 5, 1979.

121. **Szabo, P., Kunkel, L., Yu, L. C., George, D., and Smith, K. K.,** Chromosomal distribution of DNA sequences derived from the human-Y chromosome in human and higher primates, in *Committee Reports, Human Gene Mapping,* 5, 1979.

121a. **Shafit-Zagardo, B., Maio, J. J., and Brown, F. L.,** KpnI families of long interspersed repetitive DNAs in human and other primate genomes, *Nucleic Acids Res.,* 10, 3175, 1982.

122. **Warburton, D. and Pearson, P. L.,** Report of the committee on comparative mapping, *Human Gene Mapping,* 3, 75, 1975.

122a. **Mitchell, A. R., Gosden, J. R. and Ryder, O. A.,** Satellite DNA relationships in man and the primates, *Nucleic Acids Res.,* 9, 3235, 1981.

123. **Prosser, J., Moar, M., Bobrow, M., and Jones, K. W.,** Satellite sequences in chimpanzee *(Pan troglodytes), Biochim. Biophys. Acta,* 319, 122, 1978.

124. **King, M. C. and Wilson, A. C.,** Evaluation of two levels in humans and chimpanzees, *Science,* 188, 107, 1975.

124a. **Goldstein, D. J., Rogers, C. and Harris, H.,** Evolution of alkaline phosphatases in primates, *Proc. Natl. Acad. Sci. U.S.A.,* 79, 879, 1982.

125. **Benveniste, R. E. and Todaro, G. J.,** Evolution of type C viral genes: evidence for an Asian origin of man, *Nature (Lond.)* 261, 101, 1976.

126. **Szalay, S. F. and Delson, E.,** in *Evolutionary history of the primates,* Academic Press, New York, 1980.

127. **Bonnot, T. J., Heinemann, R., and Todaro, G. J.,** Evolution of DNA sequences has been retarded in Malagasy primates, *Nature (Lond.),* 286, 420, 1980.